PHYSICS AND MECHANICS OF NEW MATERIALS

Synthesis, Processing, and Emerging Applications

PHYSICS AND MECHANICS OF NEW MATERIALS

Synthesis, Processing, and Emerging Applications

Edited by

Devrim Balkose, PhD

Ana Cristina Faria Ribeiro, PhD

Nandakumar Kalarikkal, PhD

Ann Rose Abraham, PhD

A. K. Haghi, PhD

AAP | APPLE ACADEMIC PRESS

First edition published 2025

Apple Academic Press Inc.
1265 Goldenrod Circle, NE,
Palm Bay, FL 32905 USA

760 Laurentian Drive, Unit 19,
Burlington, ON L7N 0A4, CANADA

CRC Press
2385 NW Executive Center Drive,
Suite 320, Boca Raton FL 33431

4 Park Square, Milton Park,
Abingdon, Oxon, OX14 4RN UK

Library and Archives Canada Cataloguing in Publication

CIP data on file with Canada Library and Archives

Library of Congress Cataloging-in-Publication Data

CIP data on file with US Library of Congress

ISBN: 978-1-77491-573-8 (hbk)
ISBN: 978-1-77491-574-5 (pbk)
ISBN: 978-1-00348-776-0 (ebk)

About the Editors

Devrim Balkose, PhD

Professor, Chemical Engineering, Izmir Institute of Technology, Izmir, Turkey

Devrim Balkose, PhD, is currently a retired faculty member in the Chemical Engineering Department at Izmir Institute of Technology, Izmir, Turkey. She graduated from the Middle East Technical University in Ankara, Turkey, with a degree in chemical engineering. She received her MS and PhD degrees from Ege University, Izmir, Turkey, in 1974 and 1977, respectively. She became an Associate Professor in macromolecular chemistry in 1983 and Professor in process and reactor engineering in 1990. She was the Head of Chemical Engineering Department at Izmir Institute of Technology, Izmir, Turkey, between 2000 and 2009. She is now a retired faculty member in the same department. Her research interests are in polymer reaction engineering, polymer foams and films, adsorbent development, and moisture sorption. Her research projects are on nanosized zinc borate production, ZnO polymer composites, zinc borate lubricants, antistatic additives, and metal soaps.

Ana Cristina Faria Ribeiro, PhD

Researcher, Department of Chemistry, University of Coimbra, Portugal

Ana C. F. Ribeiro, PhD, is a researcher in the Department of Chemistry at the University of Coimbra, Portugal. Her area of scientific activity is physical chemistry and electrochemistry. Her main areas of research interest are transport properties of ionic and nonionic components in aqueous solutions. She has experience as a scientific adviser and teacher of different practical courses. Dr. Ribeiro has supervised master's degree theses as well as some PhD theses, and has been a theses jury member. She has been a referee for various journals as well as an expert evaluator of some of the research programs funded by the Romanian government through the National Council for Scientific Research. She has been a member of the organizing committee of scientific conferences and she is an editorial member of several journals. She has received several grants, consulted for a number of major corporations and is a frequent speaker to national and international audiences. She is a member of the Research Chemistry Centre, Coimbra, Portugal.

Nandakumar Kalarikkal, PhD
Senior Professor, School of Pure and Applied Physics, Mahatma Gandhi University and School of Nanoscience and Nanotechnology; Director, International and Inter University Center for Nanoscience and Nanotechnology, Mahatma Gandhi University, India

Nandakumar Kalarikkal, PhD, is the Senior Professor, School of Pure and Applied Physics, Mahatma Gandhi University and School of Nanoscience and Nanotechnology, and Director of the International and Inter University Center for Nanoscience and Nanotechnology, Mahatma Gandhi University, India. His current research interests include synthesis, characterization, and applications of various nanostructured materials, laser plasma, and phase transitions. He has published more than 260 research articles in peer-reviewed journals, holds six patents, and has edited over 30 books. Dr. Kalarikkal obtained his master's degree in Physics with a specialization in Industrial Physics and his PhD in Semiconductor Physics from Cochin University of Science and Technology, Kerala, India. He was a postdoctoral fellow at the National Institute for Interdisciplinary Science and Technology (NIIST), Trivandrum, and later joined Mahatma Gandhi University, Kerala, India. He has received several research grants and supervised more than 22 PhD candidates. He has carried out several international and national project works and has received many international fellowships and honors.

Ann Rose Abraham, PhD
Assistant Professor, Sacred Heart College (Autonomous), Thevara, Kochi, Kerala, India

Ann Rose Abraham, PhD, is currently an Assistant Professor at the Department of Physics, Sacred Heart College (Autonomous), Thevara, Kochi, Kerala, India. She has expertise in the field of condensed matter physics, nanomagnetism, multiferroics and polymeric nanocomposites, etc. She has research experience at various reputed national institutes, including Bose Institute, Kolkata, India; SAHA Institute of Nuclear Physics, Kolkata, India; UGC-DAE CSR Centre, Kolkata, India; and she has collaborated with various international laboratories. She is a recipient of a young researcher award in physics and best paper awards–2020, 2021. She served as an Assistant Professor and Examiner at the Department of Basic Sciences, Amal Jyothi College of Engineering, under APJ Abdul Kalam Technological University, Kerala, India. Dr. Abraham is a frequent speaker at national and international conferences. She has a good number of publications to her credit in many peer-reviewed high-impact journals of international repute. She has authored many book chapters and edited more than 10 books with Taylor and Francis, Elsevier, etc.

A. K. Haghi, PhD

Research Associate, Department of Chemistry, University of Coimbra, Portugal

A. K. Haghi, PhD, is a retired professor and has written, co-written, edited, or co-edited more than 1000 publications, including books, book chapters, and papers in refereed journals with over 3800 citations and h-index of 32, according to the Google Scholar database. He is currently a research associate at the University of Coimbra, Portugal. Professor Haghi has received several grants, consulted for several major corporations, and is a frequent speaker to national and international audiences. He is Founder and former Editor-in-Chief of the *International Journal of Chemoinformatics and Chemical Engineering* and *Polymers Research Journal*. Professor Haghi has served as an editorial board member of many international journals. He has served as a member of the Canadian Research and Development Center of Sciences and Cultures. He has supervised several PhD and MSc theses at the University of Guilan (UG) and co-supervised international doctoral projects. Professor Haghi holds a BSc in urban and environmental engineering from the University of North Carolina (USA) and holds two MSc, one in mechanical engineering from North Carolina State University (USA) and another one in applied mechanics, acoustics, and materials from the Université de Technologie de Compiègne (France). He was awarded a PhD in engineering sciences at Université de *Franche-Comté* (France). He is a persistent reviewer for leading international journals.

Contents

Contributors

G. Anjali
International and Inter University Centre for Nanosciences and Nanotechnology, Mahatma Gandhi University, Kottayam, Kerala, India

L. S. Archana
Department of Physics, Govt. College for Women, University of Kerala, Thiruvanandapuram, Kerala, India

Mohammad Ashfaq
Department of Biotechnology & University Centre for Research & Development (UCRD), Chandigarh University, Gharaun, Mohali, Punjab, India

Pius Augustine
Material Research Laboratory, Sacred Heart College (Autonomous), Thevara Kochi, Kerala, India
Department of Physics, Sacred Heart College (Autonomous), Thevara Kochi, Kerala, India
Material Research Centre, Indian Institute of Science, Bangalore, Karnataka, India

Saji Augustine
Department of Physics, Deva Matha College, Kuravilangad, Kottayam, Kerala, India

Devrim Balköse
İzmir Institute of Technology, Urla İzmir, Turkey

M. C. F. Barros
CQC-IMS, Department of Chemistry, University of Coimbra, Coimbra, Portugal

Brahmananda Chakraborty
High Pressure and Synchroton Radiation Physics Division, Bhabha Atomic Research Centre, Trombay, Mumbai, India
Homi Bhabha National Institute, Mumbai, India

Divya Chauhan
Department of Chemical and Biomedical Engineering, University of South Florida, Tampa, FL, USA

Amal Ajith Cherian
Department of Physics, Marian College Kuttikkanam (Autonomous), Kuttikkanam, Peerumade, Idukky, Kerala, India

Gizem Cihanoğlu
İzmir Institute of Technology, Urla İzmir, Turkey

Jincemon Cyriac
Department of Physics, Marian College Kuttikkanam (Autonomous), Kuttikkanam, Peerumade, Idukky, Kerala, India

Jisty Ann Jiji
Department of Physics, Marian College Kuttikkanam (Autonomous), Kuttikkanam, Peerumade, Idukky, Kerala, India

Saritha K. Jose
Department of Physics, Deva Matha College, Kuravilangad, Kottayam, Kerala, India

Shilpa Joy
Sri Shakthi Institute of Engineering and Technology, Coimbatore, Tamilnadu, India

Nandakumar Kalarikkal
School of Pure and Applied Physics, Mahatma Gandhi University,Kottayam, Kerala, India
International and Inter University Centre for Nanoscience and Nanotechnology, Mahatma Gandhi University, Kottayam, Kerala, India
School of Nanoscience and Nanotechnology, Mahatma Gandhi University, Kottayam, Kerala, India

Merve Karabıyık
İzmir Institute of Technology, Urla İzmir, Turkey

Sema Kırköse
İzmir Institute of Technology, Urla İzmir, Turkey

J. S. Jyothish Kumar
International and Inter University Centre for Nanosciences and Nanotechnology, Mahatma Gandhi University, Kottayam, Kerala, India

M. V. Manjusha
Department of Physics, the Cochin College, Kochi, Ernakulam, Kerala, India

Sunny Mathew
Department of Physics, Marian College Kuttikkanam (Autonomous), Kuttikkanam, Peerumade, Idukky, Kerala, India

A. Mialdun
Université libre de Bruxelles, Brussels, Belgium

S. L. Navya
Department of Physics, Sacred Heart College (Autonomous), Thevara Kochi, Kerala, India

Aybike Nil Olcay
İzmir Institute of Technology, Urla İzmir, Turkey

Dhanya Raj
Department of Physics, the Cochin College, Kochi, Ernakulam, Kerala, India

Kavya Ravindran
Material Research Laboratory, Sacred Heart College (Autonomous), Thevara Kochi, Kerala, India
Department of Physics, Sacred Heart College (Autonomous), Thevara Kochi, Kerala, India

A.C.F. Ribeiro
CQC-IMS, Department of Chemistry, University of Coimbra, Coimbra, Portugal

C. I. A. V Santos
CQC-IMS, Department of Chemistry, University of Coimbra, Coimbra, Portugal

Gopal Sanyal
Mechanical Metallurgy Division, Bhabha Atomic Research Centre, Trombay, Mumbai, India

S. Sashank
Department of Physics, Marian college Kuttikkanam (Autonomous), Kuttikkanam, Peerumade, Idukky, Kerala, India

Merin Mary Sebastian
School of Pure and Applied Physics, Mahatma Gandhi University,Kottayam, Kerala, India

Tina Sebastian
Department of Physics, Deva Matha College, Kuravilangad, Kottayam, Kerala, India

Nidhi Sharma
Department of Physics, Dhanauri P.G. College, Dhanauri, Roorkee Haridwar, Uttarakhand, India

V. Shevtsova
Mechanical and Manufacturing Department, Mondragon University, Loramendi, Mondragon, Spain
Ikerbasque, Basque Foundation for Science, Bilbao, Spain

Vidhya Sivan
Material Research Laboratory, Sacred Heart College (Autonomous), Thevara Kochi, Kerala, India
Department of Physics, Sacred Heart College (Autonomous), Thevara Kochi, Kerala, India

Mridula Sreedharan
International and Inter University Centre for Nanosciences and Nanotechnology,
Mahatma Gandhi University, Kottayam, Kerala, India

L. R. Sriramachandru
Department of Physics, Marian college Kuttikkanam (Autonomous), Kuttikkanam, Peerumade, Idukky, Kerala, India

Neetu Talreja
Department of Science, Faculty of Science and Technology, Alliance University, Anekal, Bengaluru, Karnataka, India

Nebu G. Thomas
Pushpagiri College of Dental Sciences, Pushpagiri Institute of Medical Sciences, Kerala, India

Sabu Thomas
International and Inter University Centre for Nanosciences and Nanotechnology,
Mahatma Gandhi University, Kottayam, Kerala, India

Ravi Kumar Trivedi
High Pressure and Synchroton Radiation Physics Division, Bhabha Atomic Research Centre, Trombay, Mumbai, Maharashtra, India

Fadeela Chundekatt Ummer
International and Inter University Centre for Nanoscience and Nanotechnology,
Mahatma Gandhi University, Kottayam, Kerala, India

T. V. Venkidesh
Department of Physics, the Cochin College, Kochi, Ernakulam, Kerala, India

Abbreviations

2D	two-dimensional
2D-NM	two-dimensional nanomaterials
2D-ZnO-NS	2D-ZnO nanosheets
2-EHA	2-ethylhexyl acrylate
AAD	average absolute deviation
AC	activated carbon
AlACA	agent aluminum acetylacetonate
ASCs	asymmetric supercapacitors
$BaTiO_3$	barium titanate
BDS	breakdown strength
BESS	battery energy storage systems
BF-BT	bismuth ferrite-barium titanate
BFO	bismuth ferrite
BP	black phosphorus
CC	carbon cloth
CFP	carbon fiber papers
CNCs	cellulose nanocrystals
CNFs	carbon nanofibers
CNTs	carbon nanotubes
CPs	conducting polymers
CV	cyclic voltammetry
CVD	chemical vapor deposition
DCMIX	diffusion coefficients in mixtures
DFT	density functional theory
DM	Dzyaloshinskii–Moriya
EDLCs	electrochemical double-layer capacitors
EES	electrical energy storage
EES	electrochemical energy storage
EMC	electromagnetic compatibility
EMI	electromagnetic interference
FEPV	ferroelectric photovoltaic effect
FeRAM	ferroelectric random-access memories
FFT	fast Fourier transform
FNPs	ferrite nanoparticles
FTIR	Fourier transform infrared spectra
FTJs	ferroelectric tunnel junctions
GHG	greenhouse gas

HDPE	high-density polyethylene
HER	hydrogen evolution reaction
HOR	hydrogen oxidation reaction
HR-TEM	high-resolution transmission electron microscopy
HS	hybrid supercapacitors
IP	intercalation pseudocapacitance
ISS	International Space Station
LDHs	layered-double hydroxide
LDPE	low density polyethylene
LIB	lithium–ion battery
LSV	linear sweep voltametric
ME	magnetoelectric
MEMS	microelectromechanical systems
MFNPs	magnetic ferrite nanoparticles
MLCC	multilayer ceramic capacitors
MnO_2	manganese oxide
MNPs	magnetic nanoparticles
MOFs	metal–organic frameworks
MPB	morphotropic phase boundary
MRI	magnetic resonance imaging
MSN	mesoporous silica nanoparticles
MTX	methotrexate
MW	megawatts
MX2	metal dichalcogenides
MX3	metal trichalcogenides
NACC	AC covered carbon cloth
NaDS	sodium dodecyl sulfate
NIR	near-infrared radiation
NM	nanomaterials
NMP	N-methyl-2-pyrrolidone
NPs	nanoparticles
OBD	optical beam deflection technique
OCP	open circuit potential
ODI	optical digital interferometry
OER	oxygen evolution reaction
OHP	outer Helmholtz plane
ORR	oxygen reduction reaction
PA	polyaniline
PACA	polyalkyl cyano acrylates
PANI	polyaniline
PB	polybutadiene
PCE	power conversion efficiency
PCL	polycaprolactone

PCs	polarizable probes
PDT	photodynamic therapy
PEDOT	poly(3,4-ethylenedioxythiophene)
PEG	polyethylene glycol
PEMF	pulsed electromagnetic field
PFH	perfluorohexane
PGA	polyglycolide
PHA	polyhydroxyalkanoates
PHB	polyhydroxybutyrate
PL	photoluminescence
PLA	polylactide
PLGA	polylactide coglycolide
PPy	polypyrrole
Pr	polarization
PS	polystyrene
PSA	pressure-sensitive adhesives
PTA	photothermal therapeutic agent
PTT	photothermal therapies
PV	photovoltaic
PVA	polyvinyl alcohol
PVP	polyvinylpyrrolidone
RES	reticuloendothelial system
rGO	reduced graphene oxide
RK	Redlich–Kister
RuO_2	ruthenium oxide
SAR	specific absorption rate
SCs	supercapacitors
SEM	scanning thermal microscopy
SERB	Science and Engineering Research Board
SF	spinel ferrites
SPION	superparamagnetic iron oxide
SQUID	superconducting quantum interference device
SRP	surface redox pseudocapacitance
SSA	specific surface area
SSCs	symmetric supercapacitors
SWNT	single-walled nanotube
TARE	Teacher Associate for Research Excellence
TEM	transmission electron microscopy
TM	transition metals
TMDCs	transition-metal dichalcogenides
UD	underpotential deposition
VACNT	vertically aligned carbon nanotube
VS	vermiculite sheets

XPS	X-ray photoelectron spectroscopy
XRD	X-ray diffraction
ZHMSC	zinc hybrid microsupercapacitor

Preface

This new volume covers new advancements in materials science and engineering technology for scientists and postgraduate research students and provides a basic overview with a comprehensive background in material physics. It also presents chapters on the practical aspects rather than theoretical for professional engineers.

The first section of this book delivers a comprehensive overview of ferrite materials with special attention to their structure, types, and properties. It presents a comprehensive survey about ferrite materials with discussions of selected applications of modern ferrite materials in emerging technologies.

In the second section, the impact of technological change upon the application of selected materials and energy resources are presented in detail.

In the last section of this volume, case studies in thermophysics and applied mechanics for engineering technology are covered, which provide a thorough examination of various physical, chemical, and material properties.

In each chapter, the explanations and also the selection of the references are sufficiently detailed to provide the essential background for further studies and explore relevant literature for postgraduate research students.

This research-oriented book is an essential reference source intended for scientific researchers, lecturers of universities, postgraduates, and senior students who specialize in the field of the material science and applied mechanics of materials. The results presented are useful for the design, fabrication, testing, and industrial applications of advanced materials and structures.

—***Editors***

SECTION 1

Modern Ferrites: Emerging Technologies and Applications

CHAPTER 1

Synthesis and Mechanism of Functionalized Ferrite Nanoparticles for Water Splitting Applications

MERIN MARY SEBASTIAN[1], FADEELA CHUNDEKATT UMMER[2], and NANDAKUMAR KALARIKKAL[1,2,3]

[1]*School of Pure and Applied Physics, Mahatma Gandhi University, Kottayam, Kerala, India*

[2]*International and Inter University Centre for Nanoscience and Nanotechnology, Mahatma Gandhi University, Kottayam, Kerala, India*

[3]*School of Nanoscience and Nanotechnology, Mahatma Gandhi University, Kottayam, Kerala, India*

ABSTRACT

Properties of ferrite nanoparticles (NPs) are a swiftly arising field for numerous applications like hydrogen evolution via water splitting. NPs of metal ferrites and other binary/ternary/military compounds can act as apromising catalyst in the field of water splitting application. In catalysis, generally highly exposed active sites enhance the catalytic activity of the material. NPs are engineered to have a large surface/volume ratio so that even a small amount can provide high catalytic activity. The electrochemical water splitting is predicted as the most dominating and promising future tool for the production of clean hydrogen which is considered the future fuel. When the ferrites are combined or anchored with any two-dimensional materials like graphene or TMD, it will enhance the surface properties and thus will lead to best catalytic activity. Therefore, the selection of the substrate and method of synthesis for functionalization

Physics and Mechanics of New Materials: Synthesis, Processing, and Emerging Applications. Devrim Balkose, Ana Cristina Faria Ribeiro, Nandakumar Kalarikkal, Ann Rose Abraham, A. K. Haghi (Eds.)

are significant for catalyst NPs. Applications of catalytic processes of ferrite particles include oxidation, reduction, fuel cell, batteries, etc. Moreover, functionalized ferrite NPs catalysts can be effectively used without losing catalytic activity, and also it may enhance the catalytic activity through the synergic effect of hybridization with the substrate. Synthesis of NPs must be environment-friendly, faster, and economic to accommodate bulk preparation. This chapter covers the synthesis methods and how differently functionalized ferrite NPs affect various water splitting applications

1.1 INTRODUCTION

Since long, hydrogen has been considered an alternative renewable energy source as well as a perfect replacement for fossil-fuel-based energysources, with zero CO_2 emission. The electrocatalytic water splitting, which includes an oxygen evolution reaction (OER), oxygen reduction reaction (ORR), hydrogen oxidation reaction (HOR), and a hydrogen evolution reaction (HER), where the HER is considered promising for the generation of hydrogen gas.

Electrocatalytic water splitting is a very flexible experimental way of the production of clean hydrogen. The water splitting reactions have very sluggish kinetics and it needs a potential of 1.23 V for the reaction to get initiated. To overcome the sluggish kinetics and thereby produce hydrogen with lower overpotential, metal-based catalysts are chosen to be catalysts due to their low-cost and earth abundance which leads to the large-scale industrial applications.[1] Here, we are focused on the spinel ferrites with a general formula AFe_2O_4, they are a prime class of semiconductor materials, primarily due to their electrical, optical, magnetic, and catalytic properties.[2,3] Kalarikkal et al. and his co-workers had developed different spinel ferrites and its heterostructures for optical limiting applications and the multiferroic heterostructures which contributes for the wireless communication and information storage applications. They even developed spinel hybrid catalyst for OER and ORR bifunctional activity in alkaline medium.[4–11] The various properties of ferrites include superior ion transport, rapid electrode kinetics, and high electrochemical behavior, which are very favorable for water splitting applications.[12,13] More than that, the spinel ferrites with chemical composition, low band gap, valence states, and good stability which makes the spinel ferrites more interesting and noticeable.

Spinel ferrites, such as $NiFe_2O^4$, $CuFe_2O_4$, $ZnFe_2O_4$, and $CoFe_2O_4$, are composed of low cost easily accessible materials. Studies had indicated their superior application as catalysts for electrochemical water splitting.[14] The crystal structure of ferrites is classified into normal spinel, inverse spinel, and complex spinel as shown in Figures 1.1 and 1.2. As mentioned, the general formula of the spinel's AFe_2O_4 is where the "A" represents the divalent ions. For normal spinel materials, A(II) and Fe (III) occupy tetrahedral and octahedral sites, respectively, while for inverse spinel type ferrites, half of the Fe (III) locates at tetrahedral sites, and the A(II) along with the other half Fe (III) locates at octahedral sites, for example, $NiFe_2O_4$. Finally, complex spinel structures, Fe (III) and A(II), occupy the octahedral sites and tetrahedral randomly.[15,16] The production of hydrogen is directly related to the efficiency of hydrogen produced immaculately and also its utilization in energy-converting devices like fuel cells, batteries, etc. For instance, the electrochemical water electrolysis method can deliver an eco-friendly method to produce clean hydrogen by breaking the bonds between water. Generally, the alkaline condition is preferred for the transition metal-based catalysts for better performance. The above-mentioned catalysts can pave way for the production of large quantities and thus can be converted into storable combustible fuel in H_2/O_2 fuel cells and metal-air batteries to produce energy.[17–19]

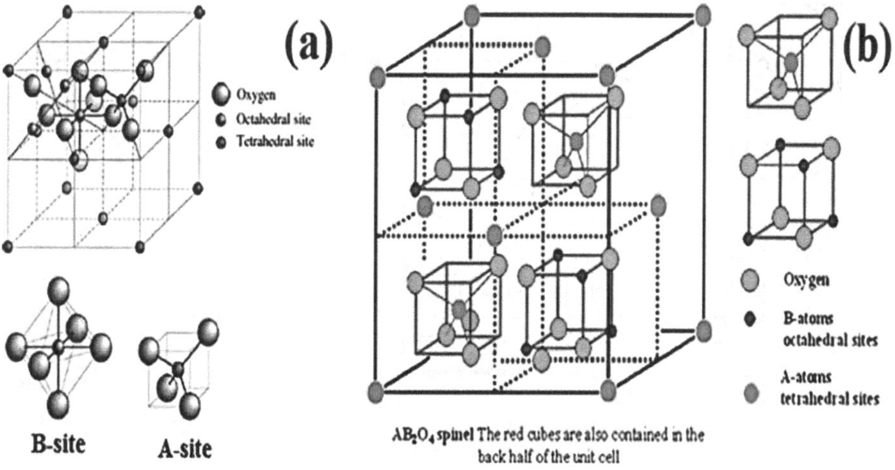

FIGURE 1.1 Spinel ferrite cell structure as reported by: (a) Ozgur et al. and (b) Issa et al. © 2020 Elsevier B.V. All rights reserved.[12]

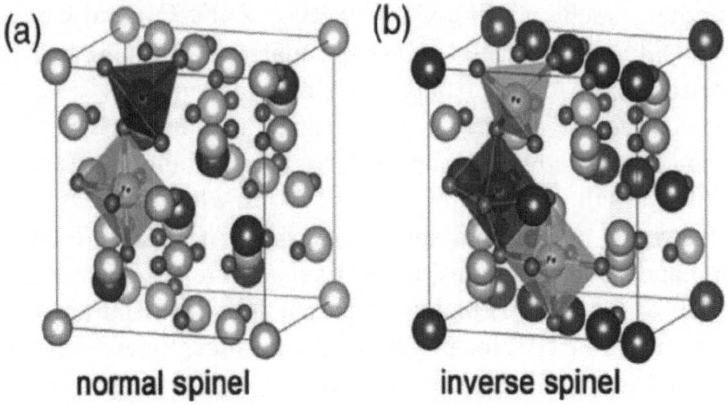

FIGURE 1.2 Unit cell structure of: (a) normal spinel ferrite and (b) inverse spinel ferrite, Taffa et al. © 2020 Elsevier B.V. All rights reserved.

Source: Reproduced with permission from Ref. [20]

1.2 SYNTHESIS OF FERRITE NANOPARTICLES FOR CATALYSIS

1.2.1 HYDROTHERMAL METHOD

The hydrothermal or solvothermal synthesis method is one of the most common and effective synthesis methods to synthesize the different types of nanomaterials of different morphologies and sizes for different applications. More than that, the hydro/solvothermal method is efficient in anchoring one type of compound with the other in-virtually. In this method, the precursors were transferred into an autoclave which contains Teflon-lined beaker filled with de-ionized water or organic compound and the reaction is carried out reaction under high temperature and pressure conditions. If the non-aqueous solvents are used as a medium for the reaction, it is known as solvothermal method or if the preparation is carried using water as medium, then it is known as a hydrothermal process. Teflon-lined beaker-contained autoclaves were commonly used as they are capable of working at high temperature and pressure, it also sustains alkaline media. Therefore, a Teflon-lined autoclave was chosen as best container for the reaction under desired temperature and pressure conditions. This method can facilitate and accelerate the reaction among the reactants, promote hydrolysis, and lead to uniform crystal growth, which results in self-assembly and a variety of morphologies of nanomaterials. Moreover, the properties, morphology, size, and structure of nanomaterials can be varied and controlled easily or tuned accordingly by varying the different reaction parameters, such as time of reaction, temperature given to the reaction,

reaction medium, pressure, pH of the medium, and concentration of the reactants and filled volume of the autoclave. This method also facilitates low cost and low environmental pollution.[21–23] The schematic illustration is added in Figure 1.3.

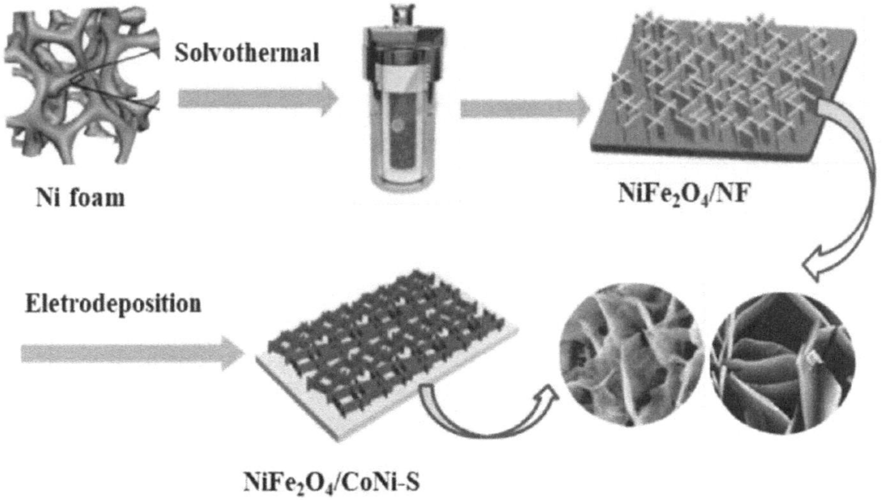

FIGURE 1.3 The schematic illustrations of the fabricated process of $NiFe_2O_4$/CoNi–S. Copyright © 2020 Hydrogen Energy Publications LLC. Published by Elsevier.

Source: Reproduced with permission from Ref. [24]

1.2.2 SOL-GEL METHOD

Sol-gel method is one of the well-established, inexpensive chemical synthetic approaches which involves the preparation of nanoparticles as well as nanoparticle composites. The sol-gel method includes the following steps—hydrolysis, condensation, and drying process, which has excellent control over the surface properties of the materials. The formation of nanoparticles involves different reaction steps. In the first step, the corresponding precursors undergo rapid hydrolysis to produce thesolution, followed by immediate condensation which leads to the formation of gels; then the gel is subjected to the drying process which leads to the drive-off of residues or water content from the samples. The sol-gel synthesis route can be classified into two, one is aqueous sol-gel and the other is non-aqueous sol-gel depending on the nature of the solvent used in the reaction. If water is used as medium, it is aqueous sol-gel method, and if the medium is organic solvent, then it is non-aqueous sol-gel method.[25–27]

1.2.3 CO-PRECIPITATION ROUTE

The co-precipitation method has been widely used for the synthesis of ferrite nanoparticles. In this synthesis route, the nanometer-sized reactors have been used for the formation of homogeneous nanoparticles of different ferrite. To protect the samples from oxidation when exposed to the atmospheric oxygen along with the agglomeration of the particles, the particles are coated with a surfactant like sodium dodecyl sulfate (NaDS) or oleic acid, etc., and then dispersed in some carrier liquids like ethanol, methanol, ammonia, etc., depending upon the nature of the nanoparticles we are synthesizing. Generally, the reaction is performed under an inert atmosphere like N_2 or Ar along with degassed solutions to avoid the uncontrollable oxidation of Fe^{2+} into Fe^{3+}. The advantage of this method is that uniform-size distribution can be obtained more easily than other synthesis methods without giving extra heat treatments. In this method, Fe^{2+} and Fe^{3+} ions are generally precipitated in alkaline solutions, such as ammonium hydroxide, potassium hydroxide, or sodium hydroxide. In most cases of ferrite synthesis, it is carried out in the range of 70–80 °C temperature or higher temperatures. Several parameters like effects of mixing methods, stirring rate, digestion time, initial pH, and the presence or absence of a magnetic field can affect the nature of ferrites prepared.[28–30]

1.2.4 SONOCHEMICAL METHOD

The sonochemical method of synthesis deals with the synthesis or the chemistry in the application of high-energy ultrasound to a reaction mixture where we can avoid the high-temperature, high pressure, and long-run reactions. Ultrasound is cyclic mechanical vibration within a frequency range of 20 kHz and 15 MHz. As the wavelength is much larger than the molecular size scale, there is no possibility of direct interaction between ultrasound and molecules which may lead to a chemical reaction. Instead of that, a physical phenomenon called acoustic cavitation occurs. Acoustic cavitation is when liquid and ultrasound interact, there is compression and rarefactions are formed. Bubbles will be generated from pre-existing impurities, which undergo strongly to the acoustic field and undergo a rapid inertial overgrowth during expansion, Once the bubbles reach their maximum size, it will get followed by a collapse. This process of bubble formation, growth, and collapse together is called cavitation and this leads to the rapid release of energy with heating and cooling rates.[31,32]

1.3 ELECTROCHEMICAL WATER SPLITTING

Water electrolysis is considered as a sustainable way to produce clean hydrogen gas, which is going to be the future fuel without the production of CO_2 and other global warming-causing gases. Water splitting reaction includes anodic OER and cathodic HER. For both the reactions to happen, it requires a minimum energy of 1.23 eV to break the bond and thereby the evolution of hydrogen to happen. In this scenario, we need catalysts that are efficient to enhance the catalytic activity. The best catalysts are Ir/Ru oxides for the OER and Pt-based catalysts for HER, which are the state-of-the-art in this field. The scarcity and high cost of the catalysts hinder its large-scale adoption.[33]

Total reaction

$$2H_2O \rightarrow 2H_2 + O_2 \tag{1.1}$$

Acidic condition

$$\text{Cathode: } 2H^+ + 2e^- \rightarrow H_2 \; Ec = 0 \text{ V} \tag{1.2}$$

$$\text{Anode: } H_2O \rightarrow O_2 + 4H^+ + 4e^- \; Ea = 1.23 \text{ V} \tag{1.3}$$

Alkaline condition

$$\text{Cathode: } 2H_2O + 2e^- \rightarrow H_{2+} 2OH^- \; Ea = -0.83 \text{ V} \tag{1.4}$$

$$\text{Anode: } 4OH^- \rightarrow O_2 + 2H_2O + 4e^- \; Ea = -0.40 \text{ V} \tag{1.5}$$

1.3.1 HER MECHANISM

HER is an electrocatalytic cathodic reaction in electrochemical water splitting which is often studied. The theoretical thermodynamical potential for water splitting is 1.23 V. However, extra potentials are always required for the HER to happen. The acidic conditions are more favorable for hydrogen evolution. In the HER mechanism, the hydrogen molecule tends to get adsorbed on the surface of the electrode depending on the electrode material. There are two types of hydrogen adsorptions, underpotentially deposited hydrogen and overpotential deposited hydrogen.

As we already mentioned, HER in acidic medium is more preferable because the reduction of an H^+ is energetically more beneficial than that of a neutral/buffer or in alkaline electrolytes. The mechanism of HER is defined and can be explained from the Tafel slope. Depending on the Tafel value, we can predict the possible mechanism which caused the HER. From the experiment, the obtained Tafel slope is around 30 mV/dec then the occurred mechanism is considered as Volmer–Tafel mechanism. Otherwise, it is considered a Volmer–Heyrovsky mechanism.

In the first step, the Volmer step is common in both mechanisms. For that, the Tafel equation is

$$\eta = a + b \log j \tag{1.6}$$

where η, j, b, and a are the overpotential, current density, Tafel slope, and constant, respectively. If the Tafel slope is small, then it is more conducive to the practical application since it will lead to a faster increase in HER rate.

The Volmer reaction,

$$H+ + e- \rightarrow H_{ads} \tag{1.7}$$

which is the "discharge step", that is, which involves in the discharging of protons on the surface of the electrode to form the adsorbed hydrogen "H_{ads}."

After the hydrogen gets adsorbed, the possibility of two mechanisms for the Hydrogen evolution

The Heyrovsky reaction

$$H_{ads} + H_2 + e- \rightarrow\rightarrow H_2 \tag{1.8}$$

which is the electrochemical desorption step, or,Tafel reaction

$$H_{ads} + H_{ads} \rightarrow H_2 \tag{1.9}$$

which is the recombination reaction step that leads to hydrogen evolution, where H_{ads} indicates the adsorption of atomic hydrogen on the active sites of an electrocatalyst.

The rate determining the Tafel slope value is about 120 mV dec^{-1}; while for the Heyrovsky and Tafel mechanism will occur, if the Tafel slopes are approximately 40 and 30 mV dec^{-1}, respectively, then the possible HER mechanism will be Volmer–Tafel mechanism. The production of molecular hydrogen requires the either Volmer–Heyrovsky reactions or Volmer–Tafel reactions during the HER process.[34,35]

1.3.2 OER MECHANISM

Electrochemical water splitting is the cleanest way to produce hydrogen gas. Both the HER and OER are occurring in the cathodic and anodic sides of the cell where the whole reaction inherently depends on the nature of the electrocatalyst used. Considering the sluggish and the hydrogen fuel production, there is a slow four-electron transfer process involved in OER which is the critical step in the cell system. As the requirement for commercial inexpensive electrolyzes increases, the essentiality for the development of cost effective OER catalysts is inevitable so that it can reduce the overpotential and increase the activity.

In an electrochemical reaction, the HER follows the reaction pathway and thus the $4e^-$ pathway which is followed by the OER makes the reaction more complex. The reaction steps in alkaline conditions are explained below.

$$M + OH^- \rightarrow MOH + e^- \quad (1.10)$$

$$MOH + OH^- \rightarrow MO + H2O + e^- \quad (1.11)$$

$$2MO \rightarrow 2M + O_2 \quad (1.12)$$

$$MO + OH^- \rightarrow MOOH + e^- \quad (1.13)$$

$$MOOH + OH^- \rightarrow M + O2 + H2O + e^- \quad (1.14)$$

Here, M is the surface metal active cations, which are the active sites for OER. The two routes of OER, that is, following the reactions and by the direct dissociation of MO and thereby the production of oxygen. On the other side, the active site M will form oxyhydroxide by choosing the association and dissociation of the oxyhydroxide intermediate and thus producing the oxygen gas eqs (1.13) and (1.14). The M–O bond strength has a critical role which decides the activity of the catalyst in OER, an M–O bond that is too strong or too weak leads to the hindering of the desorption of products, or a weak bond that prevents the intermediate binding on the surface of the anode. An optimal bond strength causes the higher electrocatalytic activity.[36–38]

1.4 PARAMETERS FOR EVALUATING CATALYSTS FOR WATER SPLITTING

1.4.1 OVERPOTENTIAL

The overall water splitting theoretically requires a cell voltage of 1.23 V to begin the reaction and production of hydrogen and oxygen. In practicality, there are so many reaction hindrances which in both the reactions need additional potentials to initiate the reactions. The additional potential other than the cell potential that requires to initiate the reaction is called overpotential (η). The overpotential is an important parameter that evaluates the activity and the efficiency of the catalyst. In general, all the research work articles report the overpotential of different catalysts at a fixed current density of 10 mA cm^{-2} and then compare the value. Smaller the value of overpotential, the better the catalyst and its activity.

1.4.2 TAFEL SLOPE AND EXCHANGE CURRENT DENSITY

Tafel slope calculation is an important parameter that determines the hydrogen evolution pathway. The Tafel slope is obtained from the linear sweep voltametric (LSV) data where the slope of the linear portion in the graph of overpotential vs. log of current density in the graph is taken. The graph is fitted using eq (1.6), where η is the overpotential, b is the Tafel slope, and j is the current density. Specifically, the Tafel slopes of the Volmer, Heyrovsky, and Tafel reactions are calculated to be 118.2, 39.4, and 29.6 mV dec^{-1}, respectively.[39] Depending on the "b" value, the rate-determining step can be found. The Volmer reaction is the rate-determining step when the B value is close to 118.2 mV dec^{-1}, and the kinetics for adsorption of H atoms are sluggish. If b is around 39.4 mV dec^{-1}, it is a Heyrovsky reaction. If b approaches close to 29.6 mV dec^{-1}, the hydrogen gas is produced by the combination of adsorbed H atoms.

Another key descriptor of electrocatalytic activity is the exchange current density, which profoundly affects the whole electrochemical reaction. By extrapolating the linear part of Tafel plots where the overpotential is zero, the exchange current density (j_0), can be obtained. The intrinsic properties like electrocatalysts and experimental conditions will affect the current density value. An ideal efficient catalyst have a low b and a high j_0.[34]

1.4.3 STABILITY

Stability is another important factor that determines the practicability of an electrocatalyst in commercial applications. There are mainly three popular methods to record the long-term stability of the catalyst. The methods are running the cyclic voltammetry for many cycles (>2000) and recording the initial and final LSV data and then comparing the degradation. The shift in the onset and overpotential parameters at a constant current density indicates durability. The second and third methods are chronopotentiometry and chronoamperometry. In chronopotentiometry, the decrease in the current density at a fixed potential for a long time of greater than 10 h isnoted. In chronoamperometry, the rise in potential at a constant current density, commonly 10 mA cm^{-2} is taken as standard and reported. Less shift in data denotes the better stability of the material in the reaction.

1.5 FERRITE NANOPARTICLES FOR WATER SPLITTING A PPLICATIONS

1.5.1 NICKEL FERRITE ON WATER SPLITTING APPLICATION

When we design a catalyst, the ideal catalyst should meet the following criteria's such as: (1) more exposed active sites which enables the catalyst to attain desirable catalytic activity, which can be attained by significantly enhancing the surface area and thus increasing the active sites. (2) The designed catalyst should promote more electron transfer, by incorporating the catalyst with some conducting material or making the catalyst itself more conducting can improve the activity and reaction. (3) Most importantly, when it comes to commercialization, the stability of the catalyst should be excellent and catalyst materials should be low-cost, earth-abundant, and synthesized using facile techniques for large-scale applications. While considering all the requirements, the Ni-Fe-based compounds had attracted the electrochemical society because of their catalytic performance.

$NiFe_2O_4$ is an inverse-spinel-structured oxide that is low-cost, earth-abundant, and environmentally benignancy was being investigated for many years as an electrocatalyst for water splitting. Ni-Fe-based compounds are known for their catalytic activity in OER and are thus recognized as the best catalysts for instance, $NiFe_2O_4$-based catalyst activity is shown in Figure 1.4. Fe and Ni are attractive non-noble metals where the rational Ni-Fe-based catalysts are mostly stable in alkaline conditions. Wu et al. synthesized the $NiFe_2O_4$/NiFe-LDH array by solvothermal method, which had an activity toward the OER with a small overpotential of 213 mV at 100 mA cm^{-2} and an overpotential of 101 mV to attain 10 mA cm^{-2} for HER.[40] Functionalized $NiFe_2O_4$– CoS on the active nickel form substrate exhibits a three-dimensional self-supported model which offers good conductivity and abundant active sites which paves way for the both HER and OER bifunctional catalytic performance with an overpotential of 116 mV and 227 mV at a current density of 10 mA cm^{-2} in 1 M alkaline medium.[41] Liu et al. showed that $NiFe_2O_4$–Sulphur composite strongly binds on the nickel form requires 267 mV of overpotential for OER for a current density of 10 mA cm^{-2} with the lowest OER Tafel slope of 36.7 mV dec^{-1}, due to its ultra-small particle size close to 2 nm and 3D-structure which implies appreciable kinetics for the OER.[42] Similarly, as-fabricated $NiFe_2O_4$ anchored on CoNi-S heterostructure array exhibits remarkable bifunctional catalytic activity and a noticeable stability of 72 h toward both the OER and HER in alkaline media. It presents a small overpotential of 219 and 149 mV for OER and HER for a standard current density of 10 mA cm^{-2}.[24] Selvasundarasekar et al. had recently proved that palladium decorated nickel ferrite subjected to the HER process in an alkaline medium showed activity with a minimum overpotential of 61 mV @ 10 mA cm–2. Hence, this showed the synergic effect in HER activity by the incorporation of Pd nanosheets, that is Pd nanosheets on $NiFe_2O_4$-Nickel foam help to reduce catalytic poising and improves hydrogen adsorption on the catalytic surface and thus improve the activity with low overpotential.

Many studies had proved that oxygen vacancy can enhance the electronic conductivity and then thereby increase the OER performance. Zhao et al. reported in his work that that the oxygen vacancies created can efficiently reduce the energy barrier of OH* to O* in OER thus improving the performance of the catalyst[43] (Cao et al.). The resultant oxygen deficient–enriched $NiFe_2O_4$@N/rGO-800 exhibits a low overpotential of 252 mV at 20 mA cm^{-2} for OER and 157 mV at 10 mA cm^{-2} for HER in1 M KOH.[44] Balasubramaniyan et al. showed that engineered oxygen vacancies in $NiFe_2O_4$ composite achieved a current density of 100 mA cm^{-2} at an overpotential of 230 mV with the smallest Tafel slope 42 mV/dec.[45] Similarly, another oxygen vacancy-rich NiO/$NiFe_2O_4$ nanosheets have an overpotential of 279 mV at 10 mA cm^{-2} and the same Tafel slope of 42 mV /dec just as the previous catalyst.[46] $NiFe_2O_4$ incorporated with Co_3O_4 heterostructure nanoarrays with abundant oxygen vacancies, providing low overpotentials of 251 and 77 mV, respectively, to achieve 10 mA·cm^{-2} for catalyzing OER and HER in an alkaline medium. The synergistic approach enhances the electrochemical activity in both reactions of the catalyst by improving its mass diffusion efficiency, conductivity, and increased active sites.[47]

The Ni-Fe-based oxides had shown better conductivity because of the electron hopping between valence bands in oxygen sites and metal centers which are active in redox and contribute to the adsorption and activation of catalyst, the substrate also contributes to the catalytic activity by providing good electrical contact between catalyst and substrate which enhances the interactions and promotes the catalytic reaction for long term.

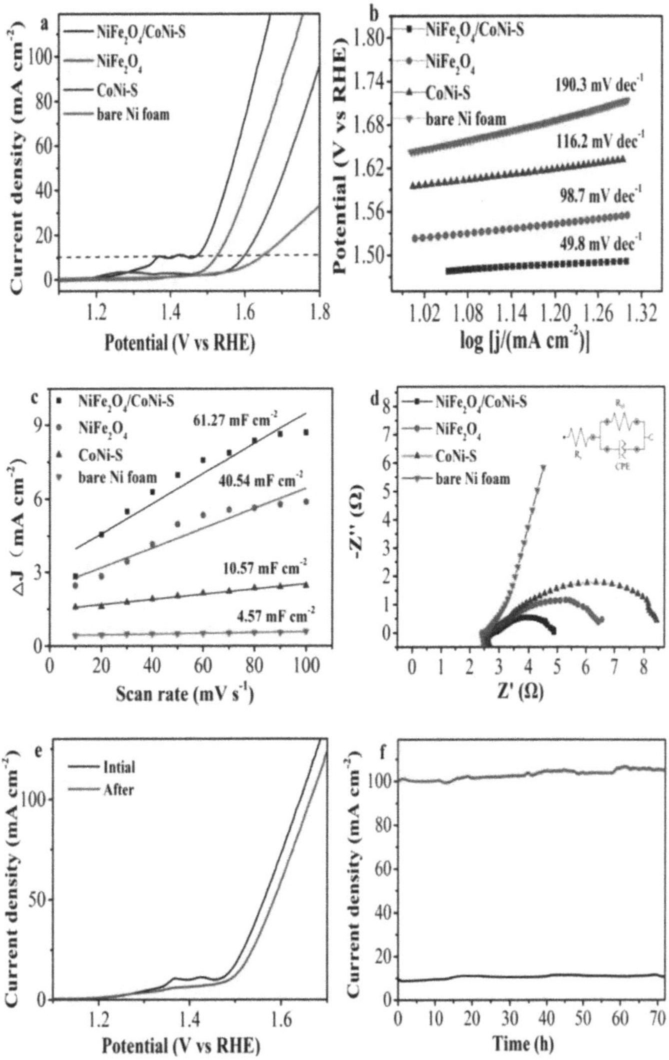

FIGURE 1.4 (a) OER polarization curves of $NiFe_2O_4$/CoNi-S, $NiFe_2O_4$, CoNi-S, and bare NF; (b) The corresponding Tafel plots; (c)The ECAS for different samples in 1 M KOH; (d) The Nyquist plots; (e) Polarization curves before and after 2000 CV sweeps; and (f) Chronoamperometric curves of $NiFe_2O_4$/CoNi-S at 1.45 and 1.57 V for 72 h copyright © 2020 Hydrogen Energy Publications LLC. Published by Elsevier.

Source: Reproduced with permission from Ref. [24]

Another trend with nickel ferrite composite with MXenes[48] and carbon which proved to be an excellent bifunctional catalyst attracted the electrochemical community. Shinde et al. experimented between the two-dimensional graphene oxide and one-dimensional carbon nanotube (CNT) combined with the Nickel ferrite. It is found that the two-dimensional graphene oxide with Nickel ferrite had shown better activity with a low overpotential of 327 mV to reach up to 10 mA cm^{-2} with a Tafel slope of 103 mV/dec for OER.[49] Another carbon-coated mesoporous Ni-$NiFe_2O_4$ nanosheet had performed as a bifunctional catalyst in alkaline electrolyte, which requires an overpotential of 212 mV for OER and 217 mV for HER to reach 10mAcm^{-2} with Tafel slope of 51 and 96 mV /dec respectively.[50] Xu et al. studied the integration of $NiFe_2O_4$ on vertically aligned carbon nanotube (VACNT) arrays hybrid as the bi-functional catalyst. The integrated electrocatalyst of $NiFe_2O_4$/VACNT showed noticeable catalytic activities with overpotentials of only 240 and 150mV to achieve 10 mA cm^{-2} OER and HER.[51] Similarly, $NiFe_2O_4$ integrated with CNT composites by Yang's and Yu's group had shown OER activity. Analyzing the integrated nickel ferrite particles on carbon matrix provides the synergic effect due to several reasons, which are, first, the morphology of ferrite particles with carbon network which promotes the reactions. Second, synergistic interactions between nickel and iron metal species together with the carbon network which is conductive. Third, encourages surface electronic state modification of metals when combined with a conductive carbon network. Finally, the heterostructure with a large surface area and exposed catalyst sites promotes electron transfer and thereby high catalytic action.[52,53] Combining metal oxide with graphene oxide/carbonaceous material can provide a better platform with a high surface area, more exposed catalytic active sites, good conductivity, and stable catalytic activity.

Other $NiFe_2O_4$-based catalysts are given in the Table 1.1.

1.5.2 COBALT FERRITE HYBRIDS FOR WATER SPLITTING APPLICATION

Cobalt ferrite nanoparticles are another category of spinel ferrite which has fascinating research interest for water splitting applications. Cobalt ferrite (CFO) nanoparticles have desirable characteristics, such as a small band gap (<2 eV), which favors the solar water splitting reaction in visible light.[61] However, the cobalt ferrite suffers from low conductivity, and a

TABLE 1.1 $NiFe_2O_4$-based Catalysts.

Catalyst description	Function/ feature	Overpotential/Tafel slope	References
$NiFe_2O_4$/N-graphene	Bifunctional HER a OER catalyst	HER: 184 mV reach 10 mA cm^{-2} Tafel slope: 8 mV dec^{-1} OER: 340 mV reach 10 mA cm^{-2} Tafel slope: 9 mV dec^{-1}	[54]
Oxygen vacancies-rich nickel ferrite	Bifunctional HER a OER catalyst	HER: 102.9 mV reach 10 mA cm^{-2} Tafel slope: 4 mV dec^{-1} OER: 448.8 mV reach 10 mA cm^{-2} Tafel slope: 5 mV dec^{-1}	[55]
$NiFe_2O_4$-graphene	HER catalyst	259 mV at a curr density of 10 cm^{-2}/121.4 mV d_1	[56]
Carbon paste modified $NiFe_2O_4$	HER catalyst	97.75 $mVdec^{-1}$	[57]
$NiFe_2O_4$ embedded in reduced graphene oxide	HER catalyst	HER: 42 mV reach 10 mA cm^{-2} Tafel slope: 58 dec^{-1}	[58]
Tubular $NiFe_2O_4$	Bifunctional HER a OER catalyst	HER: 201 mV reach 10 mA cm^{-2} Tafel slope: 6 mV dec^{-1} OER: 340 mV reach 10 mA cm^{-2} Tafel slope: 12 mV dec^{-1}	[59]
$NiFe_2O_4$-MoS_2	OER catalyst	OER: 290 mV reach 10 mA cm^{-2} Tafel slope:69.2 dec^{-1}	[60]

limited number of active sites causes poor OER and ORR results.[62] As a result, in most of the studies, hybrid nanostructures with carbon-based nanomaterials or other supportive functional materials are explored toovercome the above-mentioned limitation. Carbon-based materials can effectively reduce the particle size of ferrites and thereby can improve the number of active sites in electrocatalysts.[62] Kashyap et al. explored the efficiency of a hybrid of cobalt ferrite with nitrogen-doped graphene oxide for ORR. The nanoparticle hybrid is prepared by a solvothermal process which facilitates the fine dispersion of CFO over N-dopped graphene oxide. N-doping results in the buckling of graphene layers and facilitates the anchoring of nanoparticles. The hybrid catalyst supported with microporous carbon (Vulcan) shows enhanced electrochemical ORR performance. The microporous carbon prevents the aggregation of CFO decorated nanosheets, this high conductivity of it prevents the drop in iR and it facilitates the reactant distribution over the active sites on N-dopped reduced graphene oxide. As a result, the Vulcan-supported catalytic system outperforms the Vulcan-free counterpart by 17 mV.[63]

Yan et al. prepared a hybrid catalyst of monodispersed porous cobalt ferrite microspheres (CFO-ns) on reduced graphene oxide (rGO) using a solvothermal process. Both the components act in a synergetic way to the characteristic structure of the CFO-ns/rGO composite; the incorporation of rGO prevents the agglomeration of CFO nanoparticles and the incorporation of CFO prevents the restacking of graphene layers. The structure of the hybrid improves the active sites and facilitates the oxygen support during ORR and OER bifunctional reaction.[64]

Xu et al. showed the electrocatalytic activity of $CoFe_2O_4$ anchored on walnut shell-derived microporous carbon prepared via a sustainable and cost-effective route. The electrocatalyst shows trifunctional catalytic activity toward ORR, OER, and HER. The catalyst exhibits better performance toward the OER with a small potential of 1.683 V (453 mV) to reach at 10 mA cm^{-2} and the HER performance with a small potential (− 0.6505 V) at 10 mA cm^{-2}.[65]

The stability of the electrocatalyst is an important criterion which is essential for the catalyst. In the study by Kargar et al., $CoFe_2O_4$ isdecorated on three-dimensional carbon fiber papers (CFP) using spin coating and studied the OER and HER performance. $CoFe_2O_4$/CFP shows improved OERcompared to bare $CoFe_2O_4$. The $CoFe_2O_4$/CFP shows better stability without any change in morphology and composition. Over continuous cycles for 15 h and operating with high current density at a fixed potential, the catalyst was stable for more than 40 h. The CFP-supported catalyst also shows better HER performance compared to bare $CoFe_2O_4$.[66] Ding et al. showed the bifunctional electrocatalytic activity in using single-walled nanotube (SWNT) wrapped $CoFe_2O_4$. The incorporation of SWNT improved the electrical conductivity of the hybrid and improves the overall water splitting performance by providing a large surface area and a greater number of active sites. In addition to this both the component synergistically favors the stability of the hybrid. The hybrids exhibit good OER with a lower overpotential of 310 mV and HER performance with a low overpotential 263 mVto reach 10 mA cm^{-2} in 1.0 M KOH.[67]

Srinivas et al. showed superior electrocatalytic activity of metal-organic framework derived $CoFe_2O_4$/CoO catalyst embedded in CNT network toward water splitting applications. The excellent performance of the catalyst is due to the factors (I) MOF-derived porous carbon with large number of oxygen vacancies, (II) large number ofcatalytic sites in $CoFe_2O_4/CoO$ heterostructure. The catalyst performs with low over potential 246 mV and 164 mV at 10 mA/cm^{-2} with low Tafel slopes of 45.27 and 86.38 mV/dec for OER and HER, respectively.[68] Wu et al. designed an efficient and stable electrocatalyst by growing $CoFe_2$ on $CoFe_2O_4$ porous hollow nanofibers which is having plenty ofoxygen vacancy. The catalyst is prepared by electrospinning andsubsequent two step-heating treatment process in 10% H_2/Ar atmosphere. The porosity of the composite structure increases the contact area between electrocatalysts and electrolyte and also the nanoparticles in the nanofiber provide abundant grain interspaces which in turn provides larger interface between electrocatalysts and electrolyte which causes a decrease the ion diffusion length and kinetics of mass diffusion. The oxygen vacanciesincrease the electrical conductivity and provide active sites for OER. The electrocatalyst shows low overpotential for OER performance say 300 mVat 10 mA cm^{-2} with a small Tafel slope of 73.34 mV/ dec along with stable durability in 1 M KOH.[69] Zhang et al. prepared V_2O_3-decorated $CoFe_2O_4$ with carbon-encapsulated mesoporous nanosheets for OER and HER performance. The catalyst requires lower overpotentials for HER, that is 61 mV and OER 226 mV to reach up to 10 mA cm^{-2}.[70] Jiang et al. studies the electrocatalytic performance of $CoFe/CoFe_2O_4$ electrocatalyst with graphitic shells in which cationic sites are modified by ruthenium ($Ru\text{-}GC@CoFe/CoFe_2O_4$) prepared by self-assembling and vapor-reduction strategies. The electrocatalyst shows excellent OER performance at a low overpotential of 237 mV at 10 mA cm^{-2} and a small Tafel slope of 47 mV dec^{-1} in alkaline electrolyte.[71] Li et al. prepared a completely porous bifunctional hierarchical electrocatalyst $CoNi/CoFe_2O_4$ overnickel form which exhibits an excellent OER performance with over potential 360 mV to achieve a current density 1000 mA cm^{-2} in an alkaline medium. The catalyst also shows excellent HER performance with low over potential 82 ±3 and 189 ±3 mV to achieve current densities 10 and 100 mA cm^{-2}.[72]

Some other cobalt ferrite-supported catalysts with their electrocatalytic performance are shown in Table 1.2.

TABLE 1.2 Other Cobalt Ferrite-Supported Catalysts with Their Electrocatalytic Performance.

Catalyst description	Function/ feature	Overpotential/Tafel slope	References
Spinel $CoFe_2O_4$/CNT	OER catalyst	OER: 1.62 V (vs. RHE) to reach 10 mA cm^{-2} Tafel slope: 82 mV dec^{-1}	[73]
Co_3O_4/$CoFe_2O_4$ decorated on nickel foam	OER catalyst	OER: 215 mV (vs. RHE)) to reach 10 mA cm^{-2} Tafel slope: 90 mV/dec	[74]
$FeSe_2$/$CoFe_2O_4$ heterojunction electrocatalyst	HER catalyst	HER: 231 mV at a current density of 10 mA cm^{-2} Tafel Slope:88.76 mV/ dec	[75]
Sulfur Incorporated $CoFe_2O_4$/Multi-walled CNT	OER catalyst	OER: 0.36 V to reach current density 10 $mAcm^{-2}$ Tafel Slope:43 mV/dec	[76]
P-Ni_3S_2/$CoFe_2O_4$/NF	OER catalyst	OER: 254 mV at 50 mA cm^{-2} Tafel Slope: 54.43 mV/dec	[77]
NiO–$CoFe_2O_4$/NF	Bifunctional HER and OER catalyst	HER: 97 mV to reach 10 mA cm^{-2} Tafel slope: 89 mV/dec OER: 157 mV to reach 10 mA cm^{-2} Tafel slope: 82 mV/dec	[78]
Ni-S-$CoFe_2O_4$	OER catalyst	OER: 228 mV to reach 10 mA cm^{-2} Tafel slope: 72 mV/dec	[79]

1.5.3 OTHER FERRITES FOR WATER SPLITTING

Other ferrites and bimetal oxides are there which can act as a better catalyst for water splitting. For instance, Zn, Cu, and Mn, etc., can induce HER or OER in the electrolyte. Belhadj et al. took different ferrites like MFe_2O_4 (M=Co, Ni, Zn, and Cu) in an alkaline medium and comparedit's HER activity and its overpotential. The cobalt ferrite showed better catalytic activity compared to other catalysts which required 270 mV to reach up to a current density of 10 mA cm^{-2}. Other catalyst required 320, 420, and 520 mV for $CuFe_2O_4$, $NiFe_2O_4$, and $ZnFe_2O_4$.[80] The copper ferrite can further act as a bifunctional catalyst with high stability. Pragmatically designed tetragonal copper ferrite and porous bulk copper ferrite spinel act as an HER and OER bifunctional catalyst in an alkaline medium where the first one shows activity with 298 and 103 mV for OER and HER. The porous copper ferrite is also active in OER (290 mV) and HER (241 mV) in 6 M KOH.[81,82] Manganese ferrite composites which are grown on elm-money-derived biochar[83] composite show the same overpotential of 156 mV for both HER and OER and its composite on graphene showed its HER activity in an acidic medium with an overpotential of 313 mV.[84] Doped or bimetallic ferrites on carbon matrix also show interesting overpotentials where the carbon matrixprovides good conductivity and, for instance, in the case of cobalt ferriteduring the reaction transforms to Co(oxy)hydroxide which can act as an active site.[84–86] While considering the bimetallic oxides/hydroxides which also act as efficient catalysts, as Gua et al. mentioned in their work, the CoFe-LDH/ $CoFe_2O_4$ composite has good electrical conductivity and more state density, which results in the rapid transfer of electrons and perfectly beneficial to improve the electrocatalytic performance. Lee and Xu et al. studied the electrocatalytic activity of bimetallic oxides which were found to be efficient in both HER. Well, the bimetallic oxide composite by Xu's group proved its bifunctionality thus $NiCo_2O_4$ nanoneedle/Mo_2C-coated carbon cloth as the catalyst used as air cathode in the application of metal-airbatteries.[87–89]

TABLE 1.3 Other Ferrites for Water Splitting.

Catalyst description	Function/feature	Overpotential/T afel slope	References
MgFe2O4 nanospheres	Bifunctional HER and OER catalyst	HER: 402 mV to reach 10 mA cm^{-2} Tafel slope: 241 mV dec^{-1} OER: 1.09 V Tafel slope: 317 mV dec^{-}	[90]
1D hollow CuFe2O4 nanospheres	OER catalyst	OER: 367 mV to reach 10 mA cm^{-2} Tafel slope: 82 mV dec^{-1}	[91]
NiCo2O4 nanostructure	Bifunctional HER and OER catalyst	OER Tafel slope:46.74 mV dec^{-1} HER Tafel slope: 90.63 mV dec^{-1}	[92]

1.6 CONCLUSION

Several interesting ferrites and their composites had been explored in the area of water splitting. Their catalytic activity and abundance make the ferrites more suitable as a catalyst and direct it's for large-scale commercialization.

However, the ferrites suffer from low conductivity and a limited number of active sites causes poor activity in the water splitting reaction. To overcome the limitations, the ferrites are functionalized, doped, or incorporated with conducting materials or carbon matrix which helps to overcome the conductivity limitation. $NiFe_2O_4$ and $CoFe_2O_4$ are the most studied catalysts in the area due to their catalytic activity. In addition, charge transfer within A(II) and Fe (III) bimetallic centers in AFe_2O_4 may create oxygen vacancies and mediate the redox reaction of water oxidation. Spinel-type oxides with Ni^{2+} cations occupying the octahedral sites and cobalt cations exist in both the octahedral and tetrahedral sites. The existence of redox couples $Ni^{3+/2+}$ and $Co^{3+/2+}$ can endow the catalysts with a unique electronic structure.[93] The proximity of the bimetallic center to the ionically conductive substrate/matrix facilitates cooperative, effective adsorption of the reactants and ions from the electrolyte thus improving the electron transfer and thus leads to successive redox reactions. Engineering with the morphology and arrangement of the catalyst increases the exposure of active sites in the electrolyte and thus improves the catalytic activity. In previous years, this area and ferrite catalysts had been explored for water splitting, but still, many improvements are needed to implement a perfect ferrite catalyst for commercialization. Many spinel compounds are used to facilitate HER, OER, and ORR. OER/ORR, HER/OER-like bifunctional catalytic activity is the key process for energy conversion and storage devices, thus used as anodes and cathodes in metal-air batteries.

This chapter addresses the basic mechanism of water splitting reactions and different synthesis methods of ferrite nanoparticles. Along with that, the recent progress of ferrite and its composite catalysts acts in the reactions were also discussed.

KEYWORDS

- **mechanism**
- **environmental-friendly**
- **synthesis**
- **functionalized**
- **ferrite**
- **water splitting**

REFERENCES

1. Li, X.; Hao, X.; Abudula, A.; Guan, G. Nanostructured Catalysts for Electrochemical Water Splitting: Current State and Prospects. *J. Mater. Chem. A* **2016,** *4* (31), 11973–12000. DOI: 10.1039/c6ta02334g.
2. Rani, R.; Kumar, G.; Batoo, K. M.; Singh, M. Electric and Dielectric Study of Zinc Substituted Cobalt Nanoferrites Prepared by Solution Combustion Method. *Am. J. Nanomater.* **2013,** *1* (1), 9–12. DOI: 10.12691/ajn-1-1-3.
3. Shobana, M. K.; Kim, K.; Kim, J. H. Impact of Magnesium Substitution in Nickel Ferrite: Optical and Electrochemical Studies. *Phys. E Low-Dimensional Syst. Nanostruct.* **2019,** *108*, 100–104. DOI: 10.1016/j.physe.2018.12.013.
4. Thomas, J. J.; Krishnan, S.; Sridharan, K.; Philip, R.; Kalarikkal, N. A Comparative Study on the Optical Limiting Properties of Different Nano Spinel Ferrites with Z-Scan Technique. *Mater. Res. Bull.* **2012,** *47* (8), 1855–1860. DOI: 10.1016/j.materresbull.2012.04.067.
5. Thomas, J. J.; Shinde, A. B.; Krishna, P. S. R.; Kalarikkal, N. Temperature Dependent Neutron Diffraction and Mössbauer Studies in Zinc Ferrite Nanoparticles. *Mater. Res. Bull.* **2013,** *48* (4), 1506–1511. DOI: 10.1016/j.materresbull.2012.12.058.
6. Thomas, J. J.; Shinde, A. B.; Krishna, P. S. R.; Kalarikkal, N. Cation Distribution and Micro Level Magnetic Alignments in the Nanosized Nickel Zinc Ferrite. *J. Alloys Compd.* **2013,** *546*, 77–83. DOI: 10.1016/j.jallcom.2012.08.011.
7. Abraham, A. R. et al. Magnetic Performance and Defect Characterization Studies of Core-Shell Architectured $MgFe_2O_4$@$BaTiO_3$ Multiferroic Nanostructures. *Phys. Chem. Chem. Phys.* **2019,** *21* (17), 8709–8720. DOI: 10.1039/c8cp04946g.
8. Abraham, A. R. et al. Realization of Enhanced Magnetoelectric Coupling and Raman Spectroscopic Signatures in 0-0 Type Hybrid Multiferroic Core-Shell Geometric Nanostructures. *J. Phys. Chem. C* **2017,** *121* (8), 4352–4362. DOI: 10.1021/acs.jpcc.6b12461.
9. Abraham, A. R.; Raneesh, B.; Nambissan, P. M. G.; Sanyal, D.; Thomas, S.; Kalarikkal, N. Defects Characterisation and Studies of Structural Properties of Sol–Gel Synthesised $MgFe_2O_4$ Nanocrystals Through Positron Annihilation and Supportive Spectroscopic Methods. *Philos. Mag.* **2020,** *100* (1), 32–61. DOI: 10.1080/14786435.2019.1668576.
10. Abraham, A. R.; Raneesh, B.; Sanyal, D.; Thomas, S.; Kalarikkal, N.; Nambissan, P. M. G. Defect-Focused Analysis of Calcium-Substitution-Induced Structural Transformation of Magnesium Ferrite Nanocrystals. *New J. Chem.* **2020,** *44* (4), 1556–1570. DOI: 10.1039/c9nj04068d.
11. Sebastian, M. M.; Velayudham, P.; Schechter, A.; Kalarikkal, N. Spinel Nickel Ferrite Nanoparticles Supported on a 1T/2H Mixed-Phase MoS2Heterostructured Composite as a Bifunctional Electrocatalyst for Oxygen Evolution and Oxygen Reduction Reactions. *Energy Fuels* **2022,** *36* (14), 7782–7794. DOI: 10.1021/acs.energyfuels.2c01191.
12. Yuan, C.; Bin Wu, H.; Xie, Y.; Lou, X. W. Mixed Transition-Metal Oxides: Design, Synthesis, and Energy-Related Applications. *Angew. Chemie—Int. Ed.* **2014,** *53* (6), 1488–1504. DOI: 10.1002/anie.201303971.
13. Teh, P. F.; Sharma, Y.; Pramana, S. S.; Srinivasan, M. Nanoweb Anodes Composed of one-Dimensional, High Aspect Ratio, Size Tunable Electrospun $ZnFe_2O_4$ Nanofibers for Lithium Ion Batteries. *J. Mater. Chem.* **2011,** *21* (38), 14999–15008. DOI: 10.1039/c1jm12088c.

14. Malima, N. M. et al. Solventless Synthesis of Nanospinel Ni1−xCoxFe2O4(0 ≤x≤ 1) Solid Solutions for Efficient Electrochemical Water Splitting and Supercapacitance. *RSC Adv.* **2021,** *11* (49), 31002–31014. DOI: 10.1039/d1ra04833c.
15. Qin, H. et al. Spinel ferrites (MFe2O4): Synthesis, improvement and catalytic application in environment and energy field. *Adv. Colloid Interface Sci.* **2021,** *294*, 102486. DOI: 10.1016/j.cis.2021.102486.
16. Zhao, Q.; Yan, Z.; Chen, C.; Chen, J. Spinels: Controlled Preparation, Oxygen Reduction/Evolution Reaction Application, and Beyond. *Chem. Rev.* **2017,** *117* (15), 10121–10211. DOI: 10.1021/acs.chemrev.7b00051.
17. Xu, Y.; Sumboja, A.; Zong, Y.; Darr, J. A. Bifunctionally Active Nanosized Spinel Cobalt Nickel Sulfides for Sustainable Secondary Zinc-Air Batteries: Examining the Effects of Compositional Tuning on OER and ORR Activity. *Catal. Sci. Technol.* **2020,** *10* (7), 2173–2182. DOI: 10.1039/c9cy02185j.
18. Y. W. Li et al. Fe-MOF-Derived Efficient ORR/OER Bifunctional Electrocatalyst for Rechargeable Zinc-Air Batteries. *ACS Appl. Mater. Interfaces* **2020,** *12* (40), 44710–44719. DOI: 10.1021/acsami.0c11945.
19. Zhan, Y.; Lu, M.; Yang, S.; Liu, Z.; Lee, J. Y. The Origin of Catalytic Activity of Nickel Phosphate for Oxygen Evolution in Alkaline Solution and Its Further Enhancement by Iron Substitution. *ChemElectroChem* **2016,** *3* (4), 615–621. DOI: 10.1002/celc.201500511.
20. Narang, S. B.; Pubby, K. Nickel Spinel Ferrites: A review. *J. Magn. Magn. Mater.* **2021,** *519* (June 2020. DOI: 10.1016/j.jmmm.2020.167163.
21. Bellamkonda, S.; Chakma, C.; Guru, S.; Neppolian, B.; Rao, G. R. Rational Design of Plasmonic Ag@CoFe2O4/g-C3N4 p-n Heterojunction Photocatalysts for Efficient Overall Water Splitting. *Int. J. Hydrogen Energy* **2022,** *47* (43), 18708–18724. DOI: 10.1016/j.ijhydene.2022.04.059.
22. Zhang, X.; Ma, G.; Wang, J. Hydrothermal Synthesis of Two-Dimensional MoS_2 and Its Applications. *Tungsten* **2019,** *1* (1), 59–79. DOI: 10.1007/s42864-019-00014-9.
23. Lassoued, A.; Lassoued, M. S.; Dkhil, B.; Ammar, S.; Gadri, A. Synthesis, Photoluminescence and Magnetic Properties of Iron Oxide (α-Fe2O3) Nanoparticles Through Precipitation or Hydrothermal Methods. *Phys. E Low-Dimensional Syst. Nanostructures* **2018,** *101* (January), 212–219. DOI: 10.1016/j.physe.2018.04.009.
24. Shi, Y.; Feng, X.; Guan, H.; Zhang, J. ScienceDirect Porous Sunflower Plate-Like $NiFe_2O_4$/CoNieS Heterostructure as Efficient Electrocatalyst for Overall Water Splitting. *Int. J. Hydrogen Energy* **2021,** *46* (12), 8557–8566. DOI: 10.1016/j.ijhydene.2020.12.062.
25. Xu, J. et al. Preparation and Magnetic Properties of Magnetite Nanoparticles by Sol-Gel Method. *J. Magn. Magn. Mater.* **2007,** *309* (2), 307–311. DOI: 10.1016/j.jmmm.2006.07.037.
26. Parashar, M.; Shukla, V. K.; Singh, R. Metal Oxides Nanoparticles via Sol–Gel Method: A Review on Synthesis, Characterization and Applications. *J. Mater. Sci. Mater. Electron.* **2020,** *31* (5), 3729–3749. DOI: 10.1007/s10854-020-02994-8.
27. Sen, R.; Jain, P.; Patidar, R.; Srivastava, S.; Rana, R. S.; Gupta, N. Synthesis and Characterization of Nickel Ferrite (NiFe2O4) Nanoparticles Prepared by Sol-Gel Method. *Mater. Today Proc.* **2015,** *2* (4–5), 3750–3757. DOI: 10.1016/j.matpr.2015.07.165.
28. Maaz, K.; Karim, S.; Mumtaz, A.; Hasanain, S. K.; Liu, J.; Duan, J. L. Synthesis and Magnetic Characterization of Nickel Ferrite Nanoparticles Prepared by Co-precipitation

Route. *J. Magn. Magn. Mater.* **2009,** *321* (12), 1838–1842. DOI: 10.1016/j.jmmm.2008.11.098.

29. Ding, Y. et al. Preparation of Nano-Structured LiFePO4/Graphene Composites by Co-Precipitation Method. *Electrochem. Commun.* **2010,** *12* (1), 10–13. DOI: 10.1016/j.elecom.2009.10.023.
30. Mascolo, M. C.; Pei, Y.; Ring, T. A. Room Temperature Co-Precipitation Synthesis of Magnetite Nanoparticles in a Large ph Window with Different Bases. *Materials* **2013,** *6* (12), 5549–5567. DOI: 10.3390/ma6125549.
31. Shahangi Shirazi, F.; Akhbari, K. Sonochemical Procedures; The Main Synthetic Method for Synthesis of Coinage Metal Ion Supramolecular Polymer Nano Structures. *Ultrason. Sonochem.* **2016,** *31*, 51–61. DOI: 10.1016/j.ultsonch.2015.12.003.
32. Hangxun, X.; Zeiger, B. W.; Suslick, K. S. Sonochemical Synthesis of Nanomaterials. *Chem. Soc. Rev.* **2013,** *42* (7), 2555–2567. DOI: 10.1039/c2cs35282f.
33. Lasia, A. Mechanism and Kinetics of the Hydrogen Evolution Reaction. *Int. J. Hydrogen Energy* **2019,** *44* (36), 19484–19518. DOI: 10.1016/j.ijhydene.2019.05.183.
34. Chen, Z.; Duan, X.; Wei, W.; Wang, S.; Ni, B. J. Recent Advances in Transition Metal-Based Electrocatalysts for Alkaline Hydrogen Evolution. *J. Mater. Chem. A* **2019,** *7* (25), 14971–15005. DOI: 10.1039/c9ta03220g.
35. Joo, J.; Kim, T.; Lee, J.; Il Choi, S.; Lee, K. Morphology-Controlled Metal Sulfides and Phosphides for Electrochemical Water Splitting. *Adv. Mater.* **2019,** *31* (14), 1–23. DOI: 10.1002/adma.201806682.
36. Feng, C.; Faheem, M. B.; Fu, J.; Xiao, Y.; Li, C.; Li, Y. Fe-Based Electrocatalysts for Oxygen Evolution Reaction: Progress and Perspectives. *ACS Catal.* **2020,** *10* (7), 4019–4047. DOI: 10.1021/acscatal.9b05445.
37. Burke, M. S.; Enman, L. J.; Batchellor, A. S.; Zou, S.; Boettcher, S. W. Oxygen Evolution Reaction Electrocatalysis on Transition Metal Oxides and (Oxy)hydroxides: Activity Trends and Design Principles. *Chem. Mater.* **2015,** *27* (22), 7549–7558. DOI: 10.1021/acs.chemmater.5b03148.
38. Rossmeisl, J.; Qu, Z. W.; Zhu, H.; Kroes, G. J.; Nørskov, J. K. Electrolysis of Water on Oxide Surfaces. *J. Electroanal. Chem.* **2007,** *607* (1–2), 83–89. DOI: 10.1016/j.jelechem.2006.11.008.
39. Zhang, R. et al. Ternary NiCo2Px Nanowires as pH-Universal Electrocatalysts for Highly Efficient Hydrogen Evolution Reaction. *Adv. Mater.* **2017,** *29* (9. DOI: 10.1002/adma.201605502.
40. Wu, Z.; Zou, Z.; Huang, J.; Gao, F. $NiFe_2O_4$ Nanoparticles/NiFe Layered Double-Hydroxide Nanosheet Heterostructure Array for Efficient Overall Water Splitting at Large Current Densities, 2018. DOI: 10.1021/acsami.8b07835.
41. Meng, L. et al. Rational Construction of Uniform CoS/$NiFe_2O_4$ Heterostructure as Efficient Bifunctional Electrocatalysts for Hydrogen Evolution and Oxygen Evolution Reactions. *Electrochim. Acta* **2022,** *404*, 139596. DOI: 10.1016/j.electacta.2021.139596.
42. Liu, J.; Zhu, D.; Ling, T.; Vasile, A.; Qiao, S. Nano Energy S-$NiFe_2O_4$ Ultra-Small Nanoparticle Built Nanosheets for Efficient Water Splitting in Alkaline and Neutral pH. **2017,** *40*, 264–273. DOI: 10.1016/j.nanoen.2017.08.031.
43. Asnavandi, M.; Yin, Y.; Li, Y.; Sun, C.; Zhao, C. Promoting Oxygen Evolution Reactions through Introduction of Oxygen Vacancies to Benchmark NiFe-OOH Catalysts. *ACS Energy Lett.* **2018,** *3* (7), 1515–1520. DOI: 10.1021/acsenergylett.8b00696.

44. Cao, L.; Li, Z.; Su, K.; Zhang, M.; Cheng, B. Rational Design of Hollow Oxygen Deficiency-Enriched $NiFe_2O_4$ @ N/rGO as Bifunctional Electrocatalysts for Overall Water Splitting. *J. Energy Chem.* **2021,** *54*, 595–603. DOI: 10.1016/j.jechem.2020.06.053.
45. Balasubramanian, P.; He, S.; Jansirani, A.; Deng, H. Engineering of Oxygen Vacancies Regulated Core-Shell *N*-Doped Carbon @ $NiFe_2O_4$ Nanospheres : A Superior Bifunctional Electrocatalyst for Boosting the Kinetics of Oxygen and Hydrogen Evaluation Reactions. *Chem. Eng. J.* **2021,** *405*, 126732. DOI: 10.1016/j.cej.2020.126732.
46. Zhong, H. et al. Ion Irradiation Inducing Oxygen Vacancy-Rich NiO/$NiFe_2O_4$ Heterostructure for Enhanced Electrocatalytic Water Splitting. *2103501*, 1–10. DOI: 10.1002/smll.202103501.
47. Zhu, J. et al. NiFe2O4 @Co3O4 Heterostructure with Abundant Oxygen Vacancies as a Bifunctional Electrocatalyst for Overall Water Splitting. *J. Alloys Compd.* **2022,** *918*, 165705. DOI: 10.1016/j.jallcom.2022.165705.
48. Shinde, P. V.; Mane, P.; Chakraborty, B.; Sekhar, C. Spinel $NiFe_2O_4$ Nanoparticles Decorated 2D Ti_3C_2 MXene Sheets for Efficient Water Splitting : Experiments and Theories. *J. Colloid Interface Sci.* **2021,** *602*, 232–241. DOI: 10.1016/j.jcis.2021.06.007.
49. Shinde, P. V.; Samal, R.; Sekhar, C. Comparative Electrocatalytic Oxygen Evolution Reaction Studies of Spinel—$NiFe_2O_4$ and Its Nanocarbon Hybrids. *Trans. Tianjin Univ.* **2022,** *28* (1), 80–88. DOI: 10.1007/s12209-021-00310-x.
50. Zhang, J.; Jiang, Y.; Wang, Y.; Yu, C.; Cui, J.; Wu, J. Ultrathin Carbon Coated Mesoporous Ni-$NiFe_2O_4$ Nanosheet Arrays for Efficient Overall Water Splitting. *Electrochim. Acta* **2019,** *321*, 134652. DOI: 10.1016/j.electacta.2019.134652.
51. Xu, Y. et al. Supercritical CO_2—Assisted Synthesis of $NiFe_2O_4$/Vertically-Aligned Carbon Nanotube Arrays Hybrid as a Bifunctional Electrocatalyst for Efficient Overall Water Splitting. *Carbon NY* **2019,** *145*, 201–208. DOI: 10.1016/j.carbon.2019.01.011.
52. Srinivas, K. et al. Metal–Organic Framework-Derived Fe-Doped Ni_3 Fe/$NiFe_2O_4$ Heteronanoparticle-Decorated Carbon Nanotube Network as a Highly Efficient and Durable Bifunctional Electrocatalyst. **2020**. DOI: 10.1021/acsami.0c13836.
53. Yu, X. et al. Hierarchical Coupling Effect in Hollow Ni/$NiFe_2O_4$-CNTs Microsphere via Spray-Drying for Enhanced Oxygen Evolution Electrocatalysis. *13* (2), 437–446.
54. Navadeepthy, D.; Rebekah, A.; Viswanthan, C.; Ponpandian, N. Boosting the Kinetics of Oxygen and Hydrogen Evolution in Alkaline Water Splitting Using Nickel Ferrite /*N*-Graphene Nanocomposite as a Bifunctional Electrocatalyst. *Int. J. Hydrogen Energy* **2021,** *46* (41), 21512–21524. DOI: 10.1016/j.ijhydene.2021.03.244.
55. Wang, C. et al. A Novel Self-Activation Strategy for Designing Oxygen Vacancies-Rich Nickel Ferrite and Cobalt Ferrite Microspheres with Large Specific Surface Area for Overall Water Splitting. *Int. J. Hydrogen Energy* **2022,** *47* (58), 24343–24357. DOI: 10.1016/j.ijhydene.2022.05.185.
56. Nivetha, R.; Chella, S.; Kollu, P.; Jeong, S. K.; Bhatnagar, A.; Andrews, N. G. Cobalt and Nickel Ferrites Based Graphene Nanocomposites for Electrochemical Hydrogen Evolution. *J. Magn. Magn. Mater.* **2018,** *448*, 165–171. DOI: 10.1016/j.jmmm.2017.05.083.
57. Abbaspour, A.; Mirahmadi, E. Electrocatalytic Hydrogen Evolution Reaction on Carbon Paste Electrode Modified with Ni Ferrite Nanoparticles. *Fuel*, *104*, 575–582. DOI: 10.1016/j.fuel.2012.07.016.

58. Mukherjee, A.; Chakrabarty, S.; Su, W. N.; Basu, S. Nanostructured Nickel Ferrite Embedded in Reduced Graphene Oxide for Electrocatalytic Hydrogen Evolution Reaction. *Mater. Today Energy* **2018,** *8*, 118–124. DOI: 10.1016/j.mtener.2018.03.004.
59. Yuan, F.; Cheng, X.; Wang, M.; Ni, Y. Controlled Synthesis of Tubular Ferrite MFe2O4 (M = Fe, Co, Ni) Microstructures with Efficiently Electrocatalytic Activity for Water Splitting. *Electrochim. Acta*, *324*, 134883. DOI: 10.1016/j.electacta.2019.134883.
60. Karpuraranjith, M. et al. Hierarchical Ultrathin Layered MoS2@NiFe2O4 Nanohybrids as a Bifunctional Catalyst for Highly Efficient Oxygen Evolution and Organic Pollutant Degradation. *J. Colloid Interface Sci.* **2021,** *592*, 385–396. DOI: 10.1016/j.jcis.2021.02.062.
61. Vinosha, P. A. et al. Review on Recent Advances of Synthesis, Magnetic Properties, and Water Treatment Applications of Cobalt Ferrite Nanoparticles and Nanocomposites. *J. Supercond. Nov. Magn.* **2021,** *34* (4), 995–1018. DOI: 10.1007/s10948-021-05854-6.
62. Alshehri, S. M.; Alhabarah, A. N.; Ahmed, J.; Naushad, M.; Ahamad, T. An Efficient and Cost-Effective Tri-Functional Electrocatalyst Based on Cobalt Ferrite Embedded Nitrogen Doped carbon. *J. Colloid Interface Sci.* **2018,** *514*, 1–9. DOI: 10.1016/j.jcis.2017.12.020.
63. Kashyap, V.; Singh, S. K.; Kurungot, S. Cobalt Ferrite Bearing Nitrogen-Doped Reduced Graphene Oxide Layers Spatially Separated with Microporous Carbon as Efficient Oxygen Reduction Electrocatalyst. *ACS Appl. Mater. Interfaces* **2016,** *8* (32), 20730–20740. DOI: 10.1021/acsami.6b05416.
64. Yan, W.; Cao, X.; Ke, K.; Tian, J.; Jin, C.; Yang, R. One-Pot Synthesis of Monodispersed Porous $CoFe_2O_4$ Nanospheres on Graphene as an Efficient Electrocatalyst for Oxygen Reduction and Evolution Reactions. *RSC Adv.* **2016,** *6* (1), 307–313. DOI: 10.1039/C5RA23306B.
65. Xu, J. et al. Anchoring CoFe2O4 Nanospheres on Two-Dimensional Microporous Carbon from Walnut Shell as Efficient Multifunctional Electrocatalyst. *J. Solid State Chem.* **2021,** *299*, 122106. DOI: 10.1016/j.jssc.2021.122106.
66. Kargar, A. et al. Solution-Processed CoFe2O4 Nanoparticles on 3D Carbon Fiber Papers for Durable Oxygen Evolution Reaction. *ACS Appl. Mater. Interfaces* **2015,** *7* (32), 17851–17856. DOI: 10.1021/acsami.5b04270.
67. Ding, Y. et al. Single-Walled Carbon Nanotubes Wrapped CoFe2O4 Nanorods with Enriched Oxygen Vacancies for Efficient Overall Water Splitting. *ACS Appl. Energy Mater.* **2019,** *2* (2), 1026–1032. DOI: 10.1021/acsaem.8b01338.
68. Srinivas, K. et al. Heterostructural CoFe2O4/CoO Nanoparticles-Embedded Carbon Nanotubes Network for Boosted Overall Water-Splitting Performance. *Electrochim. Acta* **2022,** *404*. DOI: 10.1016/j.electacta.2021.139745.
69. Wu, L.; Shi, L.; Zhou, S.; Zhao, J.; Miao, X.; Guo, J. Direct Growth of $CoFe_2$ Alloy Strongly Coupling and Oxygen-Vacancy-Rich $CoFe_2O_4$ Porous Hollow Nanofibers: An Efficient Electrocatalyst for Oxygen Evolution Reaction. *Energy Technol.* **2018,** *6* (12), 2350–2357. DOI: 10.1002/ente.201800298.
70. Zhang, H. et al. V2O3-Decorated Spinel CoFe2O4with Carbon-Encapsulated Mesoporous Nanosheets for Efficient Water Splitting. *ACS Sustain. Chem. Eng.* **2021,** *9* (2), 980–986. DOI: 10.1021/acssuschemeng.0c08477.
71. Jiang, M. et al. Ru-Optimized Geometric Sites of Cations in CoFe/CoFe2O4 Electrocatalysts with Graphitic Carbon Shells for Boosting Water Oxidation. *Electrochim. Acta 2022, 425* , 140665. DOI: 10.1016/j.electacta.2022.140665.

72. Li, S. et al. Bifunctional CoNi/$CoFe_2O_4$/Ni Foam Electrodes for Efficient Overall Water Splitting at a High Current Density. *J. Mater. Chem. A 2018*, *6* (39), 19221–19230. DOI: 10.1039/C8TA08223E.
73. Huang, Q.; Li, C.; Tu, Y.; Jiang, Y.; Mei, P.; Yan, X. Spinel $CoFe_2O_4$/Carbon Nanotube Composites as Efficient Bifunctional Electrocatalysts for Oxygen Reduction and Oxygen Evolution Reaction. *Ceram. Int.* **2021,** *47* (2), 1602–1608. DOI: 10.1016/j.ceramint.2020.08.276.
74. Nozari-Asbemarz, M. et al. In Situ Synthesis of Co3O4/CoFe2O4Derived from a Metal-Organic Framework on Nickel Foam: High-Performance Electrocatalyst for Water Oxidation. *ACS Appl. Energy Mater.* **2021,** *4* (3), 2951–2959. DOI: 10.1021/acsaem.1c00429.
75. Zhang, H.; Chao Nengzi, L.; Li, B.; Cheng, Q.; Gou, J.; Cheng, X. Successfully Synthesis of $FeSe_2$/$CoFe_2O_4$ Heterojunction with High Performance for Hydrogen Evolution Reaction. *Renew. Energy* **2020,** *155*, 717–724. DOI: 10.1016/j.renene.2020.03.171.
76. Sun, X. et al. Sulfur Incorporated CoFe2O4/Multiwalled Carbon Nanotubes Toward Enhanced Oxygen Evolution Reaction. *Electrochim. Acta* **2017,** *247*, 843–850. DOI: 10.1016/j.electacta.2017.07.091.
77. Duan, J. J.; Zhang, R. L.; Feng, J. J.; Zhang, L.; Zhang, Q. L.; Wang, A. J. Facile Synthesis of Nanoflower-Like Phosphorus-Doped Ni_3S_2/$CoFe_2O_4$ Arrays on Nickel Foam as a Superior Electrocatalyst for Efficient Oxygen Evolution Reaction. *J. Colloid Interface Sci.* **2021,** *581*, 774–782. DOI: 10.1016/j.jcis.2020.08.005.
78. Wang, X.; Liu, Z.; Guo, Z.; Ge, L.; Liu, Z. NiO–CoFe2O4 Electrocatalyst Prepared on Ni foam by One-Step Hydrothermal Method for Efficient Overall Water Splitting. *J. Mater. Sci.* **2021,** *56* (14), 8575–8587. DOI: 10.1007/s10853-020-05763-4.
79. Li, Z. et al. Magnetic-Field Guided Synthesis of Highly Active Ni–S–$CoFe_2O_4$ Electrocatalysts for Oxygen Evolution Reaction. *Renew. Energy*, *165*, 612–618. DOI: 10.1016/j.renene.2020.11.083.
80. Belhadj, H.; Messaoudi, Y.; Khelladi, M. R.; Azizi, A. A Facile Synthesis of Metal Ferrites (MFe_2O_4, M = Co, Ni, Zn, Cu) as Effective Electrocatalysts Toward Electrochemical Hydrogen Evolution Reaction. *Int. J. Hydrogen Energy* **2022,** *47* (46), 20129–20137. DOI: 10.1016/j.ijhydene.2022.04.151.
81. Sharma, V.; Narayan Pan, U.; Ibomcha Singh, T.; Kumar Das, A.; Hoon Kim, N.; Hee Lee, J. Pragmatically Designed Tetragonal Copper Ferrite Super-Architectures as Advanced Multifunctional Electrodes for Solid-State Supercapacitors and Overall Water Splitting. *Chem. Eng. J.* **2021,** *415* , 127779. DOI: 10.1016/j.cej.2020.127779.
82. Tan, J.; Xu, S.; Zhang, H.; Cao, H.; Zheng, G. Preparation of a Porous Bulk Copper Ferrite Spinel with High Performance in the Electrolysis of Water. *Electrochim. Acta* **2021,** *381*, 138199. DOI: 10.1016/j.electacta.2021.138199.
83. Liu, X.; Huo, Y. Q.; Yan, L. K.; Fan, N.; Cai, K. Z.; Su, Z. M. Hollow Porous $MnFe_2O_4$ Sphere Grown on Elm-Money-Derived Biochar Towards Energy-Saving Full Water Electrolysis. *Chem.—Eur. J.* **2020,** *26* (63), 14397–14404. DOI: 10.1002/chem.202002134.
84. Nivetha, R.; Grace, A. N. Manganese and Zinc Ferrite Based Graphene Nanocomposites for Electrochemical Hydrogen Evolution Reaction. *J. Alloys Compd.* **2019,** *796*, 185–195. DOI: 10.1016/j.jallcom.2019.05.021.
85. Chamani, S.; Khatamian, M.; Peighambardoust, N. S.; Aydemir, U. Microwave-Assisted Auto-Combustion Synthesis of Binary/Ternary $CoxNi_{1-x}$ Ferrite for Electrochemical

Hydrogen and Oxygen Evolution. *ACS Omega* **2021,** *6* (48), 33024–33032. DOI: 10.1021/acsomega.1c05183.

86. Kashyap, V.; Kurungot, S. Zirconium-Substituted Cobalt Ferrite Nanoparticle Supported N-Doped Reduced Graphene Oxide as an Efficient Bifunctional Electrocatalyst for Rechargeable Zn-Air Battery. *ACS Catal.* **2018,** *8* (4), 3715–3726. DOI: 10.1021/acscatal.7b03823.
87. Guo, Z.; Wang, X.; Yang, F.; Liu, Z. Synergistic Effect of Co and Fe Bimetallic Oxides/Hydroxides Composite Structure as a Bifunctional Electrocatalyst for Enhancing Overall Water Splitting Performance. *J. Alloys Compd.* **2022,** *895*, 162614. DOI: 10.1016/j.jallcom.2021.162614.
88. Xu, C.; Wang, Q.; Zhao, S.; Wang, S. NiCo2O4 Nanoneedle/Mo2C-Coated Carbon Cloth as Efficient Catalyst for Water Splitting and Metal-Air Battery. *Synth. Met.* **2021,** *280*, 116894. DOI: 10.1016/j.synthmet.2021.116894.
89. Lee, J. et al. Electrochemical Behavior of the Flower Shaped $CoMn_2O_4$ Spinel Structure Assembled for Effective HER from Water Splitting. *Electrochim. Acta 2021*, *379*, 138168. DOI: 10.1016/j.electacta.2021.138168.
90. Maitra, S.; Mitra, R.; Nath, T. K. Investigation of Electrochemical Performance of Sol-Gel Derived $MgFe_2O_4$ Nanospheres as Aqueous Supercapacitor Electrode and Bi-Functional Water Splitting Electrocatalyst in Alkaline Medium. *Curr. Appl. Phys.* **2021,** *27* (March), 73–88. DOI: 10.1016/j.cap.2021.04.009.
91. Silva, V. D.; Ferreira, L. S.; T. A. Simões, Medeiros, E. S.; Macedo, D. A. 1D Hollow MFe_2O_4 (M = Cu, Co, Ni) fibers by Solution Blow Spinning for Oxygen Evolution Reaction. *J. Colloid Interface Sci.* **2019,** *540*, 59–65. DOI: 10.1016/j.jcis.2019.01.003.
92. Debata, S.; Patra, S.; Banerjee, S.; Madhuri, R.; Sharma, P. K. Controlled Hydrothermal Synthesis of Graphene Supported $NiCo_2O_4$ Coral-Like Nanostructures: An Efficient Electrocatalyst for Overall Water Splitting. *Appl. Surf. Sci.* **2018,** *449*, 203–212. DOI: 10.1016/j.apsusc.2018.01.302.
93. Cao, X.; Wang, T.; Jiao, L. Transition-Metal (Fe, Co, and Ni)-Based Nanofiber Electrocatalysts for Water Splitting. *Adv. Fiber Mater.* **2021,** *3* (4), 210–228. DOI: 10.1007/s42765-021-00065-z.

CHAPTER 2

Biomedical Applications of Magnetic Nanomaterials/Ferrites

MRIDULA SREEDHARAN[1], J. S. JYOTHISH KUMAR[1], G. ANJALI[1], NEBU G. THOMAS[2], SHILPA JOY[3], NANDAKUMAR KALARIKKAL[1], and SABU THOMAS[1]

[1]*International and Inter University Centre for Nanosciences and Nanotechnology, Mahatma Gandhi University, Kottayam, Kerala, India*

[2]*Pushpagiri College of Dental Sciences, Pushpagiri Institute of Medical Sciences, Kerala, India*

[3]*Sri Shakthi Institute of Engineering and Technology, Coimbatore, Tamilnadu, India*

ABSTRACT

Various biomaterial-based multifunctional nanoparticles have been evaluated in recent years for targeted therapeutics and biomedical imaging. The main characteristics of multifunctional nanoparticles are specific target stability, intracellular internalization, bioimaging, and good biocompatibility. Multifunctional nanoparticles can combine magnetic, electrical, optical, and catalytic properties apart from being biocompatible. In this chapter, we are describing the applications of magnetic nanomaterials in various branches of medicine. In the second half of the chapter, magnetic ferrite nanoparticles are highlighted, explaining the advances in the field of medicine where magnetic ferrite nanoparticles have been applied, their synthesis route and functionalization strategies to be used as imaging agents and therapeutic agents. This chapter highlights surface modifications, the challenges and solutions for the use of ferrites in drug delivery and as contrast agent in MRI imaging,

Physics and Mechanics of New Materials: Synthesis, Processing, and Emerging Applications. Devrim Balkose, Ana Cristina Faria Ribeiro, Nandakumar Kalarikkal, Ann Rose Abraham, A. K. Haghi (Eds.)

in therapy for various diseases such as cancer by hyperthermia method, as magnetofection agent, and in tissue engineering. Recent works in this area of drug delivery and treatment are also explained. Finally, the clinical translation of magnetic nanoparticles (MNPs) and their as sensing agent are discussed.

2.1 INTRODUCTION

The use of nanomaterials to treat diseases have been practised from ancient times. The behavior of a material changes when it is made into nanoform. This creates novel properties needed for treating various diseases and combating with pathogens in the body. Nanomedicine is a rapidly growing field with new inventions being translated into clinics that change the face of treatment regimen. The field of imaging and disease diagnosis has grown enormously with the use of nanoparticles especially with the use of magnetic nanoparticles (MNPs). The most difficult-to-treat diseases like cancer has been defeated with the help of nanomedicine. The type of treatment, the dosage of drugs, the side effects, and toxicity caused by over dosage has been overcome with the emergence of nanomaterials into clinics.[1] For drug delivery, the type of conjugation between the nanoparticle surface and the drug should be well understood. Multifunctional nanoparticles can attach more than one drug, contrast agent, and targeting ligands on its surface. The type of surface coatings on the nanoparticles determine their efficiency. In order to enhance the concentration of the drug in a specific area, active and passive targeting can be utilized to modify the nanoparticles. The targeted nanoparticles release the drug when changes in the environment, like temperature, pH, enzymatic activity, etc., happen

MNPs are considered versatile and their unique features has made them a potential candidate for various applications in medicine and health. MNPs intended to use in biomedical application are synthesized using pure metals such as iron, copper, nickel, or alloys such as alnico, FeCo, or ferrites with high saturation magnetizations. Among the various types of MNPs, iron oxide-based nanoparticles are used in medicine due to their stability and biocompatibility. The nanoparticles that rapidly change their magnetic state upon applying an external magnetic field and showing super paramagnetism are only preferred in medicine. Biofunctionalization of MNPs are very crucial in determining their efficiency in vivo.[2] This chapter mainly focusses on the various aspects of utilizing magnetic nanomaterials in medicine. The first part of the chapter explains the areas where magnetic nanomaterials are applied and then we are focussing on magnetic ferrite nanoparticles with recent advances in its use in MRI, targeted drug delivery, and therapy.

2.2 BIOCOMPATIBLE MULTIFUNCTIONAL NANOMATERIALS

Biocompatible nanomaterials have an inevitable role in preventive and therapeutic healthcare, designed novel medical devices, prostheses, bioassays, tissue engineering, and controlled drug delivery and diagnostics. The nanocomposition shows different properties for the particles when compared to its macro composition. Functional nanomaterials have noticeable and potential applications in biomedical field because of their excellent biocompatibility.[1] Multifunctional nanomaterials consist of more than one therapeutic agent, biomolecules, or antibodies for cell recognition or targeting along with imaging signal amplification with the aid of contrast agents. Multifunctional nanoparticles provide interesting promise for a myriad of in vivo applications as in medical diagnostics, therapy, and targeted drug delivery. For these functions to be viable in the clinic, these nanoparticles should be biocompatible and non-toxic. Core-shell nanoparticles are a potent system where their properties can be modified by changing the core-to-shell ratio or by changing the constituting materials. Some examples of nanomaterials used as multifunctional systems are liposomes, quantum dots, dendrimers, carbon nanotubes, MNPs, gold nanoparticles, and polymeric micelles. To summarize, the major properties that multifunctional nanomaterials should have are

1. Stability
2. Biocompatibility
3. Specific targeting
4. Intracellular penetration
5. Imaging
6. Stimulus sensitive drug release.[2]

2.3 NANOMATERIALS IN TISSUE ENGINEERING

The scope of nanoparticles and necessity for control over nanomaterials have had relevant impact on regenerative medicine and tissue engineering. Nanoparticles with low toxicity, biocompatible property, detection tags, targeted delivery potential, and precise control over external magnetic fields have made it most promising therapeutic element in overcoming tissue engineering obstacles.[3] The three-dimensional nano scaffolds mimic natural extracellular matrix, with appropriate mechanical strength, ease of monitoring cellular activities, and delivering of bioactive agents. This is efficiently achieved through nanoscale approach rather than a macroscopic one. Nanoparticles can provide high control over scaffolding features such as regulating their mechanical strength and favoring controlled release of bioactive molecules. Metallic nanoparticles of gold or silver facilitate conjugation of various ligands including polypeptide sequences, antibodies, and proteins with various functional groups such as phosphines, amines, and thiols, as their strong affinity to gold/silver is clearly understood. The gold nanoparticles (Au) are therapeutically used in bone regeneration, as it supports cell adhesion and differentiation. The Au nanoparticles sprinkled in electrospun silk fibroin hastens wound healing because of the ability of scaffold to support the cell differentiation and repair. The metallic silver nanoparticles can be conjugated with natural polymers like chitosan to develop antibacterial scaffold. This scaffold has significant role in wound healing.[4]

2.4 NANOMATERIALS IN DRUG DELIVERY

Key challenges faced in delivering drugs to infected sites are biodistribution and intracellular trafficking. The major portion of the drug which we ingest orally or through intravenous gets distributed all over the body and their efficiency is thus altered due to non-targeted delivery. In order to compensate for this issue, we take in a higher dose of the drug than needed. These challenges could be answered with the development of nanotechnology where nanoparticles carry the drugs to destined sites and release them at those infected sites, thus improving the efficiency, safety, and distribution. The nanocarriers enhance the drug solubility and stability along with their transport across membranes. In this era of precision medicine, engineered nanoparticles which can overcome biological barriers to delivery drugs efficiently have a key role in determining the fate of precision therapies and thus their designing needs more importance.[5]

MNPs, mainly made of iron oxide, is less toxic and is widely employed in tissue engineering, nowadays. One of the exciting applications of tissue engineering is the neuroregeneration, for example, the superparamagnetic iron oxide—Au core-shell nanoparticles bound with nerve growth factor with low toxicity have been developed for neuron growth and differentiation.

Several polymeric nanomaterials like chitosan, gelatine, alginate, etc., are conjugated with modifiers like PEG to improve the cell-nanoparticle interaction. Presently, modern medicine is linked with tissue engineering to enhance tissue reconstruction, to avoid organ transplantation difficulties, and to highlight ease in treatment.[6]

2.4.1 CLASSES OF NANOMATERIALS

The nanomaterials for drug delivery can be broadly classified into inorganic, polymeric, and lipid-based nanoparticles. In this chapter, we will be detailing the inorganic nanoparticle-based delivery.

2.4.1.1 LIPID-BASED NANOPARTICLES

Lipid nanoparticles have been in clinical use as drug delivery system due to its non-toxicity, ease of preparation, biocompatibility, easy scale up, and targeted delivery. The most recent example of lipid-based drug delivery is mRNA COVID-19 vaccine developed by Pfizer and Moderna. The main classes of lipid nanoparticles are liposomes, solid lipid nanoparticles, and nanostructured lipid carriers where they are very effective in delivering nucleic acid apart from hydrophilic and hydrophobic drugs. One of the greatest breakthroughs in nanomedicine was the approval of the cancer drug Doxil, which is a PEGylated liposome with doxorubicin drug. The first siRNA loaded drug carrier which entered clinical use is patisiran, which is a lipid nanoparticle. Due to their unique features and efficiency in delivering drugs for various diseases, lipid nanoparticles are considered the next generation drug carriers in clinics. Advanced technologies like micro-fluidics, supercritical fluid technology, and coacervation methods are used to synthesise lipid nanoparticles where more studies are needed to understand the structure-function relationship of newly designed lipid nanoparticles to enter into targeted drug delivery systems.[7]

2.4.1.2 POLYMERIC NANOPARTICLES

Compared to other classes of nanoparticles, polymeric nanoparticles as drug delivery system is advantageous as (1) it shows good stability in biological fluids, (2) it can be functionalized with ease so as to control polymer degradation and leaking of the drug, and (3) the availability of various natural and synthetic polymers. Polymeric nanoparticles have been shown effective in cancer therapeutics where functionalized biodegradable polymeric systems are used. Poly lactide (PLA), poly glycolide (PGA), and poly lactide co glycolide (PLGA) are commonly used nanoparticles (NPs) for cancer drug encapsulation. The in vivo degradation of such polymeric NPs can be controlled according to the conditions which enhances the delivery and release of drug at target sites. Other degradable polymers used are poly hydroxy alkanoates (PHA); poly hydroxy butyrate (PHB); polycaprolactone (PCL); polyalkyl cyano acrylates (PACA); animal-based polymers like gelatin, hyaluronic acid, and albumin; marine-based polymers like chitosan, alginate, and carageenan; and plant polymers like starch, zein, and cellulose.[8]

2.4.1.3 INORGANIC NANOPARTICLES

Iron oxide, gold, silica, and quantum dots are some examples of inorganic nanoparticles used in drug delivery which have exceptional magnetic, electrical, and optical properties. Majority of the FDA-approved nanomedicines have iron oxide nanoparticles as the carrier. Examples of such nanomedicines are Injectafer, Feraheme, Venofer, and Ferrlecit used for treatment of iron deficiency in many diseases. Magnetite or maghemite containing iron oxide nanoparticles show superparamagnetism at particular sizes and is used in drug delivery and as contrast agents in imaging.[9]

Most of the inorganic nanoparticles release drugs in a controlled fashion and also protect the drugs from degradation and reduce the frequency and dose of drug intake. All these factors make inorganic nanoparticles a potent candidate for drug delivery.[10]

2.4.2 GOLD NANOPARTICLES

The most studied inorganic nanoparticles used are gold and silver oxide, carbon nanotubes, silica, and calcium phosphate. Gold and silver nanoparticles are the most widely studied and applied as they can be synthesized into a variety of shapes and numerous kinds of coatings can be attached for efficient drug delivery. Gold nanoparticles are a potent and very promising drug carrier system for cancer as they have high perfusion rate and thus the dose of drugs needed for treatment can be reduced. Gold NPs can also be used as therapeutic agent for destroying tumor cells when irradiated by laser by making it sensitive to stimuli like pH and temperature. Silica gold nanoshell with temperature-sensitive hydrogel can modulate drug delivery when irradiated by laser. Targeted drug delivery can be achieved by gold nanoparticles when coated with anti-EGFR—epidermal growth factor receptor—where the size range would be around 30 nm. Polymeric coatings like polyethylene glycol (PEG) on gold nanoparticles make them an efficient drug carrier and contrast agent with a size range significant for tumor uptake. Tumour necrosis factor (TNF) coupled with colloidal gold can be in the size range sufficient to pass through tumor blood vessels but not through normal blood vessels.[10] The major challenge in using metal NPs is the formation of protein-corona.

2.4.3 MESOPOROUS SILICA NANOPARTICLES

Mesoporous silica nanoparticles (MSN) due to their porous structure is chosen for efficient and targeted drug delivery carrier. They can incorporate large dose of drugs, can be functionalized with a wide variety of molecules due to its active surface, and release it in a controlled fashion thus widely used in many disease therapies like infections and cancer. Any type and quantity of drugs can be loaded since its pores are modifiable. They can also form hybrid systems with other NPs and polymers forming more efficient drug delivery system.[11] MSN can incorporate up to 600 mg drug per 1 g silica and their stability and biocompatibility is high when compared to other drug carriers. MSN can be synthesized by various methods like sol-gel process, evaporation-induced self-assembly, and synthesis with solution.[12]

2.4.4 IRON OXIDE NANOPARTICLES

Due to their unique features like stability, biocompatibility, and magnetic property, iron oxide nanoparticles are used clinically for various applications as contrast agents, targeted drug delivery system, and in cancer therapy. Their magnetic property and surface area varies with the size, range from 1 to 100 nm, which makes them protean, robust, and reactive. Magnetic iron oxide nanoparticles when shelled with materials like ceramics or polymers make them stable, allow conjugation with various proteins in the blood and tissues, and make them potent anticancer drug delivery systems.[13]

The main objectives to be achieved when using iron oxide NPs as drug delivery system are

1. Carry the pharmaceutical drug

2. Enhance the dissolution rate
3. Direct it to target tissue
4. Increase the local concentration in the affected area
5. Controlled release of therapeutic agent
6. Minimal systemic toxicity

Magnetic directed drug delivery is an effective treatment for cancer. The MNPs carrying the anticancer drug can be targeted to the tissue using image guided targeting by applying an external magnetic field. But the major limitations are the strength of external magnetic field applied and the tissue penetration. Tumors located deep into the tissues cannot be targeted in such cases. Most of the clinically approved MNPs are synthesized through coprecipitation synthetic method.[14] The major biomedical applications of MNPs are shown in Figure 2.1.

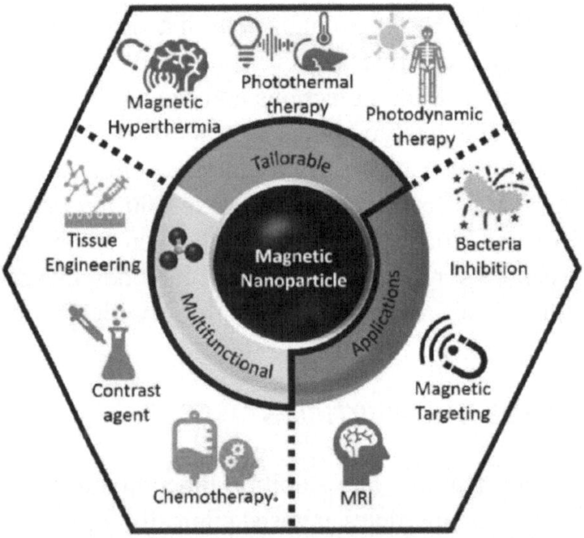

FIGURE 2.1 Various biomedical applications of magnetic nanoparticles. Reproduced with permission from Ref. [13].

2.4.5 MAGNETIC FERRITE NANOPARTICLES (MFNP)

Ferrites are ceramics made up of iron oxide with metallic elements. Ferrites can mediate external magnetic field which has made them useful in biomedicine. Ferrites are classified into spinel (cubic), orthogonal, garnet, and hexagonal according to their crystal structure.[15] In nature, we can find MNPs in certain magnetotactic bacteria where they produce MNPs and move according to the earth's magnetic field. MFNP are used in clinics as MRI contrast agents, drug carriers, hyperthermic treatment agents, and targeted drug delivery for cancer,[16] which are explained in detail in this chapter.

2.5 FERRITES IN MAGNETIC RESONANCE IMAGING

Magnetic resonance imaging (MRI) is regarded as the most powerful diagnostic tool in biomedicine due to its extreme imaging versatility, greater client acceptance, the capacity to evaluate anatomical and physiological parameters, and the capability to collect unique clinical information among different medical fields. The first clinical MRI studies appeared in 1979, and MRI scanners became commercial in 1980. Even for commercial clinical MR imaging systems, 1.5 T (tesla) magnets were available as early as 1985.[17,18] The advanced capacity of cellular MRI to spot, track, and evaluate cells in vivo over time will allow for the study of cellular responses related to disease processes and can be used as a screening tool for treatment monitoring.

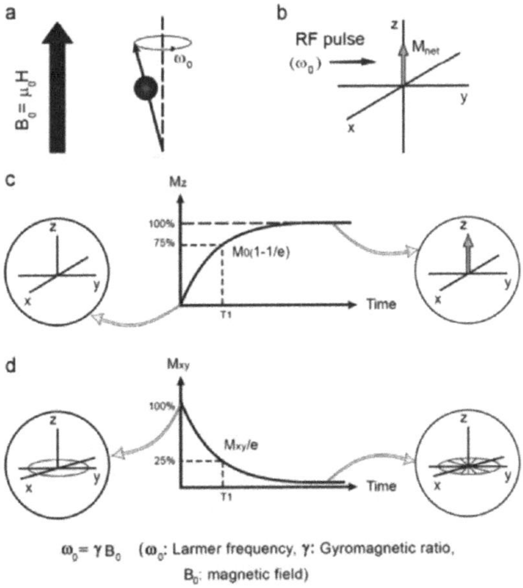

FIGURE 2.2 Magnetic resonance imaging: (a) spins align to the magnetic field and process under Larmor frequency and (b) after induction of RF pulse, magnetization of spins changes. Spins take relaxation process of: (c) T1 relaxation and (d) T2 relaxation. Reproduced with licence no 5382341151877 from Ref. [17].

The MRIs require strong magnetic field which aligns the protons (nuclei precision) in the body along it, at the same time a radio frequency current is given through the patient body to stimulate the protons and to strain against the external field. As the radiofrequency field is removed, the MRI sensors can monitor the energy released by protons when they realign. The relaxation times T_1 (longitudinal: spin-lattice relaxation) and T_2 (transverse: spin-spin relaxation), during which protons return to their initial condition, can be helpful in creating an MRI image after the application of RF pulses.[19] The energy released and time taken by the protons to realign, depends on the environment as well as the chemical nature of the molecules; this will be detected by the MRI sensors and distinguishes different tissues. Prior to the MRI, a patient may receive intravenous contrast agents to enhance the rate at which protons realign with the magnetic field. The image becomes brighter as the protons realign more quickly. Figure 2.2 represents the steps in MRI. A MRI contrast agent based on nanoparticles has three parts: a core that enhances signal quality[20]; a shell (water dispersing) which is biocompatible with environment; and a bioactive substance that targets.[21] Gadolinium (Gd) was initially utilized as an MRI contrast agent, but its toxicity restricted its employment in the medical industry. The low toxicity and other benefits of superparamagnetic ferrite nanoparticles led to its replacement.

Ferrites, with spinel structures, are homogenous material with chemical formula MFe_2O_4, where M=Mn, Fe, Co, Ni, Co, and Zn, has special attention among other ferrites. Figure 2.3 shows the characteristics of spinel ferrites (SF) which makes it applicable for the biomedical applications. Because of their enhanced relaxation time, the spinel ferrites have found use as contrast agents for MRI, in ferrofluid technology for MRI monitoring in hypothermia and in cancer detection via superconducting quantum interference device (SQUID) magnetometry.[22,23] The fundamental composition, the presence of minor impurities, and the characteristics of the microstructure, such as grain size, porosity, grain borders, and crystallographic texture, are the factors that ultimately influence the physical qualities of ferrite material. By properly tweaking their characteristics, magnetic NPs can be created for biomedical applications that have the requisite magnetic properties. Ferrites properties are tuned to suit the MRI application with size and shape, chemical composition and coating (ferrofluids).

Before designing a synthesis method for ferrite-based application, the right composition, size, and the dispersity of the NPs must be chosen based on the intrinsic magnetic properties, such as saturation magnetic moment, magnetic anisotropy, and magnetostriction.

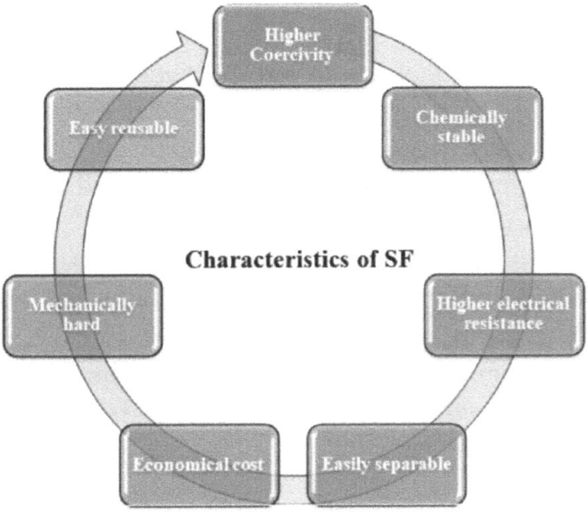

FIGURE 2.3 Characteristics of spinel ferrites made it applicable in biomedical applications.

Spinel ferrites with divalent metal ions as Co, Li, Mg, Mn, and Ni ferrite clearly indicate the variety of applications, based on the high saturation magnetization at room temperature. Comparison of various ferrite NPs has shown that manganese ferrite NPs are a great option for MRI contrast agents due to their higher magnetic susceptibility and magnetic saturation than conventional iron oxide and other spinel ferrite NPs. $MnFe_2O_4$ NPs had the strongest MR contrast effect, according to the magnetization data of all manufactured SFNPs. Investigators found that $MnFe_2O_4$ NPs, which have around 5 μB magnetic spins, exhibited the highest magnetic susceptibility of the synthetic spinel ferrite NPs.[24,25] The physical stability and characteristics of SFNPs, such as residual magnetizations, and coercivity M_s are largely determined by the size and form of the particles. Individual NPs' magnetic performance is dominated by finite-size and surface effects, which give rise to the special features of SFNPs.

2.5.1 DOPING IN SPINEL FERRITES

The chemical compositions, the kind of cations, and the distribution of the cations over tetrahedral and octahedral sites all directly affect the magnetic characteristics of SF NPs. The type of cations and their number distribution at the octahedral and tetrahedral sites of the spinel structure play a very important role in the overall change in magnetic characteristics. Because of the variations in composition or cation distribution over the tetrahedral and octahedral sites, SF NPs that are almost the same size might have distinct M_s values.[26] The M_s value of the nanoparticles has a direct relation on the contrast enhancement effects in MRI. By doping different elements with spinel ferrites enhances the magnetic properties and thereby increasing its efficiency in various application. For example, Zn^{2+} doped to $CoFe_2O_4$ causes magnetic tuning. Since the antiferromagnetic coupling interactions between tetrahedral and octahedral sites may be adjusted by utilizing Zn^{2+} dopants, the magnetism tuning of nanoparticles is successful. With the use of this magnetism engineering idea, high-performance MNPs could significantly advance present treatments, diagnostics, and other biomedical research projects like cell actuation.[27]

2.5.2 COATING IN SPINEL FERRITES (FERROFLUIDS)

It is crucial to coat the surface of SF cores with a layer of hydrophilic molecules or polymers for biomedical applications, to enhance better interactions with or binding to a variety of functional biological entities, including polypeptides, antibodies, enzymes, DNA, and RNA, among others, as well as better biocompatibility and colloid stability. Surface tension causes the uncoated SF NPs to be unstable and to clump together; consequently, surfactant polymers are added throughout the manufacturing process to reduce the surface tension. The polymers polyvinyl alcohol (PVA), PEG, polyaniline (PA), polystyrene (PS), and polyvinylpyrrolidone (PVP) are utilized to reduce the aggregation of inorganic SF NPs.[28] They are usually referred to as ferrofluids, which can enhance the reflectivity. These results are significant in a broader sense since they may be used to the efficient conjugation of NPs with a variety of macromolecules, giving them viable options for improving MRIs signal-to-noise ratio and producing more accurate anatomical pictures.

Till now we discussed how the SF can be utilized as contrast for MRI. There are some reasons to be noticed in MRI at the opposite end of the spectrum, even if the trends in MRI will probably continue toward larger fields. The SQUID-based ultra-low field MRI developments during the coming years will certainly be crucial. For instance, if the method can be demonstrated to add significant value to MRI in terms of cost and/or accuracy, this may prove motivation to advance the field. The spinel ferrite will play an important role in this advancement in the scientific field.

2.6 TARGETED DRUG DELIVERY USING MAGNETIC FERRITE NANOPARTICLES

Impressive advancements in the field of nanotechnology over the years has revolutionized treatment methods in medical field. Nowadays, scientists are able to design multi-functional nanoparticles (NPs), with properties like biocompatibility, non-toxicity, and target specificity. Among them iron-oxide-based NPs have been successfully employed as drug delivery vehicles for the treatment of medical complications like cancer. Attractive properties like ease of synthesis, size, and magnetic tenability of substituted iron-oxides nanoparticles support them in their utilization in drug delivery applications. MNPs consist of metal nanoparticles, iron oxide-based NPs, and alloys. Iron oxide NPs include magnetite (Fe_3O_4), maghemite (γ-Fe_2O_3), and ferrites (e.g., $CoFe_2O_4$, $MnFe_2O_4$, and $ZnFe_2O_4$). Magnetite NP with ferromagnetic properties and exhibiting reverse cubic spinel structure is the dominant variant among these three and possess different interesting properties due to the two valence states, Fe^{2+} and Fe^{3+}.[29] Ferrites are the combination of one or more different type of metal and iron oxide. The metals can be cobalt, manganese, nickel, and zinc. Ferrites are formed by the reaction between these metals and ferric oxide. Ferrite nanoparticles (FNPs) can be denoted as MFe_2O_4 where M stands for metals mentioned above. FNPs exhibit magnetic properties due to the existence of both octahedral and tetrahedral sites in their single-phase cubic spinel structure, with magnetic moments aligned in antiparallel direction. This results in the high magnetism of FNPs. Eventually the FNPs exhibit superior magnetic nature, large surface area, and biocompatibility which are suitable for biomedical applications.[30]

Ferrite materials hold high electrical resistivity, low dielectric losses and eddy current, adequate saturation magnetization, equitable permeability, acceptable permittivity, and drug/gene binding capacity which makes it practical for different applications across various fields. These properties of ferrite nanoparticles can be tailored by synthesis methods, dopant incorporation, inclusion of impurities, cationic distribution over available sites, adjusting particle size, distribution, shape and crystallinity, and also by changing sintering conditions.[31,32] When it comes to biomedical field nanocrystalline, ferrites have applications like MRI, hyperthermia, and magnetically guided drug delivery. Delivery of cargos to specific sites by magnetic ferrite nanoparticles (MFNPs) can be achieved through passive targeting, active targeting, and magnetic targeting. Figure 2.4 demonstrates the mechanism of two types of targeting in cells. Passive targeting is accomplished by taking advantage of the distinct physiology of tumor tissue, capability of MFNPs to escape phagocytosis by the reticuloendothelial system (RES), and by having a prolonged half-life in blood circulation. Passive targeting has the advantage of lowering the accumulation of MFNPs in organs like liver and spleen which lead to increased deposition of MFNPs in tumor, and thus a limited systemic toxicity.[33] By actively targeting MFNPs to receptors or other surface membrane proteins which is overexpressed on target cells, the uptake can be achieved. The surface of the MFNPs can be decorated with targeting ligands which will allow delivery of drug loaded nanoparticles to specific cells or even subcellular sites. This will help to reduce the systemic exposure of the cytotoxic drug. The ligand-receptor interactions facilitate nanoparticle internalization through receptor-mediated endocytosis. Magnetic targeting is another competing drug delivery technique which is very close to clinical execution. It is based on intra-arterial administration of drug-loaded MFNPs and enrichment by externally applied magnetic field that is generated by a strong electromagnet.[34]

2.6.1 PASSIVE TARGETING WITH MFNP

The major problem associated with bare MFNPs is that these NPs get cleared from blood circulation when passing through the biological defence system and vascular barriers. The accumulation of bare MFNPs in different organs like liver, spleen, or lymph nodes is also high. The factors like particle size and shape, surface chemistry, and functionalization have greater influence in prolonging half-life in blood circulation, promoting the EPR effect and enhancing colloidal stability of the MFNPs. Fe_3O_4 NPs with a size of ~10 nm are considered more for biomedical applications due to their excellent magnetic properties, low cost, and low toxicity.[35] These superparamagnetic NPs become magnetic in the presence of an external magnetic field and return to a non-magnetic state with the withdrawal of external field. This behavior of the MNPs helps to reduce the aggregation and thereby from a table suspension after removal of the external magnetic field which is very useful during the in vivo applications.[36] Similarly, the hydrodynamic diameter indicating the level of hydration around MFNPs also influence its injectability, biodistribution, and clearance of hyperthermia or MRI agents.[37] The properties of MNPs depend upon its shape also. In many studies, it is found that MNPs with a spherical morphology showed less aggregation due to its superparamagnetic nature.

The surface of the MFNPs can be tailored using organic biopolymers or inorganic layers to enhance the properties like biocompatibility, solubility, stability, and size distribution. The nature of the coating can be considered as the key factor in deciding the behavior of MFNPs in biological system which include cellular uptake and accumulation, circulation, and clearance from the body. Both natural and synthetic polymers have been used as the protective coating for coating MFNPs. Surface coating with PEG is commonly used to prevent the phagocytosis by RES and for having a prolonged circulation time in blood. Jun Xie and co-workers synthesized Mn-Zn ferrite MNCs coated with PEG-phospholipids (1,2-distearoyl-snglycero-3-phosphoethanolamine-N-[methoxy(PEG copolymers, DSPE-PEG2000) with an average diameter of 48.6 nm exhibiting high magnetism of 98 emu g^{-1}Fe with minimum recognition and phagocytosis of macrophages and an improved biocompatibility.[33] The intravenous administrated MNCs exhibited effective passive targeting ability and an increased accumulation in tumor tissues via the EPR effect. D. Ramnandan and co-workers functionalized $MgFe_2O_4$ with the polymers chitosan (CHI), PVA, and PEG for the delivery of anticancer drug doxorubicin (DOX). They were able to conclude that CHI-coated MFNPs showed more stability and drug loading efficiency than the other two combinations.[38] When a combination of PEG and chitosan which is palmitoylated was used to coat $Fe_{0.3}Mn_{0.7}Fe_2O$, the MNP showed an encapsulation efficiency of 92.8% for anticancer drug methotrexate (MTX) with the magnetization value of 19.7 emu/g and proved to be potential candidate for cancer therapy.[39] Chitosan composed of chitin contains reactive functional groups such as amines, which will improve biocompatibility and biodegradability of MFNPs and also a way to conjugate therapeutics, imaging agents, and targeting ligands.[40] PVA is a water-soluble synthetic organic polymer possessing emulsifying and adhesive properties which helps in reducing the aggregation of MNPs.

2.6.2 ACTIVE TARGETING WITH MFNP

Active targeting of the MFNPs is carried out through strong affinity of surface conjugated ligands with specific target molecules at the surface of the cells of interest. Antigen-antibody or receptor-ligand interactions are mostly preferred for the delivery of the components to cells and the specific binding will trigger cell internalization of NPs through ligand induced endocytosis. Common targeting ligands that are implemented to the design of MFNPs include small organic molecules, peptides, proteins, antibodies, and aptamers. The density of active targeting molecules on the surface, size, and shape of the MFNPs have great influence in the targeting capacity of the NPs.

Magnetite Fe_3O_4 is one of the most commonly used spinel ferrite nanoparticles with a unique inverse spinel structure. Huong et al. had synthesized folic acid conjugated MNP Fe_3O_4 based curcumin loading nanosystem to study the role of folic acid as a targeting factor.[41] They were able to find that the small sized folate attached Fe_3O_4-based curcumin loading nanosystem exhibited better targeting effect compared to the counterpart without folate. The cytotoxicity assay on MCF-7 cells of folic acid conjugated bovine serum albumin-calcium ferrite nanohybrid synthesized for targeted delivery of natural cytotoxic agents hesperidin and eugenol toward breast cancer cells showed the capability of targeting FA receptor positive breast cancer cells.[40] Gong et al. had prepared peptide CL 1555, conjugated manganese ferrite nanomicelles in which poly(ε-caprolactone)-block-PEG amphiphilic copolymer shell, and dozens of manganese ferrite ($MnFe_2O_4$) nanoparticle.[42] This nanoprobe can be used to specifically target tumor vascular endothelial cells with CD105 expression.

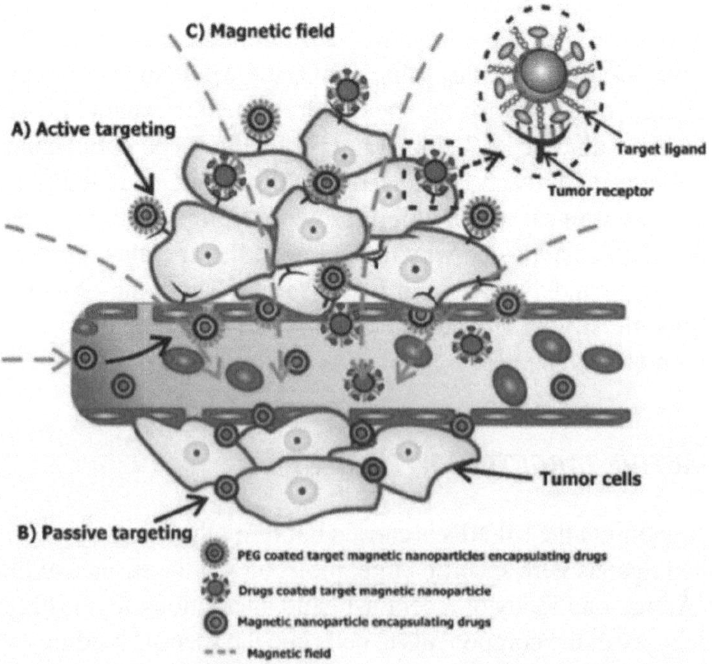

FIGURE 2.4 Schematic diagram showing the types of targeting by magnetic nanoparticles. Reproduced with permission from Ref. [66]. http://creativecommons.org/licenses/by/4.0/.

2.6.3 MAGNETIC TARGETING WITH MFNP

Owing to the economic aspect and functional ease, magnetic targeting system is widely explored in the field of drug delivery. In magnetic targeting, the drug-loaded MNPs are injected into the artery that delivers essentials to tumor and permanent dipole magnet is placed directly over the patient for the accumulation of MNPs in the tumor sites. The magnet is withdrawn after MNPs are leeched out into the surrounding tissues and the particles will get trapped in the tumour cells where the drug is released. Magnetic targeting of the MNP is possible due to their strong response to time-modulated magnetic fields that can penetrate human tissues without hindrance. Thus, by controlling the movement of MNPs carrying targeting drugs, such as anticancer or radionuclide atoms can be delivered to targeted area in the human body under the influence of external magnetic field. García-Jimeno et al. studied the effect of external magnetic field on the distribution of magnetoliposomes after injecting them intravenously and observed the accumulation of magnetoliposomes at the site of inflammation.[44] The studies conducted in mice found that the presence of magnetic field reduced the amount of iron in the liver, spleen, and plasma compared to that without any magnetic field exposure. Single nanoparticle consist of several thousands of crystal units within the structure and each crystal has a single magnetic domain which will align in the direction of the external magnetic field applied to nanoparticles creating a large magnetic momentum. This magnetic momentum is completely lost when the external magnetic field removed. This switch on mechanism of MFNPs in magnetic fields and the ability of functionalization of their surfaces at the molecular level make them good candidates for enhanced drug delivery.[34] Liu et al. developed an apparatus comprising of two magnets which are oppositely polarized that will allow deep penetration into tumours and they were able to demonstrate five-fold increase in the penetration and a three-fold increase in the accumulation of MNPs within solid tumours compared to EPR effect.[45]

2.7 NANOFERRITES IN THERAPY

The use of nanoscale materials in biological and health science has improved significantly in recent years. Magnetic ferrite nanoparticles have received a lot of attention since they could be used as efficient therapeutic tools in hyperthermia therapies; photodynamic therapy (PDT), and photothermal therapies (PTT). Due to its effectiveness in treating cancer, atherosclerotic plaques, and various bacterial infections, nanoferrite materials are becoming more prominent. The spinel ferrites and their nanocomposite forms are also prominent aspirant in this area.

2.7.1 MAGNETIC HYPERTHERMIA

Gilchrist and group first proposed this treatment in 1957, but the healing properties of heat from magnetic materials were known in China as early as 2000 BCE. The Yellow Emperor's Book of Internal Medicine, an old Chinese medical text, mentions the capability to heal cancer through the heat treatment. Faraday's electromagnetic theories paved the path for the pulsed electromagnetic field (PEMF) therapy, which attracted the interest of numerous scientists and led them to pursue the study of magnetic fields and magnetic materials. This opened the door for the use of magnetic fields in magnetic hyperthermia to cure cancer.[46]

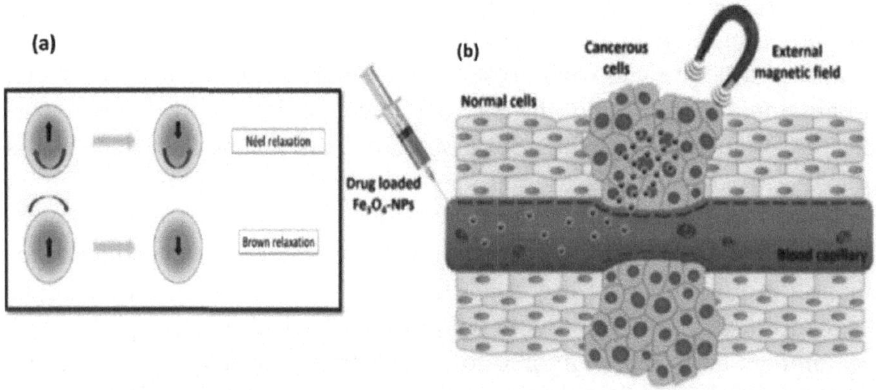

FIGURE 2.5 (a) Pictorial representation of Neel's and Brownian relaxation. Reproduced with licence 5381641445864 from Ref. [47]. (b) Systematic illustration of mechanism of hyperthermia therapy. Reproduced with permission from Ref. [48]. https://creativecommons.org/licenses/by-nc-nd/4.0/.

The fundamental idea underlying hyperthermia is when superparamagnetic nanoparticle-water dispersion is exposed to an alternating magnetic field, Néel and Brownian relaxation processes cause heat production, depicted in Figure 2.5(a). While Brownian relaxation involves the actual rotation of the particles, Neel relaxation involves the rotation of the magnetic moment.

The effectiveness of the magnetic anisotropy of the MNPs, the magnetic moment, and the viscosity of the particle dispersion all play a role in these relaxation processes. In a hyperthermia treatment, the tumor's temperature is increased beyond 44 °C for a sizable amount of time; the elevated temperature kills tumour cells but will not effect on normal cell (Fig. 2.5(b)). The specific absorption rate (SAR), which is defined as the amount of heat created by the unit weight of the MNPs during exposure per unit mass at a certain strength and frequency of the alternating magnetic field, is typically used to express the heating efficiency of nanoparticles.[49] A characteristic change in SAR values was caused by the inter-particle interactions caused by the effective magnetic anisotropy, which could be accelerated by altering the accumulation of magnetic nano-particles or regulating the core-shell structure of MNPs.

The biocompatibility of MNP with the highest SAR values may also provide difficulties. Previously reported articles still lack several of the elements that would make the therapy more effective when compared to other anticancer agents, such as critical size, enhanced biocompatibility, higher saturation magnetization, and lower toxicity. Following a comprehensive examination of MNPs previously published, iron oxide NPs produced using a one-pot synthesis approach exhibited high SAR values, showing 2452 W/g, and manganese ferrite and graphene oxide nanocomposites, getting 1541.6 W/g.[50,51] Cobalt ferrites, manganese ferrite, zinc ferrite, magnetite, and nickel ferrites are some of the spinel ferrites and their doped states that have been used in magnetic hyperthermia.

2.7.2 PHOTODYNAMIC THERAPY (PDT)

The demand for safer treatments has boosted interest in non-radiative therapies like PDT. A photosensitizer and light source are combined in PDT to create singlet oxygen ($^1O_2$) that has been proven to be harmful to cancer cells, making it particularly appealing. When one of an oxygen molecule's outermost electrons has its spin reversed, singlet oxygen, a highly reactive form of molecular oxygen, is created. Two unpaired electrons often exist individually in the outermost anti -bonded orbitals of oxygen triplet ground state. The coupling of the two electrons into one of the anti-bonding orbitals is what gives singlet oxygen its high reactivity. The spin of outer electron is flipped using an energy source to create the $^1O_2$. PDT is ideally suited for cancer therapy because it can selectively produce $^1O_2$ by activating a photosensitizer with a localized energy source. The efficiency of singlet oxygen synthesis plays a significant role in determining the efficacy of photodynamic treatment.[52] Fe ions released from magnetic ferrite nanoparticles can enhance PDT. A high concentration of hydrogen peroxide in the cancer cells act as catalyst for Fenton reaction thereby creating reactive oxygen species [ROS]. Low levels of ROS play important roles in supporting cellular survival processes like proliferation, but high concentrations have the ability to damage cellular constituents effectively.[53]

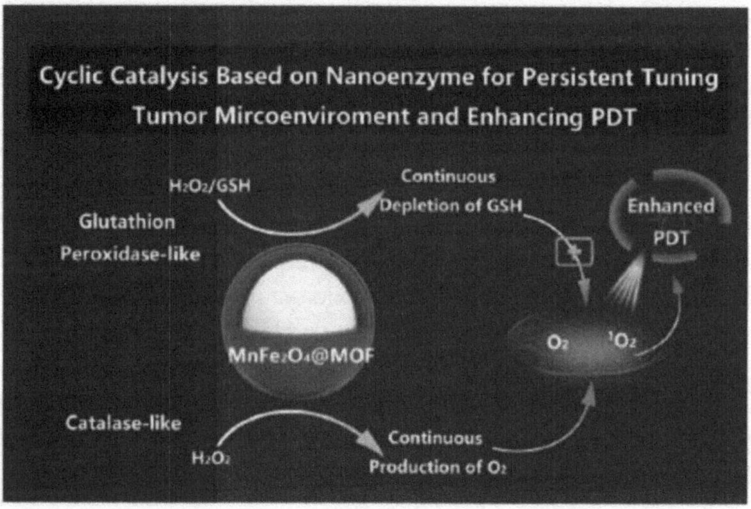

FIGURE 2.6 Schematic illustration of $MnFe_2O_4$@MOF providing O_2 and consuming GSH for enhancement of PDT. Reproduced with Licence 5384520554464 from Ref. [54].

Liposome formulation encapsulating MNPs within the liposome core and a photosensitizer m-THPC (clinical use under the name of Foscan) in the lipid bilayer showed complete tumor regression with dual therapy in which in vitro studies proved that PDT was more effective than magnetic hyperthermia.[55] Yin et al. developed a biocompatible nanoplatform ($MnFe_2O_4$@ metal-organic framework (MOF)) with porphyrin-based MOF as the photosensitizer and manganese ferrite nanoparticle ($MnFe_2O_4$) as the nanoenzyme. After incorporating into the tumor, nanoplatform was able to catalyze the production of oxygen from hydrogen peroxide through Fenton reaction. At the same time, $MnFe_2O_4$ consumed glutathione (GSH) in the presence of H_2O_2 which decreases the depletion of ROS upon laser irradiation during PDT and achieves better therapeutic efficacy in vitro and in vivo.[54] When cytotoxic and photo-cytotoxic responses of $Zn_{0.5}Co_{0.5}Fe_2O_4$ nanoparticles

(NPs) synthesized using sol-gel method were examined in a hepatocellular model under an optimum laser of wavelength of 630 nm for a short exposure time, a percentage loss of 60% was observed in cell viability of the cell model.[56]

2.7.3 PHOTOTHERMAL THERAPY

Conventional cancer therapy includes chemotherapy, radiotherapy, and surgical procedures which are not specific to cancer cells and thus is associated with many complications. In this scenario, photothermal therapy (PTT) can be a favorable technique for cancer therapy that has been explored intensively because of its minimal invasive nature and specific targeting of the tumor cells. The main two components of PTT are photothermal therapeutic agent (PTA) and near-infrared radiation (NIR) light to treat cancer cells. PTA is the component which has the ability to absorb in the NIR region of the spectrum. NIR light within the 650–900 nm range is preferred for cancer due to its ability to focus locally on a specific region, and its minimal skin and tissue absorption. These advantages will help the light to penetrate deep into the tissues. A second NIR (NIR-II) region of light within the 1000–1700 nm range is also used for PTT. But there are limitations to this region like the smaller number of stable nanoparticles that has the ability to absorb in this region and the light at this wavelength might get more absorbed due to water molecules which makes PTT with NIR-II laser less explored. A PTA is the critical component of this technique that absorb the NIR light and reaches to lowest excited singlet state and then reverts to the ground state by converting the absorbed light into heat through a non-radiative mechanism. This released heat creates hyperthermia, which functions to kill the cells where it is accumulated.[57]

Yang et al. synthesized polymeric nanosystem using manganese ferrite ($MnFe_2O_4$) and perfluorohexane (PFH) stabilized by synthetic polymer shells and conjugated to the surface of an antibody. The nanoparticles (NPs) were used as atherosclerotic plaque angiogenesis theranostic for multimodal imaging-guided photothermal therapy (PTT).[59] $MnFe_2O_4$ in this nanosystem act as the PTA which the convert NIR light to heat energy for cancer cell apoptosis. Superparamagnetic zinc ferrite spinel ($ZnFe_2O_4$)-reduced graphene oxide (rGO) nanostructures under near-infrared (NIR) irradiation resulted in an excellent destruction of the prostate cancer cells.[60] Liu et al. had prepared hemoglobin-functionalized copper ferrite nanoparticles

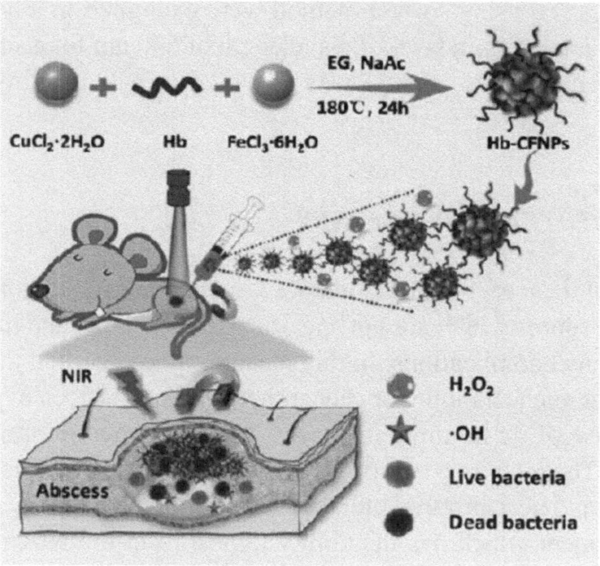

FIGURE 2.7 Schematic illustration of the synthesis of Hb-CFNPs and the corresponding antibacterial application through the combination of Fenton reaction and NIR irradiation. Reproduced with permission from Ref. [58].

(Hb-CFNPs) integrating nanocatalytic antibacterial therapy and photothermal therapy (PTT). With the assistance of NIR irradiation, hyperthermia generated by Hb-CFNPs could induce the death of the damaged bacteria like Gram-negative Escherichia coli and Gram-positive Staphylococcus aureus.[58]

2.8 MAGNETIC NANOPARTICLES IN BONE TISSUE ENGINEERING

Even though bone is a self-healing tissue, it needs the support of various factors to regenerate like external stimulus from chemical, biological factors, magnetic, electrical, and stress. The bone becomes mechanically conductive if there is a presence of paramagnetic iron. The MNPs which when engulfed by the osteocytes (bone cells) can act as a magnetic source which will enable the scaffold material/nanomaterial to engage in bone repair. When an external magnetic field is applied, the magnetic NPs become magnetized and the synergistic effect of the field and the particles will enhance the bone repair. Biomineralization process in the bone takes place in the presence of medium magnetic field. Spatial structure of cytoskeletal proteins is changed when magnetic field is applied in addition to the change in the crystallinity of the biomineralization products. The magnetic fields also promote the fusion of implant into bone cells and thus improve the density and calcium content of the bone.[61]

The magnetic field can promote attachment, proliferation, differentiation, and migration of osteoblast cells along with modulating calcium ions transport channel on the cell membrane. The magnetic field not only affects the early stage biomineralization behavior, but also controls the expression of osteocalcin genes and proteins in the later stage of bone repair.[62]

Recently, stem cell therapy is included in repairing large area bone defects. Although stem cell transplantation is successful in treating animal bone defect models, the clinical treatment results are not that satisfying. The maintenance of long-term and effective characteristics of stem cell is the key to determine the treatment success. It is necessary for bone repair and differentiation to maintain a stable cell phenotype and reduce cell necrosis at the defect site after transplantation in vivo. With the support of materials science and chemical biology, researchers tried to use iron oxide nanoparticles to control of stem cells in tissue engineering for many years. Iron oxide nanoparticles under the influence of external magnetic field affect cell adhesion, proliferation, distribution, and stem cell osteogenic differentiation.[63] The mechanism of Fe_3O_4/BSA particles uptake into stem cells, promoting the osteogenesis under the action of an external magnetic field is shown in Figure 2.8.

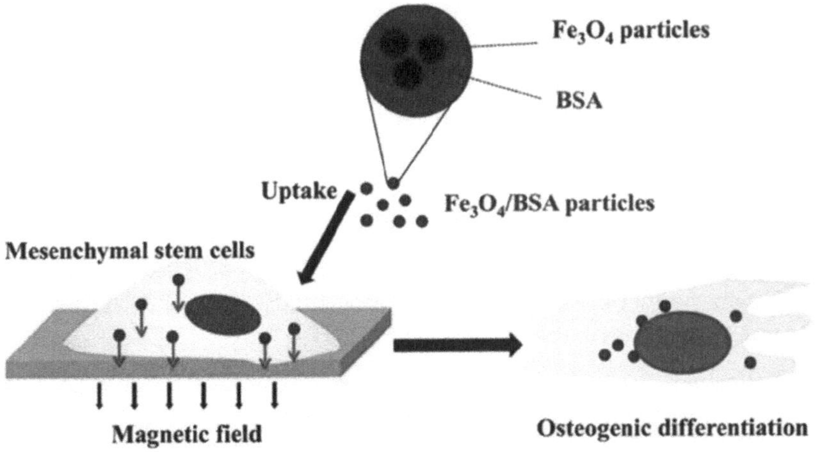

FIGURE 2.8 Fe_3O_4/BSA particles promote the osteogenesis of mesenchymal stem cells under an external magnetic field, while magnetic particles and an external magnetic field alone do not affect the differentiation of stem cells. Reproduced with licence No. 5381190665095 from Ref. [63].

2.9 CLINICAL TRANSLATION

Any kind of drug or therapy to enter into clinics undergo lengthy process of getting approval from regulatory agencies like FDA. In the nanomedicine market, as of now, MNPs are the leading category which has been successfully translated. There is a plethora of challenges both biological and chemical, which needs to be addressed before getting translated. The major challenges to be met are

1. Biodistribution of the nanocarrier
2. Imaging techniques to visualize the biodistribution and quantitatively estimate it. This can be achieved through radiolabeling.
3. Overcoming various biological barriers like endothelial and epithelial membranes in blood vessels, placenta and intestine, reticulo-endothelial system, cell/nucleus membrane, organelles, the immune reaction system, molecular pumps, interstitial pressure gradients, enzymatic degradation, and vessel fluid dynamics.
4. Understanding the basic science of nanoparticle transport by integrating basic biology, control over these biological barriers, properties such as size, shape, surface chemistry, and physical properties of the nanoparticles.
5. Developing mathematical and computer models to predict risk and benefit parameters.
6. Establishing standards, reference materials, and testing protocols for novel nanomaterials.
7. Establishing analytical toolkit, safety, biodistribution and toxicological data for manufacturing of nanomedicines.[64]

Some of the iron oxide MNP formulations used in clinics are listed below.

1. Gatromark^{TM}-Ferumoxil oral suspension—Silicone-coated superparamagnetic iron oxide nanoparticles formulation for—MRI contrast agent
2. Feridex I.V.—Ferumoxide injectable solution-superparamagnetic iron oxide (SPION) associated with dextran—MRI contrast agent
3. NanoTherm AS1—Ferrofluid-aminosilane coated SPION—Magnetic Hyperthermia Agent
4. Feraheme—Ferumoxytol Injection—Treatment of Iron deficiency.[43]

2.10 MAGNETIC NANOPARTICLES FOR PATHOGEN DETECTION DURING COVID-19

Magnetic nanosensing has taken a leap forward in the area of early disease detection. It has shown good results in sensing various pathogens and viruses even before the COVID-19 pandemic. Magnetic particle spectroscopy, nuclear magnetic imaging, and magneto resistance are some of the magneto nanosensing techniques available. During the pandemic, the whole world was demanding most sensitive, easy, low cost, and rapid SARS COV2 detection tests and magnetic nanosensors provided a good solution. Compared to other detection systems such as RT-PCR which needed trained personnel and long turnaround times, magnetic nanosensors needed no trained personnel, and a rapid detection system. These sensors could detect antigen and antibody apart from nucleic acids. Thus, they were considered versatile and advanced compared to other types of sensors. The magnetic nanosensors used for detecting SARS COV2 virus can also be extended to other viruses and bacteria if the reagents on MNPs are changed accordingly. Compared to optical sensors, magnetic nanosensors a great way ahead for detection in a controlled, highly sensitive, specific, and timely manner. Immunoassays incorporating MNPs provide rapid on-site diagnosis.[65]

2.11 CONCLUSION

Multifunctional nanomaterials have a wide range of successful applications in medicine. Ferrite nanoparticles are extensively used in MRI contrast agents and also in treating cancer with targeted delivery of drugs along with photo-sensitive treatments like PDT and PTT. All these innovations have made the clinicians move into new regimen with quick and effective results without much adverse effects. But still many of the drug delivery systems need to be revised and new advanced nanomaterial systems need to be translated into clinics. A precise and thorough understanding of the material property and their behaviour in vivo need to be practised by the researchers in order to move further in nanomedicine to combat novel pathogens and diseases.

KEYWORDS

- **multifunctional nanoparticles**
- **magnetic ferrite nanoparticles**
- **MRI**
- **cancer therapy**
- **magnetic hyperthermia**
- **targeted therapy**

REFERENCES

1. Seleci, M.; Seleci, D. A.; Joncyzk, R.; Stahl, F.; Blume, C.; Thomas, S. Smart Multifunctional Nanoparticles in Nanomedicine. *BioNanoMaterials* **2016,** *17*, 33–41.
2. Sanvicens, N.; Marco, M. P. Multifunctional Nanoparticles—Properties and Prospects for Their Use in Human Medicine. *Trends Biotechnol.* **2008,** *26* (8), 425–433.
3. Nanace, E.; Mckenna, M. Challenges and Barriers. Nanoparticles for Biomedical Applications: Fundamental Concepts. *Biol. Interact. Clin. App.* **2019,** 89–102.
4. Sakthi Devi, R.; Girigoswami, A.; Siddharth, M. et al. Applications of Gold and Silver Nanoparticles in Theranostics. *Appl. Biochem. Biotechnol.* **2022,** *194*, 4187–4219.
5. Yang, J. Jia, C.; Yang, J. Designing Nanoparticle-Based Drug Delivery Systems for Precision Medicine. *Int. J. Med. Sci.* **2021,** *18* (13), 2943–2949.
6. Yuan, M. M. S.; Wang, Y.; Xin, Y. X. Engineered Nanomedicine for Neuroregeneration: Light Emitting Diode-Mediated Superparamagnetic Iron Oxide-Gold Core-Shell Nanoparticles Functionalized by Nerve Growth Factor. *Nanomed.: Nanotechnol. Biol. Med.* **2019,** *21*, 102052.
7. Xu, L.; Wang, X.; Liu, Y.; Yang, G.; Falconer, R. J.; Zhao, C. Lipid Nanoparticles for Drug Delivery. *Adv. NanoBiomed Res.* **2022,** *2*.
8. Agnese, G.; Elena, G.; Eeda, V.; Massimo, F.; Stefania, B.; Vibhudutta, A. Biodegradable Polymeric Nanoparticles for Drug Delivery to Solid Tumors Front. *Pharmacol. Sec. Exp. Pharmacol. Drug Discov.* **2021,** *12*, 601626.
9. Mitchell, M. J.; Billingsley, M. M.; Haley, R. M. et al. Engineering Precision Nanoparticles for Drug Delivery. *Nat. Rev. Drug Discov.* **2021,** *20*, 101–124.
10. Paul, W.; Sharma, C. P. Inorganic Nanoparticles for Targeted Drug Delivery. *Biomater. Biointegr. Med. Implant Mater.* **2020,** *2*, 333–373.
11. Niculescu, V. C. Mesoporous Silica Nanoparticles for Bio-Applications. *Front. Mater. Sec. Polymeric Compos. Mater.* **2020,** *7* (36), 1–14.
12. Bharti, C.; Nagaich, U.; Pal, A. K.; Gulati, N. Mesoporous Silica Nanoparticles in Target Drug Delivery System: A Review. *Int. J. Pharm. Investig.* **2015,** *5* (3), 124–133. DOI: 10.4103/2230–973X.160844
13. Martins, P. M.; Lima, A. C.; Ribeiro, S.; Mendez, S. L.; Martins, P. Magnetic Nanoparticles for Biomedical Applications: From the Soul of the Earth to the Deep History of Ourselves. *ACS Appl. Bio Mater.* **2021,** *4* (8), 5839–5870.

14. Boubbou, K. E. Magnetic Iron Oxide Nanoparticles as Drug Carriers: Clinical Relevance. *Nanomedicine* **2018,** *13* (8), 953–971.
15. Jadhav, V. V.; Mane, R. S.; Shinde, P. V. Basics of Ferrites: Structures and Properties. In: *Bismuth-Ferrite-Based Electrochemical Supercapacitors: Springer Briefs in Materials*; Springer: Cham, 2020.
16. Wang, Y.; Miao, Y. et al. Engineering Ferrite Nanoparticles with Enhanced Magnetic Response for Advanced Biomedical Applications. *Mater. Today Adv.* **2020,** *8*.
17. Shokrollahi, H.; Khorramdin, A.; Isapour, G. Magnetic Resonance Imaging by Using Nano-Magnetic Particles. *J. Magn. Magn. Mater.* **2014,** *369*, 176–183.
18. Song, W.; Veiga, D. D.; Custódio, C. A.; Mano, J. F. Bioinspired Degradable Substrates with Extreme Wettability Properties. *Adv. Mater.* **2009,** *21* (18), 1830–1834.
19. Amiri, M.; Salavati-Niasari, M.; Akbari, A. Magnetic Nanocarriers: Evolution of Spinel Ferrites for Medical Applications. *Adv. Colloid Interface Sci.* **2019,** *265*, 29–44.
20. Shokrollahi, H. Magnetic Properties and Densification of Manganese–Zinc Soft Ferrites (Mn1-xZnxFe2O4) Doped with Low Melting Point Oxides. *J. Magn. Magn. Mater.* **2008,** *320* (3–4), 463–474.
21. Xiao, Y. D.; Paudel, R.; Liu, J.; Ma, C.; Zhang, Z. S.; Zhou, S. K. MRI Contrast Agents: Classification and Application. *Int. J. Mol. Med.* **2016,** *38* (5), 1319–1326.
22. Smolensky, E.; Park, H. Y. E. et al. Scaling Laws at the Nanosize: The Effect of Particle Size and Shape on the Magnetism and Relaxivity of Iron Oxide Nanoparticle Contrast Agents. *J. Mater. Chem.* **2013,** *1*, 2818–2828.
23. Shultz, M. D.; Calvin, S.; Fatouros, P. P.; Morrison, S. A.; Carpenter, E. E. Enhanced Ferrite Nanoparticles as MRI Contrast Agents. *J. Magn. Magn. Mater.* **2007,** *311* (1), 464–468.
24. Lee, J.-H.; Huh, Y.-M.; Jun, Y.-w.; Seo, J.-w.; Jang, J.-t.; Song, H.-T.; Kim, S.; Cho, E.-J.; Yoon, H.-G.; Suh, J.-S. Artificially Engineered Magnetic Nanoparticles for Ultra-Sensitive Molecular Imaging. *Nat. Med.* **2007,** *13* (1), 95–99.
25. Faraji, S.; Dini, G.; Zahraei, M.; Polyethylene Glycol-Coated Manganese-Ferrite Nanoparticles as Contrast Agents for Magnetic Resonance Imaging. *J. Magn. Magn. Mater.* **2019,** *475*, 137–145.
26. Lee, N.; Yoo, D.; Ling, D.; Cho, M. H.; Hyeon, T.; Cheon, J. Iron Oxide Based Nanoparticles for Multimodal Imaging and Magnetoresponsive Therapy. *Chem. Rev.* **2015,** *115* (19), 10637–10689.
27. Jang, J. T.; Nah, H.; Lee, J. H.; Moon, S. H.; Kim, M. G.; Cheon, J. Critical Enhancements of MRI Contrast and Hyperthermic Effects by Dopant-Controlled Magnetic Nanoparticles. *Angew. Chem. Int. Ed.* **2009,** *48* (7), 1234–1238.
28. Lakić, M.; Andjelković, L.; Šuljagić, M.; Vulić, P.; Perić, M.; Iskrenović, P.; Krstić, I.; Kuraica, M. M.; Nikolić, A. S. Optical Evidence of Magnetic Field-Induced Ferrofluid Aggregation: Comparison of Cobalt Ferrite, Magnetite, and Magnesium Ferrite. *Opt. Mater.* **2019,** *91*, 279–285.
29. Litewka, J. D.; Łazarczyk, A.; Hałubiec, P.; Szafrański, O.; Karnas, K.; Karewicz, A. Superparamagnetic Iron Oxide Nanoparticles-Current and Prospective Medical Applications. *Materials* **2019,** *12* (4).
30. Veiseh, O.; Gunn, J. W.; Zhang, M. Design and Fabrication of Magnetic Nanoparticles for Targeted Drug Delivery and Imaging. *Adv. Drug Deliv. Rev.* **2010,** *62* (3), 284–304.
31. Gama, L.; Diniz, A. P.; Costa, A. C.F. M.; Rezende, S. M.; Azevedo, A.; Cornejo, D. R. Magnetic Properties of Nanocrystalline Ni-Zn Ferrites Doped with Samarium. *Phys. B Condens. Matter.* **2006,** *384*, 97–99.

32. Gadkari, A. B.; Shinde, T. J.; Vasambekar, P. N. Structural Analysis of Y3+-Doped Mg-Cd Ferrites Prepared by Oxalate Co-precipitation Method. *Mater. Chem. Phys.* **2009,** *114*, 505–510.
33. Xie, J. et al. Biomaterials High-performance PEGylated MnZn Ferrite Nanocrystals as a Passive-Targeted Agent for Magnetically Induced Cancer Theranostics. Biomaterials **2014,** *35* (33), 9126–9136.
34. Amiri, M.; Salavati-niasari, M.; Akbari, A. Magnetic Nanocarriers: Evolution of Spinel Ferrites for Medical Applications. *Adv. Colloid Interface Sci.* **2019,** *265*, 29–44.
35. Ha, Y. et al. Recent Advances Incorporating Superparamagnetic Nanoparticles into Immunoassays. *ACS Appl. Nano Mater.* **2018,** *1* (2), 512–521.
36. Vallabani, N. V. S.; Singh, S. Recent Advances and Future Prospects of Iron Oxide Nanoparticles in Biomedicine and Diagnostics. *Biotech.* **2018,** *8* (6), 279.
37. Hoque, S. M.; Islam, M. K.; Hoq, A.; Haque, M. M. Comparative Study of Specific Loss Power and Transverse Relaxivity of Spinel Ferrite Nanoensembles Coated with Chitosan and Polyethylene Glycol. *Front. Nanotechnol.* **2021,** *3*, 1–13.
38. Ramnandan, D.; Mokhosi, S.; Daniels, A.; Singh, M. Chitosan, Polyethylene Glycol and Polyvinyl Alcohol Modified $MgFe_2O_4$ Ferrite Magnetic Nanoparticles in Doxorubicin Delivery: A Comparative Study In Vitro. *Molecules* **2021,** *26* (13), 3893.
39. Karimi, Z. et al. Pegylated and Amphiphilic Chitosan Coated Manganese Ferrite Nanoparticles for pH-Sensitive Delivery of Methotrexate: Synthesis and Characterization. *Mater. Sci. Eng C* **2017,** *71*, 504–511.
40. Maheswari, U. P.; Muthappa, R.; Purushothaman, B. K.; Begum, M. S. K. M. Evaluation of Folic Acid Functionalized BSA-$CaFe_2O_4$ Nanohybrid Carrier for the Controlled Delivery of Natural Cytotoxic Drugs Hesperidin and Eugenol. *J. Drug Deliv. Sci. Technol.* **2020,** *61*, 102105.
41. Huong, L. T. T.; Nam, N. H. et al. Folate Attached, Curcumin Loaded Fe_3O_4 Nanoparticles: A Novel Multifunctional Drug Delivery System for Cancer Treatment. *Mater. Chem. Phys.* **2016,** *172*, 98–104.
42. Gong, M. et al.; Targeting T1 and T2 Dual Modality Enhanced Magnetic Resonance Imaging of Tumor Vascular Endothelial Cells Based on Peptides-Conjugated Manganese Ferrite Nanomicelles. *Int. J. Nanomed.* **2016,** *11*, 4051–4063.
43. Boubbou, K. E. Magnetic Iron Oxide Nanoparticles as Drug Carriers: Clinical Relevance. *Nanomedicine* **2018,** *13* (8), 953–971.
44. García-Jimeno, S.; Escribano, E.; Queralt, J. et al. External Magnetic Field-Induced Selective Biodistribution of Magnetoliposomes in Mice. *Nanoscale Res. Lett.* **2012,** *7*, 452.
45. Liu, J. F. et al. Use of Oppositely Polarized External Magnets To Improve the Accumulation and Penetration of Magnetic Nanocarriers into Solid Tumors. *ACS Nano.* **2020,** *14* (1), 142–152.
46. Narayanaswamy, V.; Sambasivam, S.; Saj, A.; Alaabed, S.; Issa, B.; Al-Omari, I. A.; Obaidat, I. M. Role of Magnetite Nanoparticles Size and Concentration on Hyperthermia Under Various Field Frequencies and Strengths. *Molecules* **2021,** *26* (4), 796.
47. Da Silva, F.; De Campos, M. Study of Heating Curves Generated by Magnetite Nanoparticles Aiming Application in Magnetic Hyperthermia. *Braz. J. Chem. Eng.* **2020,** *37* (3), 543–553.
48. Yew, Y. P.; Shameli, K.; Miyake, M.; Khairudin, N. B. B. A.; Mohamad, S. E. B.; Naiki, T.; Lee, K. X. Green Biosynthesis of Superparamagnetic Magnetite Fe_3O_4 Nanoparticles

and Biomedical Applications in Targeted Anticancer Drug Delivery System: A Review. *Arab. J. Chem.* **2020,** *13* (1), 2287–2308.
49. Nandhini, G.; Shobana, M. Role of Ferrite Nanoparticles in Hyperthermia Applications. *J. Magn. Magn. Mater.* **2022,** *552*, 169236.
50. Peng, E.; Choo, E. S. G.; Chandrasekharan, P.; Yang, C. T.; Ding, J.; Chuang, K. H.; Xue, J. M. Synthesis of Manganese Ferrite/Graphene Oxide Nanocomposites for Biomedical Applications. *Small* **2012,** *8* (23), 3620–3630.
51. Guardia, P.; Di Corato, R.; Lartigue, L.; Wilhelm, C.; Espinosa, A.; Garcia-Hernandez, M.; Gazeau, F.; Manna, L.; Pellegrino, T. Water-Soluble Iron Oxide Nanocubes with High Values of Specific Absorption Rate for Cancer Cell Hyperthermia Treatment. *ACS Nano* **2012,** *6* (4), 3080–3091.
52. Chen, W.; Zhang, J. Using Nanoparticles to Enable Simultaneous Radiation and Photodynamic Therapies for Cancer Treatment. *J. Nanosci. Nanotechnol.* **2006,** *6* (4), 1159–1166.
53. Ranji-Burachaloo, H.; Gurr, P. A.; Dunstan, D. E.; Qiao, G. G. Cancer Treatment Through Nanoparticle-Facilitated Fenton Reaction. *ACS Nano* **2018,** *12* (12), 11819–11837.
54. Yin, S. Y.; Song, G.; Yang, Y.; Zhao, Y.; Wang, P.; Zhu, L. M.; Yin, X.; Zhang, X. B. Persistent Regulation of Tumor Microenvironment via Circulating Catalysis of MnFe2O4@ Metal–Organic Frameworks for Enhanced Photodynamic Therapy. *Adv. Funct. Mater.* **2019,** *29* (25), 1901417.
55. Di Corato, R.; Béalle, G.; Kolosnjaj-Tabi, J.; Espinosa, A.; Clement, O.; Silva, A. K.; Menager, C.; Wilhelm, C. Combining Magnetic Hyperthermia and Photodynamic Therapy for Tumor Ablation with Photoresponsive Magnetic Liposomes. *ACS Nano* **2015,** *9* (3), 2904–2916.
56. Iqbal, S.; Fakhar-e-Alam, M.; Atif, M.; Amin, N.; Alimgeer, K.; Ali, A.; Hanif, A.; Farooq, W. A. Structural, Morphological, Antimicrobial, and In Vitro Photodynamic Therapeutic Assessments of Novel Zn+ 2-Substituted Cobalt Ferrite Nanoparticles. *Results Phys.* **2019,** *15*, 102529.
57. Kumar, A. V. P.; Dubey, S. K.; Tiwari, S.; Puri, A.; Hejmady, S.; Gorain, B.; Kesharwani, P. Recent Advances in Nanoparticles Mediated Photothermal Therapy Induced Tumor Regression. *Int. J. Pharm.* **2021,** *606*, 120848.
58. Liu, Y.; Guo, Z.; Li, F.; Xiao, Y.; Zhang, Y.; Bu, T.; Jia, P.; Zhe, T.; Wang, L. Multifunctional Magnetic Copper Ferrite Nanoparticles as Fenton-Like Reaction and Near-Infrared Photothermal Agents for Synergetic Antibacterial Therapy. *ACS Appl. Mater. Interfaces* **2019,** *11* (35), 31649–31660.
59. Yang, Z.; Yao, J.; Wang, J.; Zhang, C.; Cao, Y.; Hao, L.; Yang, C.; Wu, C.; Zhang, J.; Wang, Z. Ferrite-Encapsulated Nanoparticles with Stable Photothermal Performance for Multimodal Imaging-Guided Atherosclerotic Plaque Neovascularization Therapy. *Biomater. Sci.* **2021,** *9* (16), 5652–5664.
60. Akhavan, O.; Meidanchi, A.; Ghaderi, E.; Khoei, S. Zinc Ferrite Spinel-Graphene in Magneto-Photothermal Therapy of Cancer. *J. Mater. Chem. B* **2014,** *2* (21), 3306–3314.
61. Fan, D. et al. Recent Advances of Magnetic Nanomaterials in Bone Tissue Repair. *Front. Chem.* **2020,** *8*, 745.
62. Liu, Y.; Cui, T.; Wu, T.; Li, Y.; Tong, G. Excellent Microwave-Absorbing Properties of Elliptical Fe_2O_2 Nanorings Made by a Rapid Microwave-Assisted Hydrothermal Approach. *Nanotechnology* **2016,** *27*, 165707.

63. Jiang, P.; Zhang. Y.; Zhu.; C.; Zhang. W.; Mao, Z.; Gao, C. Fe3O4/BSA Particles Induce Osteogenic Differentiation of Mesenchymal Stem Cells Under Static Magnetic Field. Acta Biomaterialia. **2016,** *46*, 141–150.
64. Sanhai, W.; Sakamoto, J.; Canady, R. et al. Seven Challenges for Nanomedicine. *Nat. Nanotech.* **2008,** *3*, 242–244.
65. Wu, K. et al. Magnetic-Nanosensor-Based Virus and Pathogen Detection Strategies before and during COVID-19. *ACS Appl. Nano Mater.* **2020,** *3* (10), 9560–9580
66. Belyanina, I.; Kolovskaya, O.; Zamay, S.; Gargaun, A.; Zamay, T.; Kichkailo, A. Targeted Magnetic Nanotheranostics of Cancer. *Molecules* **2017,** *22*, 975.

CHAPTER 3

Study on Magnetic and Magnetoelectric Properties of $BiFeO_3$ Multiferroic Systems

JINCEMON CYRIAC[1], SUNNY MATHEW[1], SAJI AUGUSTINE[2], SARITHA K. JOSE[2], L. S. ARCHANA[3], S. SASHANK[1], JISTY ANN JIJI[1], and NANDAKUMAR KALARIKKAL[4]

[1]*Department of Physics, Marian College Kuttikkanam (Autonomous), Kuttikkanam, Peerumade, Idukki, Kerala, India*

[2]*Department of Physics, Deva Matha College, Kuravilangad, Kottayam, Kerala, India*

[3]*Department of Physics, Government College for Women, University of Kerala, Thiruvananthapuram, Kerala, India*

[4]*International and Inter University Centre for Nanoscience and Nanotechnology, School of Pure and Applied Physics, Mahatma Gandhi University, Kottayam, Kerala, India*

ABSTRACT

Due to the simultaneous existence of two or more characteristics, such as ferroelasticity, ferroelectricity, and ferromagnetism, multiferroic nanomaterials are of tremendous interest. This can guarantee wide applications in eco-friendly, low-power, multifunctional products. At room temperature, bismuth ferrite ($BiFeO_3$, BFO) shows both ferroelectricity and ferromagnetism. Because of this, it has great importance in multiferroic systems. In this review, we discuss the structure, properties, and effects of doping of this system to make progress in its properties in different types of BFO, and

Physics and Mechanics of New Materials: Synthesis, Processing, and Emerging Applications. Devrim Balkose, Ana Cristina Faria Ribeiro, Nandakumar Kalarikkal, Ann Rose Abraham, A. K. Haghi (Eds.)

composites. We anticipate that this review will serve as an update and inspire additional researchers to advance the creation of BFO nanomaterials in the future.

3.1 INTRODUCTION

Multiferroic nanomaterials are substances with multiple properties, including ferroelectricity, ferromagnetism, ferroelasticity, etc. In a single phase, these materials have numerous applications in the field of spintronic devices, transducers, actuators, and sensors.[1] The magnetoelectric (ME) effect is one of the attractive phenomena that can occur when the magnetic and ferroelectric degrees of freedom in multiferroics are coupled.[2] An electric field can be used to control the effect, and the opposite is also true.

Bismuth ferrite ($BiFeO_3$; abbreviated as BFO) is a multiferroic system, which has more than one type of parameter such as ferromagnetism, ferroelectricity, and ferroelasticity.[3] The ferroelectric transition temperature (T_N) of BFO is 1103 K, and the G-type antiferromagnetic transition temperature (T_N) is 643 K.[4] BFO-based nanomaterials are classified into versatile types like BFO ceramics, thin films and nanostructures.[5] In thin films, high remnant polarization, close to the theoretical value (P_r) was produced due to the influence of the substrate, lattice strain, or buffer layer.[6] It has a rhombohedrally distorted perovskite (ABO_3) crystal structure (space group R_3c) with lattice parameters a=b=5.58 Å and c=13.87 Å and rhombohedral angle 89.45° at room temperature.[7]

It has a drawback that it suffers from high leakage current. The usefulness of undoped BFO is limited in light of the spiral-modulated spin structure (SMSS). The viable method to overcome the de-merits is the partial substitution of anions by those of foreign components.[9] By replacing Bi^{3+} with ions including Gd^{3+}, Sc^{3+}, La^{3+}, Eu^{3+}, Nd^{3+}, etc. or by simultaneously replacing Bi and Fe in $BiFeO_3$ with lanthanide and transition element ions, the parent material's multiferroicity can be increased.[10]

Defects are inherent in solid systems due to a variety of physical and thermodynamic conditions, nonstoichiometric composition, reagent purity and synthesis temperature. Factors like ionic radii mismatch, charge imbalance, and lattice parameter radiations are due to the flaws such as vacancies and interstitial defects.

BFO nanostructure exhibits highly different properties compared with BFO ceramics and thin films. Due to the vast array of morphologies and enormous surface area, the nanostructure exhibits enhanced photocatalytic activity and magnetization under visible light. BFO-based nanomaterials have a strong photovoltaic effect owing to the low band gap. At room temperature, the single phase BFO owns both ferromagnetic and ferroelectric nature. The ceramics, thin films and nanostructure are different forms of nanomaterials based on BFO.[11]

Secondary phases that can appear during the sintering process is because of the labile nature of Bi^{3+} and accumulation of oxygen vacancies. The change from Fe^{3+} to Fe^{2+} results in the generation of oxygen vacancies, which generate a significant leakage current, making it difficult to identify a saturated hysteresis loop. Bi^{3+} and Fe^{3+} are accountable for polarization and magnetization, respectively. To improve the physical properties, techniques such as ion substitution and ABO_3-containing alloys are used.[12]

The thin films can be produced with less effort than the ceramic with superior ferroelectric performance and a high degree of precision.[13] A strong correlation exists between epitaxial strain and film thickness on different substrates for epitaxial BFO films. Doping extra elements can have an impact on the structural features of thin films to produce solid solutions.

Due to the various photocatalytic behaviors and magnetic properties, numerous BFO nanostructures have been reported by a variety of techniques.[14] Their properties are highly dependent on the nanostructure's morphologies and dimensions. As particle size decreases, magnetic response increases and its nanowires exhibit higher magnetization than nanorods and nanotubes.[15] The shape and size of photocatalytic particles are determined by the bandgap and specific surface area.

3.2 STRUCTURE OF BFO

Two perovskite cells are connected in this structure in the [111] direction to form a rhombohedral cell, wherein the centers of the FeO_6 octahedra and the angular positions are occupied by Fe^{3+} and Bi^{3+}, respectively.[16,17] The displacement of the Bi^{3+} from centrosymmetric orientations as a result of the hybridization between the Bi^{3+} lone pair (6*s* orbital) and O^{2-} results in a ferroelectric polarization in the [111] direction (2*p* orbital).[6] The period of a cycloid is around 62 nm and is created by spatially modulating the spin direction of an antiferromagnetic plane.[18]

The likelihood of the dopants replacing Bi^{3+} in the A-site increases when their ionic size and electronegativity are comparable to those of Bi^{3+}.[19] Due to the fact that the majority of A-site ions have smaller radii than Bi^{3+}, they are unable to completely fill the space between the backbones of FeO_6, resulting in greater buckling between the oxygen octahedra. As a result, substitutions cause modifications to phase structure. By modifying the average ionic radius of the A-site cation, the structural transition with a double polarization hysteresis loop may be produced everywhere.[20] When 10% Ca is doped, a structural change from rhombohedral to triclinic was observed.[21] When the dopant's ionic size and electronegativity are comparable to those of Fe^{3+}, it is more commonly replaced by the dopants in the B-site; for example, by doping the B-site with Mn or Ti. When 20% Mn^{4+} is substituted, a more stable rhombohedral structure is seen in comparison to $BiFeO_3$, however, a mixed orthorhombic phase and monoclinic phase are seen when greater than 40% Fe^{3+} is substituted.[22] The structural transition from the rhombohedral phase to the orthorhombic phase takes place as a result of the distortion caused by the addition of Mn^{2+} to the crystals. When Ti^{4+} replaces greater than 33% of Fe^{3+}, the phase changes from rhombic to orthorhombic.[23]

BFO's structure can be significantly changed, and its characteristics can be effectively enhanced, by the addition of ABO_3 to form solid solutions.[24] In BFO-BTO, the Ba acts as a major blocking ion to prevent Bi displacements. Ti and Fe compete on the B-site A continuous solid solution with various crystal forms is created as a result of the competition on each site.[25] When BFO content is between 7.5 and 67%, a cubic structure is seen. When its content is beyond 67%, the structure is rhombohedral. When it is below 7.5%, it is tetragonal.[25] In BFO–BTO ceramics Lee et al. reported a morphotropic phase boundary (MPB).[13] The rhombohedral phase and tetragonal phase are believed to exist together in the MPB for 0.67BFO–0.33BTO. $BiFe_{1-x}Co_xO_3$'s crystal structure was thoroughly examined. It was found that when x is less than 0.2, it preserves the rhombohedral R_3c structure, whereas when x is greater than 0.4 results in tetragonal formations.[26] At around x is equal to 0.3, MPB, which is referred to as a monoclinic phase, is seen. While the lattice parameter c remained essentially constant, the lattice parameters $a/2$ and $b/2$ decreased linearly as x increased. With x, the monoclinic angle is also decreased. Although being grown on single-crystal substrates to produce the finest thin films, the BFO film's organization is directly tied to the epitaxial strain caused by diverse reactants. Zeches et al. revealed that epitaxial strain can be employed to maintain the structure and put it on a morphotropic phase. For tensile strain, its films have an R-like phase, however regarding heavy compressive loads, the films typically exhibit a great tetragonality (T-like). Sando et al. thoroughly encapsulated that the evolution of its crystal structure changes under different strains, from the R-like phase to the T-like phase.[27] When the tensile strain exceeds 2%, an orthorhombic phase stabilizes.

3.3 PROPERTIES OF BFO

3.3.1 MAGNETIC PROPERTIES

BFO exhibits a G-type antiferromagnetic order as a consequence of the Fe^{3+} electron spins in the rhombohedrally deformed perovskite. The ferromagnetic alignment of the Fe^{3+} spins in the (111) plane is antiparallel to each neighboring spin. Its magnetic order is distinguished by the ability of the G-type antiferromagnetic spin-lattice to generate a spin cycloid with a long periodicity of 62–64 nm along the [111] direction. Below 673 K (T_N), it exhibits poor magnetism due to the inclination of the G-type antiferromagnetic structure and the presence of an asymmetric spin cycloid structure.

When the spin cycloid's period length is less than 62 nm, the magnetic response rises dramatically.[28] Studies have shown that the particle sizes affect the theoretical values of T_C and T_N.[29] The thin films are primarily influenced by epitaxial strain, which results in additional anisotropy and impacts magnetic properties. The films have a thickness of approximately 70 nm and were developed on multiple substrates to eliminate thickness-related effects. The increasing strain suppresses the magnon phase with low energy. Chen et al.[30] demonstrated that Strain could be used to adjust the orientation of the antiferromagnetic spin with minimal effect on the polarization structure. When the strain is free, the antiferromagnetic spin axis (L) of the film degenerates in the plane that is perpendicular to the polarization direction (along [111]). Under compressive tension, L continuously points in the plane [110] as it spins. Under tensile strain, L changes gradually as the axis converges to a point roughly corresponding to [112]. Further tensile strain causes L to rotate in the direction of the out-of-plane [110], but only by approximately 15° with respect to111] within the plane (110). The consequence of this is that the polarization direction and antiferromagnetic axis are no longer parallel. Due to the Dzyaloshinskii-Moriya interaction, which is dominant at low strain, the antiferromagnetic axis and the polarization direction are perpendicular. However, at high strain, the single-ion anisotropy increases. Accordingly, these studies suggested that strain could be used to modify the bias in exchange, magnetic polarity, and the response of enormous magnetoresistive spin valves, which could be advantageous for the creation of spintronic devices.

In addition, doping substances, especially those that replace the B-site of Fe^{3+}, such as Co,[30–32] Cr [100], Ni [101], and Mn [102], can alter the magnetic properties. The interaction of the ferromagnetic elements Fe^{3+} and Cr^{3+} with three electrons in *d* orbitals was caused by the d_5–d_3 interaction. Along the *z*-axis, the magnetic contact between Fe^{3+} and Mn^{3+} ions is shown to be antiferromagnetic, whereas there is ferromagnetic exchange along the other principal axes.

The most notable feature of BFO-based nanomaterials may be their ability to support ME coupling, a mutual coupling of ferroelectric and ferromagnetic properties.[33] Zhao et al. were the first to demonstrate that electric control of antiferromagnetic domain structure exists.[34] Its film in a single phase, indicating a significant connection between the two types of room temperature arrangement. Before and after electrical poling, the structure of the antiferromagnetic domain is found to be closely connected to the structure of the ferroelectric domain. They discover that both before and after electrical poling the structure of the antiferromagnetic domain is closely connected utilizing the structure of the ferroelectric domain. Polarization only switches at 71° and 109° produce a connection between antiferromagnetic and ferroelectric domains pattern, at 180° ferroelectric polarization does not switch.

A substantial magnetization can be achieved for device applications. It has been discovered that the intersection of ferromagnetic and antiferromagnetic heterostructures exhibits magnetic anisotropy. Another advantage of thin film heterostructures is the magnetic anisotropy of the system at the interface between ferromagnetic and antiferromagnetic heterostructures. In addition, the primary advantage of thin film heterostructure is that magnetic anisotropy can be controlled via epitaxy.

Yu et al. discovered that different magnetic exchange effects are produced at various LSMO/BFO interfaces.[35] Antiferromagnetic bulk order competes with ferromagnetic Fe^{3+} and Mn^{4+}. By varying its ferroelectric polarization, they were able to achieve a reversible alternate between two states of exchange bias. This is a significant advancement for electrically manipulating magnetization and a major step toward electrically controllable spintronic devices.

Controlling a reversible electric field for ME coupling at room temperature is still challenging due to the exchange bias-associated temperature dependence in a ferromagnetic BFO heterostructure. Chu et al. utilized transition metal ($Co_{0.9}$ $Fe_{0.1}$) as the ferromagnetic layer along it to achieve reversible electric field regulation of a local magnetron at room temperature.

The magnetic anisotropy was toggled on and off using an in-plane electric field. Following the first electrical switch, it was discovered that the striped domains begin to ascend from left to right. The domains then switched back after the application of the opposing electric field. This collection of images demonstrated that the electric field can govern a switchable domain ferroelectric wall at room temperature.

It is intended for the ferroelectric polarization to be parallel to the *c* axis, the antiferromagnetic polarization to the *b*-axis, and the ferromagnetic moment to the *a*-axis. Additionally, a Co film heterostructure exists over the film. The antiferromagnetic axis of the film is perpendicular to the axis parallel to which magnetic anisotropy is aligned.

3.3.2 FERROELECTRIC PROPERTIES

The spontaneous polarization of BFO is caused by the relative movement of Bi^{3+} away from the oxygen octahedron, which generates an electric dipole moment. When making BFO, a lot of oxygen vacancies and the second phase are created due to the Bi components' erratic behavior and the fluctuating valence of the Fe element. As a result, current leakage increases dramatically, which makes it challenging to create a ferroelectric hysteresis loop which is saturated at room temperature. Its advancement as a ferroelectric/piezoelectric material has been impeded by significant leakage current.[37] To further improve the effectiveness of ferroelectric devices, major efforts have been made to reduce leakage currents. The most efficient way to reduce the high leakage current is ion replacement and creating solid solutions of it including other ABO_3-type perovskites.

Its piezoelectric and ferroelectric characteristics are susceptible to epitaxial strain.[38] As previously mentioned, the increasing compressive strain causes shifting of the rhombohedral phase to the monoclinic phase and tetragonal phase. Its polarization direction also changed at the same time. Orthorhombic BFO is seen to have its polarization in the [110] directions,[39] the MA phase has its polarization rotated toward the [001] direction,[38,40] and the T-like M_C phase has its polarization in the (010) plane slightly tilted away from the [001] direction. There is a general trend for the polarization with strain, and this pattern is compatible with the fact that the T-like phase exhibits a significantly larger spontaneous polarization. The polarization along (001) plane increases with the increased compressive strain.[27] The pseudotetragonal structure may theoretically have a P_r of 150 C cm^2.[41] A supertetragonal phase of its film on LAO substrates, with a P_r of around 130 C cm^2 has been reported.[42]

An R-like or T-like transition takes place at about 4–5% compressive strain, and it has proven to own a significantly higher response for piezoelectrics. This is particularly true in terms of the piezoelectric properties.[43] This is one of the key factors in improving the piezoelectric properties by the shifting of phase boundaries between different phases.[44] In the transformation, the third "intermediate" polymorph serves as a transitional stage. The piezoelectric response can be improved in other transitions outside the R-like/T-like transition. It has been shown that the change from the monoclinic MC to MA phase results in a large d_{33} of 100 pm V_1.[45]

3.4 TRANSITION METAL DOPED $BIFEO_3$ THIN FILMS

The following issues prevent the direct use of BFO in device applications: (a) the significant leakage current and tendency directed toward breakdown, (b) the substantial coercive field and (c) the spiral spin structure prevents the linear ME action and cancels out macroscopic magnetism. To solve these problems, both ionic substitution of transition metal dopants and the development of nanostructures with distinctive sizes less than the period of the spiral structure have been utilised.[46–56] Another intriguing idea is to use various substrates with different lattice characteristics to produce strain effects using BFO.[53–55]

Due to the various magnetic moments among the B-sites, the doping of the 3d orbit transition metal (TM = Ti, Cr, Mn, Co, Ni, Cu, and Zn) ions, the B-site in it has lately received significant attention.[56–59] Through TM doping, it is possible to produce a shift in a gradient of spin canting, a complex chiral spin structure, or a local ferromagnetic spin configuration. However, even for the same dopants across studies, the level of improvement varies.

The variation in the ionic radius of TM ions led to a little modification in the lattice constant of the films which are doped when compared to BFO. The lattice volume increases as the radius of the dopant ion decreases, which suggests that films with large doped ion radius have less local structural deformation in FeO_6. Better electrical characteristics might be obtained by reducing dense microstructure and microscopic grain boundaries.[57]

3.4.1 ELECTRICAL PROPERTIES

In comparison to BFO samples, it was discovered that the leakage current densities for BFTM samples are reduced. The presence of the oxygen vacancies which are formed as a result of Fe^{3+} electrical instability, and extrinsic influences such as porosity in microstructure, grain boundaries, and dislocations have a significant impact on the leakage current.[60–62] The growth of oxygen vacancies increases the free carrier density by introducing impurity energy levels into the band gap thereby driving electrons to the particular defect levels. By regulating the valence fluctuation of the Fe (Fe^{2+}/Fe^{3+}) atom, its B-site can be doped with TM ions to fill any oxygen vacancies.[48,56,63] Since more oxygen vacancies are created by the Ni doping compared to other divalent TM ions, the leakage current in the films is not much reduced.[62] It is interesting to notice that Mn-doped films' measured J (leakage current densities) values are roughly an order of magnitude lower than those of the other films. These findings have a strong correlation with the previously published TM-doped samples.[50,57] The reduction of Fe valence fluctuation and oxygen vacancies can be used to explain the decrease in current leakage of the TM-doped samples. On the contrary, as film grain size increases, oxygen vacancies grow as well,[17] which may have an impact on the sample leakage current.

3.4.2 MAGNETIC PROPERTIES

M-H loops of all the samples display a weak ferromagnetic activity. Due to its G-type antiferromagnetic properties, the bulk BFO sample exhibits relatively low magnetism. However, because of the strain and/or contact effects stated elsewhere, the thin film displays a modest ferromagnetic activity.[64] The magnetization rises at lower temperatures and it may be due to local spin clustering.[65] Small amounts of Mn and Co doping and substantial amounts of Cr and Ni doping both significantly raise the magnetization level of saturation. At ambient temperature, the maximum level of magnetization was found to be Ms~0.21 B per formula unit in BFCr samples, which is approximately 4 times more than Cr-doped BFO films produced using the spin coating process on the identical substrate.[56] This variation implies that the sample preparation methods might greatly influence the magnetic properties of these TM-doped films. The coercive field of a co-doped film is 162 Oe at room temperature, which is significantly larger than the other samples. The various magnetic moments of the TM ions create disruption of the cycloid spin structure in BFTM samples, which raises saturation magnetization.[49] The Fe^{3+}-O-Fe^{2+} contacts because the substantial leakage current is most likely brought on by the valence variation, the interactions are dual exchange interactions of Fe (Fe^{2+}/Fe^{3+}) atom, which may increase its saturation magnetization. It might also explain why the BFMn film has the minimum Ms among all these films, while BFCr and BFNi films hold greater Ms than BFMn and BFCo films.

3.4.3 FERROELECTRIC PROPERTIES

The distance-dependent repulsive force between adjacent domains was observed within ferroelectric thin films.[66,67] Due to the increased grain size in the Ni-doped sample, the repulsive force between adjacent domains weakens. This might make it simple for more domains to face in the direction of the field, leading to significant ferroelectric polarization. The diffusion and displacement of oxygen vacancies, as well as electronic charge trapping in the domain wall amongst the electronic switching, are the fundamental causes of fatigue in perovskite oxides.[48] The direct cause of the increased fatigue resistance in BFMn samples is the decrease in the moveable charge density.

3.4.4 MAGNETOELECTRIC PROPERTIES

The P-E loops are measured in the presence of a magnetic field *H*, to be able to ascertain the ME coupling in between the ferroelectric and magnetic orders in multiferroic samples. Its polarization *P* and magnetic field *H* applied are connected for ME effect directly by $\Delta P = \alpha \Delta H$ formula, where α is the ME coefficient. The relationship $\alpha_E^2 \propto \chi^e : \chi^m$, in which $\alpha_E = \frac{a}{\varepsilon_0 \varepsilon}$, in which ε is the dielectric constant, ε_0 is the vacuum permittivity, and χ^e is the electrical susceptibility & χ^m is the magnetic susceptibility, respectively, limits the ME voltage coefficient.[68] The samples demonstrate room temperature ME coupling, as evidenced by the polarization's increasing value when a magnetic field is present. Its magnetic field and its doped TM ion have a significant impact on the ME voltage coefficient. It is discovered that the ME voltage coefficient dips above the critical field H_c and rises with the magnetic field. At a low magnetic field, the cycloid spin structure breaks, and as H rises, it becomes depinned. The spin-flop phenomena, that is, the transition from cycloid spin structure to a homogeneously magnetic state can be used to explain the presence of H_c.[69] Due to the suppression of the cycloid spin structure, the well saturation level in magnetization was reached above H_c, which is further supported by *M-H* hysteresis loop measurements. For samples of BFO, BFCo, BFCr, BFNi, and BFMn, the ME voltage coefficient measured at H_c was approximately 1.48, 1.36, 3.79, 9.93, and 10.48 V/cm Oe, respectively. The "spin current" concept,[70] which clarifies the relationship between ferroelectricity and spin-orbit also known as Dzyaloshinskii–Moriya interaction, could be used to describe how the ME coupling in multiferroic BFO came to be physical. It demonstrates how the ferroelectricity-inducing, bond bending of the Fe-O-Fe present in it can increase the ME voltage coefficient. However, because the bond bending is affected by the magnetic field and the TM ion which is doped, it is possible that the films which is doped with TM have a larger ME coefficient than the BFO films. The magneto and electro restrictive interaction might also be used to explain the detected ME coupling,[68] it explains why the Mn-doped film has a higher ME coefficient than TM-doped films.

3.5 IMPACT ON THE MICROSTRUCTURE AND MULTIFERROIC CHARACTERISTICS OF $BIFEO_3$ CERAMICS BY DOPING WITH BA AND MN

$BiFeO_3$ ceramics with Ba, Mn single and co-dopants were made by a combination of fast sintering and solid-state reaction. Ba, Mn co-doped ceramics were found to have improved ferroelectric and ferromagnetic characteristics. Large remanent magnetization (Mr~0.834emμ/g) and coercive field (Hc~938Oe) measurements for $Bi_{0.95}Ba_{0.05}Fe_{0.9}Mn_{0.1}O_3$ and large remnant polarization (Pr~40.87μC/cm^2) measurements for $Bi_{0.9}Ba_{0.1}FeO_3$ were obtained, respectively. From the analysis of the structure and valence electron density measured by X-ray diffraction (XRD), Raman spectrum, and X-ray photoelectron spectroscopy (XPS), the doping ratio of Ba and Mn ions plays a significant role in facilitating microstructure distortion and transition, which are beneficial for enhancing the ferroelectric properties of the ceramics. Additionally, the ferromagnetic property of Ba-Mn co-doped BFO is influenced by the presence of Fe^{2+} and Mn^{4+} ions. These findings show that the room-temperature ferromagnetic and ferroelectric characteristics can be modified effectively by ions co-doping for multiferroic applications.

Candidates for multiferroic materials include the well-known compounds of BFO and $BiMnO_3$ (BMO). According to general consensus, BMO[71–73] adopts the monoclinic crystal structure of space group C2 with off-centered Bi 6S2 lone pairs rupturing the antisymmetry. However, BMO only goes through the ferromagnetic and ferroelectric transitions when it is cooled below 500 K and 100 K, respectively.[74,75] The practical applicability is greatly constrained by these modest transition temperatures. Contrarily, BFO is the most researched single-phase multiferroic material at the moment since it exhibits both ferroelectricity and magnetism above room temperature.[76,77] It is commonly accepted that the distortion of the $6s^2$ lone pair of Bi^{3+} ions at the A-site causes its ferroelectricity, while the magnetic property is connected to the remaining moment of the Fe-site spin structure. In actuality, this mild ferromagnetism results from the G type's fundamental antiferromagnetic structure, which is cycloidally modified with a long 62 nm period. Therefore, it is required to alter the electronic structure and eliminate the spatially modulated cycloid spin structures to improve the ferroelectric and magnetic properties.

The suppression of the formation of the secondary phase and the evaporation of Bi, which simultaneously facilitates the distortion and transition of crystal structure resulting in the enhancement of the ferroelectric property, and broke the spin structure by inducing the spontaneous magnetization, has recently been addressed by A- and/or B-site substitution by alkaline-earth and/or transition metals.[78–80] To enhance the ferroelectric characteristics, divalent alkaline-earth metal ions doping at its A-site results in structural distortion, transitions, and even an electrical conductor-insulator transition. On the contrary, when Bi^{3+} ions are replaced by divalent alkaline-earth metal ions, the unfavorable effect of high leakage current caused by the increasing number of oxygen vacancies as a result of charge compensation prevents the improvement of ferroelectricity.

According to reports,[81–85] Mn doping affects the multiferroic characteristics resulting in the Jahn–Teller distortion and better microstructure transition, which improves ferroelectric performance. The magnetization is improved by Mn doping, which is effective in producing double exchange (Mn^{3+}-O_2-Mn^{4+}) in it. As a result, Ba and Mn doping is a successful method for enhancing multiferroic characteristics.

Despite the fact that there have been a few reports on Mn and Ba co-doped ceramics,[86,87] the researchers are primarily interested in the effects of the same proportion of Ba and Mn ion doping on the multiferroic property or the alteration of the multiferroic property of Ba-Mn co-doped BFO on the basis of $Bi_{0.8}Ba_{0.2}FeO_3$ ceramic of heavily doped Ba ions. To further understand the causes of the improved ferroelectric and ferromagnetic capabilities, it is required to verify the precise relationship of structural characteristics with the doping content and doping ratio of Ba^{2+} and Mn^{4+} ions. Due to bismuth loss and secondary phases $Bi_2Fe_4O_9$ and $Bi_{20}FeO_{40}$, which are unavoidable at high temperatures, its single-phase synthesis is found to be challenging using standard solid-state processes. For the synthesis of a single phase of it, however, a combination of solid-state reaction and rapid sintering process offers a better solution because it takes less time and prevents the loss of Bi.

For Ba-doped BFO, as the Ba concentration rises, the intensities of the (104) and (110) diffraction peaks diminish and their widths widen. When the Ba concentration reaches 10%, the (104) and (110) diffraction peaks merge into a single peak, shifting toward a lower diffraction angle. This is due to the structural change caused by the replacement of Bi^{3+}, which has a smaller ionic radius (1.17 Å), in comparison with that of Ba^{2+} (1.42 Å). It suggests that the distortion tendency of the rhombohedral structure to a monoclinic or tetragonal form is made easier by the doping of Ba ions. The ferroelectric characteristics of the ceramic may benefit from this adjustment. All diffraction peak positions after the addition of Mn ions are still compatible with their parent diffraction peak positions. The only difference is that the (104) and (110) of its parent diffraction peaks are wider than those of (104) and (110), respectively. It is well-known that the grain size has a significant impact on the width of the diffraction peak.[88] When compared to its pure form, the doping of Ba and Mn ions results in a significant reduction of particle size. Mn ion doping in BBFO ceramics results in a deformed perovskite rhombohedral structure but does not alter its parent crystal structure.

The following irreducible representation can be used to express the Raman active modes for it with a highly deformed perovskite of the R_3c space group: $\Gamma_{R(R3c)} = 4A_1 + 9E$. Raman and infrared radiation are both active in the $4A_1$ and doubly degenerate 9E modes. The spectra show that transverse optical E(TO) phonons in it with R_3c are related with peaks at 259.3, 339.6, 366.6, 476.9, and 530.9 cm^{-1} and four peaks at 126.1, 165.5, 213, and 429 cm^{-1} can be attributed to A_1-symmetry longitudinal optical [A_1(LO)] phonons. The ceramics of the Raman spectrum were fitted in the Lorentz spectrum. The crystalline structure is not considerably altered by the substitution of Ba^{2+} for Bi^{3+}, according to the Raman spectroscopy of Ba-doped ceramics. However, the A_1-2 and A_1-3 modes' intensities substantially decline, Meanwhile, the phonon modes of A_1-1, A_1-2, A_1-3 and some E modes practically merge together along with the broadening and shifting of some A_1 and E modes and this suggests apparent alterations in phonon behaviours. These alterations are related to the insertion of Ba into its lattice, which has changed the local atomic structure. The first-principles calculation by Hermet et al.[89] suggests that Fe atoms may have contributed to higher frequency modes and are implicated in modes between 152 and 262 cm^1. In other words, the Mn ion's inclusion has no effect on the A_1 modes linked to the vibrations of the Bi-O bond. As a result, Mn doping causes the structural transition trend in the ceramic to move toward orthorhombic.

It is obvious that its pure form has a relatively large average grain size, and that when Ba content rises, the average grain size decreases significantly. The amount of oxygen vacancies present and the pace at which ions diffuse affect grain growth.[90] Since Bi is quite volatile, evaporation leaves a lot of oxygen vacancies in its pure form, which makes it simple for the ions to spread and causes the formation of larger grains. By occupying specific likely evaporated Bi sites with Ba ions and switching out smaller-ionized Bi atoms for larger-ionized Ba atoms, these phenomena may be prevented. Although the pellets' microstructures are generally thick, Ba and Mn co-doped samples show few dispersed holes. The average grain size, density, and homogeneity of the grains are all significantly impacted by the doping content of Ba and Mn ions. With more Mn ion present, the average grain size of Ba and Mn co-doped materials is shrunk by the suppression of grain growth brought on by the decrease in oxygen vacancies.[91]

With an increase in the amount of Ba atoms, the ferroelectric hysteresis loops show a more well-shaped tendency, and the remnant polarization P_r value rises gradually from 7.44 to 16.46 $\mu C/cm^2$ and then to 40.87 $\mu C/cm^2$ from $x = 0$, $x = 0.05$, and $x = 0.1$. The crystal structure distortion of the ceramic caused by the substitution of Ba^{2+} with a larger ionic radius for Bi^{3+} with a lower ionic radius in its structure is thought to be the cause of this strong ferroelectric characteristic. Ba doping, on the contrary, reduces voltage under stress, which makes it possible to apply a smaller electric field to the sample. The sample is collapsed as the voltage is increased. This is explained by the large leakage current brought on by oxygen vacancies that resulted from the replacement of Bi^{3+} with Ba^{2+}. It is generally acknowledged that a part of Fe^{3+} ions transforms into Fe^{2+} ions as a result of oxygen vacancies created during growth.[92] The high leakage of BBFO is frequently attributed to these Fe ions. As a result, it is thought that the Ba atoms added to the ceramic help to increase ferroelectricity and even lower the voltage withstanding stress. The BBFMO1 ceramic has a higher P_r value and can sustain more voltage stress than the BBFMO2 ceramic. The structural distortion and local lattice distortion of the ceramic caused by the Ba and Mn doping are nonnegligible factors in this case. The decrease in leakage current is the other crucial element. When high valence Mn^{4+} ions occupy the Fe site, they function as a donor resulting in a decrease in the oxygen vacancies, which lowers the leakage current. However, if the amounts of Mn and Ba ions differ, Mn^{3+} ions may form depending on the valence balance, which opens the door for possible electron transfers between the Mn^{4+} and Mn^{3+} ions. In contrast, this makes the leakage current stronger. It is well known that the stereochemical activity of Bi lone pairs is the root cause of the ferroelectricity of this multiferroic system. The shift in the stereochemical activity of the Bi lone electron pair and the alteration in the Bi-O covalent bonds are caused by the high Mn ion doping, which is supported by the reduced intensity of the A_1-2 and A_1-3 modes compared to the two modes of its pure form.[93] Spin canting and superexchange interactions between Fe^{3+} and Fe^{3+} on its lattices may result from the introduction of the Mn ion changing the Fe^{3+}-O-Fe^{3+} bonds. Along (001) axis, there is a decreasing relative displacement between the Fe^{3+} ion and the O^{2-} octahedron's charge center.[93] The reduction in ferroelectric polarization is then brought on by the strongly doped Mn ions.

The doping of Ba and Mn ions greatly increases the saturation magnetization (M_s), remanent magnetization (M_r), and coercive field (H_c) of the ceramics. The M_s and M_r rise in the case of Ba-doped ceramic as the Ba ion content rises. On the contrary, as the amount of Ba ion grows, the H_c decreases before increasing. Additionally, the addition of the Mn ion considerably increases its magnetization on the basis of the Ba-doped BFO. Ba and Mn co-doped BFO exhibits a strong rising tendency in the M_s, M_r, and H_c with an increase in the amount of M_n ions.

The Ba and Mn co-doped ceramics exhibit a greater improvement of room-temperature ferromagnetism for a number of reasons, including the following: First, the canting of the antiferromagnetic sublattices is what causes the weak room-temperature ferromagnetic feature of it. Second, in contrast to its pure form, all its samples which are doped by Ba show significant ferromagnetic characteristics, which can be attributed to weak ferromagnetic ordering and double exchange contact between Fe^{2+} and Fe^{3+} ions. The substitution of Bi^{3+} ions for Ba^{2+} ions resulted in a reduced bond angle of Fe-O-Fe and an increase in the spin canting angle, which distorts the rhombohedral structure of it. As a result, a weak ferromagnetic ordering ultimately develops, increasing magnetization. It is possible that the presence of Fe^{2+} ions will lead to a double exchange interaction between Fe^{2+} and Fe^{3+} ions through oxygen, which will increase ferromagnetism. Third, the following factors have a major role in the strong ferromagnetic ordering that was noticed when Mn ions were introduced into its Ba-doped sample. Due to the variations in magnetic moments among the many B0site occupants, the replacement of Mn atoms for Fe at the B-sites can result in the formation of a local ferromagnetic spin configuration. Compared to the addition of Ba atoms in its ceramics, Mn^{4+} substitution more significantly distorts the Fe-O-Fe bond angles, tilting the FeO_6 octahedrons and causing the distortion of the structure. The improvement of the canting angle of adjacent antiferromagnetically ordered planes as a result of the disturbance of Fe-O-Fe bond angles leads to a bigger increase in the magnetization values. The rise in the anisotropy constant of samples doped with Ba and Mn as well as the change in magnetic phase are both related to the increase in Hc.[94]

3.6 STRUCTURAL, DIELECTRIC AND MAGNETIC PROPERTIES OF $NI_{0.75}ZN_{0.25}FE_2O_4$–$BIFEO_3$ COMPOSITES:

The coupling between several order factors could yield a novel effect, such as the ME effect.[95] Cr_2O_3,[96] an antiferromagnetic crystal, was the first crystal family in which the linear ME effect was experimentally discovered until 1960–1961. As technology moves closer to the downsizing of devices, demand for materials with multifunctional qualities is rising. The term "multiferroic" was originally only used to describe single-phase materials, but it has since been generalized to apply to any material that displays two or more different types of long-range spontaneous orderings.[97] Only at low temperatures can the single-phase multiferroics show the coexisting order characteristics, and they also have a mild ME response at ambient temperature. Magnetoelectric composites have been produced to address the issues with single-phase multiferroics, and it has been discovered that by utilizing product characteristics, a composite of the piezoelectric phase and magnetostrictive phase is projected to achieve a much stronger ME coupling effect.[98]

With the help of an applied magnetic field and mechanical stress, the magnetostriction causes a mechanical deformation that, in turn, causes an electric field due to the piezoelectric effect.[99,100] Composite materials could exhibit product or sum characteristics.[98,101,102] Sum characteristics are produced by the weighted sum of the contributions from individual phases, corresponding to their volume fraction. Physical characteristics such as density, magnetism, and dielectric properties fall within the category of some properties. The more intriguing feature of the product is that it is present in its composite but absent in separate phases.

Suchtelen was the first to suggest the ME effect as a product tensor property in composite materials.[102] These composites have received a lot of interest in the field of ME material investigations in recent years, in addition to their considerable use in the device applications of switches, actuators, magnetic field sensors, and electronic memory devices in recent years.[103] Numerous ME composites have been created during the past few years, including $CoFe_2O_4$–$BaTiO_3$, $NiFe_2O_4$-$BaTiO_3$, $Ni_{0.8}Zn_{0.2}Fe_2O_4$–$Ba_{0.6}Sr_{0.4}TiO_3$,[104] $(x)Co_{0.5}Zn_{0.5}Fe_2O_4$–$(1–x)$PLZT,[105] and the multilayer composites of PZT/Terfenol-$_D$. To improve the ME effect, a proper ferrite and ferroelectric material combination is crucial. For this, large piezoelectric and magnetostriction coefficients of the ferroelectric phase are required.[106,107] The synthesis and characterization of $xNi_{0.75}Zn_{0.25}Fe_2O_4$–$(1–x)BiFeO_3$ composites, which contain it as a ferroelectric component and $Ni_{0.75}Zn_{0.25}Fe_2O_4$ (NZF) as a ferrite component, have been the primary emphasis of this research. Based on the previously mentioned defined criteria, these two components were chosen.

Due to the high coupling coefficient of a Jahn–Teller ion like Zn^{2+}, nickel ferrite doped with Zn^{2+} exhibits high electrical resistivity, high Curie temperature, high saturation magnetization, and high magnetostriction coefficient. Furthermore, because it contains unpaired electrons, which align when a magnetic field is applied, it displays soft ferromagnetic activity. As ferroelectric is an excellent ferroelectric and piezoelectric material, nanocomposites of Co, Ni, and Cr ferrites with their ferroelectric phase have recently been discovered on various papers as a piezoelectric phase.[108–110]

This material is important because ferromagnetism and ferroelectricity coexist in it at room temperature. It demonstrates a number of appealing optical features in addition to these intriguing ones.[111] Considering potential modifications to the various physical properties of a class of NZF–BFO composites is therefore highly intriguing. This study examines the structural, dielectric, magnetic, magnetodielectric, and optical characteristics of composites made of $xNi_{0.75}Zn_{0.25}Fe_2O_4$ and (x = 0, 0.10, 0.20, and 0.30).

3.6.1 EXPERIMENTS

Through the hybrid process method, ceramic samples of $xNi_{0.75}Zn_{0.25}Fe_2O_4$–$(1–x)BiFeO_3$ (x = 0, 0.10, 0.20, and 0.30) were created. The stoichiometry quantities of explanatory grade ZnO, NiO, and Fe_2O_3 powders were weighed and entirely mixed in an acetone medium using agate mortar for 5–6 h to create the NZF phase. After mixing the powder, it is dried in the air-filled, closed aluminum crucible for 2 h at 1200° C. Powder is once again crushed in an agate mortar after presintering to produce a uniformly fine powder of the desired chemical. The sol–gel process was used to create it in the nanoscale range. As beginning ingredients included bismuth nitrate $Bi(NO_3)^{3-}$ $5H_2O$, ferric nitrate $Fe(NO_3)^{3-}$ $9H_2O$, citric acid and ethylene glycol, metals were mixed in the proper molar ratios. Citric acid was dissolved with distilled water to create an aqueous solution. Then, with steady stirring at 60°C–70°C, ferric nitrate and bismuth nitrate were added in succession to create a homogenous solution and prevent precipitation.

Ethylene glycol is appended to the solution, with a ratio of 70:30 citric acid to ethylene glycol. The mixture was transformed into xerogel, which began to expand and fill the beaker, forming a frothy precursor made up of extremely small, nanoscale flakes. The resulting gel was then dried in a hot air oven for 12 h at 100°C, and the powder that was left over after drying was subsequently calcined for 2 h at 600°C. To get a single phase of $BiFeO_3$, the calcined powder was then drained in diluted HNO_3. To create $xNi_{0.75}Zn_{0.25}Fe_2O_4$–$(1–x)BiFeO_3$ composites with x variations of 0.10, 0.20, and 0.30, the produced powders of NZF and BFO were combined in an acetone media for 3–4 h in three distinct mass ratios, namely, 10/90, 20/80, and 30/70, respectively. The mixes were then put through a hydraulic press with a steel die and a uniaxial pressure of around 6×10^7 kg/m^2 to form pellets that were 8 mm in diameter and 1–2 mm thick. Finally, the pellets were sintered in an environment of air for 2 h at 900°C.

An X-ray diffractometer (Bruker D8 Advance) was used to perform the structural examination of the sintered samples. Using CuK_α radiation (k = 1.5418 Å), the same procedure has been carried out over a broad range of angles ($20° \leq 2\Theta \leq 80°$) at a scanning rate of 1 min^{-1} at room temperature, ensuring the synthesis and purity of the necessary compounds. With the use of an FEI Quanta 200F electron microscope running with a 20 kV voltage, field emission scanning electron microscopy images were collected to examine the distribution of grains on the surface of sintered pellets. The linear intercept approach was used to calculate the grain size.

On a Jeol model 1200 EX apparatus, transmission electron microscopy (TEM) measurement was carried out. An automated HIOKI 3532-50 Hi-Tester LCR meter was used to measure the dielectric loss (tanδ) and dielectric constant (ε) in the temperature range of 30–500°C. (1–1 MHz). Using a superconducting quantum interference device with a magnetic field in the range of (-8 kOe $\leq$ H $\leq$ 8 kOe) at 300 and 5 K, the magnetic characteristics of composite materials were determined. A modified Sawyer-Tower circuit was considered to assess the relationship between dielectric polarization and electric field (*P-E* hysteresis loops; Marine India Electronics Pvt. Ltd.).

Using a Wayne Kerr 6500 high-frequency LCR meter and a magnet with an accuracy of 10 Oe and a maximum output of 8 kOe, a magnetocapacitance study was conducted. At room temperature, the samples' photoluminescence (PL) spectra were evaluated using a fluorescence spectrophotometer (Hitachi F-4600). Using an FTIR spectrophotometer in the range of 2000–400 cm^{-1}, the dried samples embedded in KBR matrix had their Fourier transform infrared spectra (FTIR) recorded.

The XRD pattern of composites (with $x = 0.10$, 0.20, and 0.30) clearly demonstrates the presence of the mixed spinel-perovskite phase. Due to the kinematics of formation, it is challenging to obtain the pure phase of it. Along with the predominant phase of it, which is typically seen in this system, small amounts of impurity phases like $Bi_2Fe_4O_9$ and $Bi_{36}Fe_{24}O_{57}$ are also seen. The pattern corresponded with JCPDS card No. 71-2494 and the cataloguing of the diffraction peaks of the composites conform with the stated values for NZF[112] and for BFO. While it displays a rhombohedral crystal structure with lattice parameters a = 5.56 Å and c = 13.75Å, NZF stabilizes in a cubic spinel structure with a lattice parameter of 8.345 Å. Due to changes in the positions of the diffraction peaks for both phases, there is a little difference in its lattice parameter and the lattice parameter of the NZF phases in composites, which shows structural distortion in the composites. Using Scherrer's equation, $D_m = {}^{K\lambda}/_{\beta} \cos h$, where K is constant, λ is the wavelength of the X-rays employed, β is full-width half maxima, and h is the diffraction (Bragg) angle, the crystallite size (Dm) for NZF phase has been calculated from the reflection line (311).

Because each phase's thermal expansion coefficient and densification rate are different, internal stress cannot be prevented in cofired ceramics.[113] As a result, the cofired ceramics exhibit significant stress and lattice deformation due to the dense stacked grains and change in grain shape. This could be the cause of the composites' produced nonuniform microstructure with various grain sizes. Both the NZF and the BFO constituent grains of composites are seen to be tightly packed. The fact that these constituents adhere strongly to one another indicates that these composites were successfully formed.

Sharp diffraction spots in the SAED pattern suggest that its particles have a rhombohedral perovskite structure that is compatible with the earlier-shown XRD results and that well-developed single-crystalline nanoparticles are being formed. According to TEM images, the typical particle size is found to be between 38 and 40 nm.

3.6.2 DIELECTRIC PROPERTIES

For composites $x\text{Ni}_{0.75}\text{Zn}_{0.25}\text{Fe}_2\text{O}_4$–$(1-x)\text{BiFeO}_3$ (with $x = 0, 0.10, 0.20$, and 0.30) measured at two different frequencies. Near the magnetic transition temperature or its T_N, an abnormality in the dielectric constant has been noted for all the compositions.

The discovered anomalies are broad in character and exhibit slight dispersion as frequency increases.[114] The antiferromagnetic transition is thought to be the cause of the anomaly seen in its pure form.[115] The Devonshire thermodynamic theory can be used to explain the reported maximum value of the dielectric constant.[116] This form of anomaly in magnetoelectrically ordered systems was anticipated by the Landau–Devonshire theory of phase transition as the result of the vanishing magnetic order's influence on an electric order.[117,118] This type of anomaly has been discovered by several groups in its bulk and thin films[119] and has been predicted to lead to ME coupling in the system.

Up until a particular temperature, known as the transition temperature, the value of the dielectric constant rises with rising temperature; above this temperature, it falls. In the case of ferrite-based ME composites, the rise in dielectric constant may be attributed to the thermal activation of electron hopping between Fe^{2+} and Fe^{3+} ions on the octahedral sites. Furthermore, because this hopping results in local displacement in the direction of an externally applied field, it produces dielectric polarization.

Due to the effect that the concentration of Fe^{2+} ions has on the dielectric variation, it is expected that as Fe^{2+} concentration increases, so will polarization and the dielectric constant. The random vibrating motion of ions and electrons is sped up with temperature increases above the transition temperature. As a result, it is less vulnerable to field-direction orientation, which causes the dielectric constant to drop.[120]

Additionally, the result of the interaction of dipolar, electronic, ionic, and interfacial polarizations is a dielectric constant. At lower frequencies, dipolar and interfacial polarization predominate, and at higher frequencies, electronic polarization plays a crucial role.[121] As a result, the dielectric constant is found to be higher at 1 kHz than at 10 kHz. With $x = 0, 0.10, 0.20$, and 0.30, the transition temperatures for $x\text{Ni}_{0.75}\text{Zn}_{0.25}\text{Fe}_2\text{O}_4$–$(1-x)\text{BiFeO}_3$ composites were observed to be 345, 230, 285, and 420 C, respectively, at 1 kHz frequency.

According to earlier composites research,[122] the diffuse phase transition seen in composites is caused by connections between ferroelectric and nonferroelectric grains. With increasing NZF content, the value of the dielectric constant is observed to increase, and the influence of NZF concentration at $x = 0.30$ is particularly strong. As the conduction contribution of the NZF phase toward ε increases, it is discovered that both the dielectric constant (ε) and loss ($\tan\delta$) significantly rise with an increase in NZF content.[123] The polarization mechanism in ferrites and the conduction mechanism are comparable.

The conduction phenomena in ferrites, which involves the interchange of electrons between ions of the same elements residing in more than one valence state and are randomly dispersed throughout equivalent crystallographic lattice locations, is described on the basis of the Verwey de Bohr mechanism. Conduction is also caused by the exchange of electrons between Fe^{2+} and Fe^{3+} ions, and the quantity of these ions is influenced by the rate at which Fe^{3+} is reduced to Fe^{2+} at higher temperatures and during sintering.[121]

Conduction mechanisms are brought about by the electron and hole hopping between Fe^{2+}/Fe^{3+}, $Zn^{2+}/Zn^{3+}$and Ni^{2+}/Ni^{3+} ions. Along with n-type carriers, or the electronic exchange between Fe^{2+}/Fe^{3+} ions, the presence of Ni^{2+}/Ni^{3+} and Zn^{2+}/Zn^{3+} ions, which result in p-type carriers, adds to the polarization mechanism. The p-type carriers' contribution should be less than that of n-type carriers and show the opposite sign. Additionally, even at lower frequencies, the contribution of p-type carriers to polarization declines more quickly because their mobility is lower than that of n-type carriers.

As a result, the polarization initially increases and then declines with increasing frequency.[124] The space charge conduction, which is connected to the transport of defects like oxygen vacancies to the dielectric-electrode interface, is indicated by the rapid increase in the dielectric loss at high temperatures.

3.7 CONCLUSION

BFO is the perfect multiferroic system with various special properties. This chapter discusses the structure, types and various properties of BFO due to the doping of different rare earth and transition metal ions. It is mainly focused on the magnetic and ME properties of BFO nanomultiferroics. In future, the changes that occur when it is doped with other elements can be studied to enhance the magnetic and ME coupling properties. By this review, we hope to encourage the researchers to take the challenge and find the opportunity to study more details about the BFO multiferroic nanomaterial for different applications.

KEYWORDS

- **multiferroics**
- **ceramics**
- **thin films**
- **magnetoelectric coupling**
- **magnetic properties**
- **ferroelectricity**

REFERENCES

1. Eerenstein, W.; Mathur, N. D.; Scott, J. F. Multiferroic and Magnetoelectric Materials. *Nature* **2006,** *442*, 759–765. https://doi.org/10.1038/nature05023
2. Dutta, D. P.; Mandal, B. P.; Naik, R.; Lawes, G.; Tyagi, A. K. Magnetic, Ferroelectric, and Magnetocapacitive Properties of Sonochemically Synthesized Sc-Doped $BiFeO_3$ Nanoparticles. *J. Phys. Chem. C.* **2013,** *117*, 2382–2389. https://doi.org/10.1021/jp310710p
3. Catalan, G.; Scott, J. F. Physics and Applications of Bismuth Ferrite, *Adv. Mater.* **2009,** *21* (24), 2463–2485. https://doi.org/10.1002/adma.200802849
4. Yang, J.-C.; He, Q.; Yu, P.; Chu, Y.-H. $BiFeO_3$ Thin Films: A Playground for Exploring Electric Field Control of Multifunctionalities. *Annu. Rev. Mater. Res.* **2015,** *45*, 249–275. https://doi.org/10.1146/annurev-matsci-070214-020837
5. Ma, J.; Hu, J.; Li, Z.; Nan, C.-W. Recent Progress in Multiferroic Magnetoelectric Composites: From Bulk to Thin Films. *Adv. Mater.* **2011,** *23*, 1062–1087. https://doi.org/10.1002/adma.201003636
6. Wang, J.; Neaton, J. B.; Zheng, H.; Nagarajan, V.; Ogale, S. B.; Liu, B.; Viehland, D.; Vaithyanathan, V.; Schlom, D. G.; Waghmare, U. V.; Spaldin, N. A.; Rabe, K. M.; Wuttigand, M.; Ramesh, R. Epitaxial $BiFeO_3$ Multiferroic Thin Film Heterostructures. *Science* **2003,** *299*, 1719–1722. https://doi.org/10.1126/science.1080615
7. Sakar, M.; Balakumar, S.; Saravanan, P.; Bharathkumar, S.; Compliments of Confinements: Substitution and Dimension Induced Magnetic Origin and Band-Bending Mediated Photocatalytic Enhancements in $Bi_{1-x}Dy_xFeO_3$ Particulate and Fibre Nanostructures. *Nanoscale* **2015,** *7*, 10667–10679. https://doi.org/10.1039/C5NR01079A
8. Lotey, G. S.; Verma, N. K. Structural, Magnetic, and Electrical Properties of Gd-Doped $BiFeO_3$ Nanoparticles with Reduced Particle Size. *J. Nanopart. Res.* **2012,** *14*, 742. https://doi.org/10.1007/s11051-012-0742-7
9. Anjum, G.; Kumar, R.; Mollah, S.; Shukla, D. K.; Kumar, S.; Lee, C. G.; Structural, Dielectric, and Magnetic Properties of $La_{0.8}Bi_{0.2}Fe_{1-x}Mn_xO_3$ ($0.0 \leq x \leq 0.4$) Multiferroics. *J. Appl. Phys.* **2010,** *107*, 103916. https://doi.org/10.1063/1.3386527.

10. Durga Rao, T.; Asthana, S. Evidence of Improved Ferroelectric Phase Stabilization in Nd and Sc Co-substituted $BiFeO_3$. *J. Appl. Phys.* **2014,** *116*, 164102. https://doi.org/10.1063/1.4898805.
11. Guo, R.; You, L.; Zhou, Y.; Lim, Z. S.; Zou, X.; Chen, L.; Ramesh, R.; Wang, J. L. Non-Volatile Memory Based on the Ferroelectric Photovoltaic Effect. *Nat. Commun.* **2013,** *4*, 1990. https://doi.org/10.1038/ncomms2990
12. Yang, C. H.; Kan, D.; Takeuchi, I.; Nagarajan, V.; Seidel, J. Doping $BiFeO_3$: Approaches and Enhanced Functionality. *Phys. Chem. Chem. Phys.* **2012,** *14*, 15953–15962. https://doi.org/10.1039/c2cp43082g
13. Chen, L.; Ren, W.; Zhu, W. M.; Ye, Z. G.; Shi, P.; Chen, X. F.; Wu, X. Q.; Yao, X. Improved Dielectric and Ferroelectric Properties in Ti-Doped $BiFeO_3$–$PbTiO_3$ Thin Films Prepared by Pulsed Laser Deposition. *Thin Solid Films* **2010,** *518*, 1637–1640. https://doi.org/10.1016/j.tsf.2009.11.072
14. Gao, F.; Chen, X. Y.; Yin, K. B.; Dong, S.; Ren, Z. F.; Yuan, F.; Yu, T.; Zou, Z.; Liu, J. M. Visible-Light Photocatalytic Properties of Weak Magnetic $BiFeO_3$ Nanoparticles. *Adv. Mater.* **2007,** *19*, 2889–2892. https://doi.org/10.1002/adma.200602377
15. Gao, F.; Yuan, Y.; Wang, K. F.; Chen, X. Y.; Chen, F.; Liu, J. M. Preparation and Photo Absorption Characterization of $BiFeO_3$ Nanowires. *Appl. Phys. Lett.* **2006,** *89*, 102506. https://doi.org/10.1063/1.2345825
16. Moreau, J. M.; Michel, C.; Gerson, R.; James, W. J. Ferroelectric $BiFeO_3$ X-ray and Neutron Diffraction Study. *J. Phys. Chem. Solids* **1971,** *32*, 1315–1320. https://doi.org/10.1016/S0022-3697(71)80189-0
17. Royen, P.; Swarup, K. Das System Wismut-eisenberg von 0 is 55 Mol% eisenoxyd. *Angew. Chem. Int. Ed.* **1957,** *69*, 779. https://doi.org/10.1002/ange.19570 692407
18. Sosnowska, I.; Peterlinneumaier, T.; Steichele, E. Spiral Magnetic Ordering in Bismuth Ferrite. *J. Phys. C Solid State Phys.* **1982,** *15*, 4835–4846. https://doi.org/10.1088/0022-3719/15/23/020
19. Kan, D.; Palova, L.; Anbusathaiah, V.; Cheng, C. J.; Fujino, S.; Nagarajan, V.; Rabe, K. M.; Takeuchi, I. Universal Behavior and Electric-Field-Induced Structural Transition in Rare-Earth-Substituted $BiFeO_3$. *Adv. Funct. Mater.* **2010,** *20*, 1108–1115. https://doi.org/10.1002/adfm.200902017
20. Yang, C. H.; Seidel, J.; Kim, S. Y.; Rossen, P. B.; Yu, P. et al. Electric modulation of Conduction in Multiferroic Ca-Doped $BiFeO_3$ Films. *Nat. Mater.* **2009,** *8*, 485–493. https://doi.org/10.1038/nmat2432
21. Wang, N.; Luo, X.; Han, L.; Zhang, Z.; Zhang, R.; Olin, H..; Yang, Y. Structure, Performance, and Application of $BiFeO_3$ Nanomaterial. *Nano-Micro Lett.* **2020,** *12*, 81. https://doi.org/10.1007/s40820-020-00420-6
22. Khan, A. H.; Atiq, S.; Anwar, M. S.; Naseem, S.; Abbas, S. K. Optimization of Magnetodielectric Coupling in Mn Substituted $BiFeO_3$ for Potential Memory Devices. *J. Mater. Sci.* **2018,** *29*, 11812–11823. https://doi.org/10.1007/s10854-018-9281-z
23. Kumar, M.; Yadav, K. L. Strain-Relaxed Structure in (001)/ (100)-Oriented Epitaxial $PbTiO_3$ Films Grown on (100) $SrTiO_3$ Substrates by Metal Organic Chemical Vapor Deposition. *Appl. Phys. Lett.* **2007,** *91*, 112904. https://doi.org/10.1063/1.2779239
24. Zhuang, J.; Wu, H.; Ren, W.; Ye, Z. G. Local Polar Structure and Multiferroic Properties of *(1−x)* $Bi_{0.9}Dy_{0.1}FeO_3$−xPb- TiO_3 Solid Solution. *J. Appl. Phys.* **2014,** *116*, 066809. https://doi.org/10.1063/1.4891345

25. Kumar, M. M.; Srinivas, A.; Suryanarayana, S. V. Structure Property Relations in $BiFeO_3/BaTiO_3$ Solid Solutions. *J. Appl. Phys.* **2000,** *87*, 855–862. https://doi.org/10.1063/1.371953
26. Azuma, M.; Niitaka, S.; Hayashi, N.; Oka, K.; Takano, M.; Funakubo, H.; Shimakawa, Y. Rhombohedral–Tetragonal Phase Boundary with High Curie Temperature in *(1 − x)* $BiCoO_3$–*x*$BiFeO_3$ Solid Solution. *Jpn. J. Appl. Phys.* **2008,** *47*, 7579–7581. https://doi.org/10.1143/JJAP.47.7579
27. Sando, D.; Barthelemy, A.; Bibes, M. $BiFeO_3$ Epitaxial Thin Films and Devices: Past, Present and Future. *J. Phys.: Condens. Mater.* **2014,** *26*, 47. https://doi.org/10.1088/09538984/26/47/473201
28. Zavaliche, F.; Zhao, T.; Zheng, H.; Straub, F.; Cruz, M. P.; Yang, P. L.; Hao, D.; Ramesh, R. Electrically Assisted Magnetic Recording in Multiferroic Nanostructures. *Nano Lett.* **2007,** *7*, 1586–1590. https://doi.org/10.1021/nl070465o
29. Goswami, S.; Bhattacharya, D.; Choudhury, P. Particle Size Dependence of Magnetization and Non Centrosymmetry in Nanoscale $BiFeO_3$. *J. Appl. Phys.* **2011,** *109*, 07D737. https://doi.org/10.1063/1.3567038
30. Chen, Z.; Chen, Z.; Kuo, C. Y.; Tang, Y.; Dedon, L. R. et al. Complex Strain Evolution of Polar and Magnetic Order in Multiferroic $BiFeO_3$ Thin Films. *Nat. Commun.* **2018,** *9*, 3764. https://doi.org/10.1038/s41467-018-06190-5
31. Saenrang, W.; Davidson, B. A.; Maccherozzi, F.; Pod-Kaminer, J. P.; Irwin, J. et al. Deterministic and Robust Room Temperature Exchange Coupling in Mono-Domain Multiferroic $BiFeO_3$ Heterostructures. *Nat. Commun.* **2017,** *8*, 1583. https://doi.org/10.1038/s41467-017-01581-6
32. Chauleau, J. Y.; Haltz, E.; Carretero, C.; Fusil, S.; Viret, M. Multi-Stimuli Manipulation of Antiferromagnetic Domains Assessed by Second-Harmonic Imaging. *Nat. Mater.* **2017,** *16*, 803– 807. https://doi.org/10.1038/nmat4899
33. Kuo, C. Y.; Hu, Z.; Yang, J. C.; Liao, S. C.; Huang, Y. L. et al. Single-Domain Multiferroic $BiFeO_3$ Films. *Nat. Commun.* **2016,** *7*, 12712. https://doi.org/10.1038/ncomms12712
34. Zhao, T.; Scholl, A.; Zavaliche, F.; Lee, K.; Barry, M. et al. Electrical Control of Antiferromagnetic Domains in Multiferroic $BiFeO_3$ Films at Room Temperature. *Nat. Mater.* **2006,** *5*, 823–829. https://doi.org/10.1038/nmat1731
35. Yu, P.; Lee, J. S.; Okamoto, S.; Rossell, M. D.; Huijben, M. et al. Interface Ferromagnetism and Orbital Reconstruction in $BiFeO_3$–$La_{0.7}Sr_{0.3}MnO_3$ Heterostructures. *Phys. Rev. Lett.* **2010,** *105*, 027201. https://doi.org/10.1103/PhysR evLett.105.027201
36. Chu, Y. H.; Martin, L. W.; Holcomb, M. B.; Gajek, M.; Han, S. J. et al. Electric-Field Control of Local Ferromagnetism Using a Magnetoelectric Multiferroic. *Nat. Mater.* **2008,** *7*, 478–482. https://doi.org/10.1038/nmat2184
37. Rojac, T.; Bencan, A.; Malic, B.; Tutuncu, G.; Jones, J. L.; Daniels, J. E.; Damjanovic, D. $BiFeO_3$ Ceramics: Processing, Electrical, and Electromechanical Properties. *J. Am. Ceram. Soc.* **2014,** *97*, 1993–2011. https://doi.org/10.1111/jace.12982
38. Daumont, C.; Ren, W.; Infante, I. C.; Lisenkov, S.; Allibe, J. et al. Strain Dependence of Polarization and Piezoelectric Response in Epitaxial $BiFeO_3$ Thin Films. *J. Phys: Condens. Matter.* **2012,** *24*, 16. https://doi.org/10.1088/09538984/24/16/162202
39. Yang, J. C.; He, Q.; Suresha, S. J.; Kuo, C. Y.; Peng, C. Y. et al. Orthorhombic $BiFeO_3$. *Phys. Rev. Lett.* **2012,** *109*, 247606. https://doi.org/10.1103/PhysRevLett.109.247606

40. Jang, H. W.; Baek, S. H.; Ortiz, D.; Folkman, C. M.; Das, R. R. et al. Strain-Induced Polarization Rotation in Epitaxial (001) $BiFeO_3$ Thin Films. *Phys. Rev. Lett.* **2008,** *101*, 107602. https://doi.org/10.1103/PhysRevLett.101.107602
41. H. Béa, B. Dupé, Fusil, S.; Mattana, R.; Jacquet, E. et al., Evidence for Room-Temperature Multiferroicity in a Compound with a Giant Axial Ratio. *Phys. Rev. Lett.* **2009,** *102*, 217603. https://doi.org/10.1103/PhysRevLett.102.217603
42. Zhang, J. X.; He, Q.; Trassin, M.; Luo, W.; Yi, D. et al., Microscopic Origin of the Giant Ferroelectric Polarization in Tetragonal-Like $BiFeO_3$. *Phys. Rev. Lett.* **2011,** *107*, 147602. https://doi.org/10.1103/PhysRevLett.107.147602
43. Wojdeł, J. C.; Íñiguez, J. Ab Initio Indications for Giant Magnetoelectric Effects Driven by Structural Softness. *Phys. Rev. Lett.* **2010,** *105*, 037208. https://doi.org/10.1103/PhysR evLett.105.037208
44. Beekman, C.; Siemons, W.; Ward, T. Z.; Chi, M.; Howe, J. et al. Phase Transitions, Phase Coexistence, and Piezoelectric Switching Behaviour in Highly Strained $BiFeO_3$ Films. *Adv. Mater.* **2013,** *25*, 5561–5567. https://doi.org/10.1002/adma.201302066
45. Liu, H. J.; Liang, C. W.; Liang, W. I.; Chen, H. J.; Yang, J. C. et al. Strain-Driven Phase Boundaries in $BiFeO_3$ Thin Films Studied by Atomic Force Microscopy and X-ray Diffraction, *Phys. Rev. B* **2012,** *85*, 014104. https://doi.org/10.1103/PhysRevB.85.014104
46. Kartopu, G.; Lahmar, A.; Habouti, S.; Solterbeck, C. L.; Elouadi, B.; Es-Soun, M. *Appl. Phys. Lett.* **2008,** *92*, 151910. https://doi.org/10.1016/j.ssc.2013.07.032
47. Li, B.; Wang, C.; Liu, W.; Ye, M.; Wang, N. *Mater. Lett.* **2013,** *90*, 45. https://doi.org/10.1016/j.matlet.2012.09.012
48. Cheng, Z.; Wang, X.; Dou, S.; Kimura, H.; Ozawa, K. *Phys. Rev. B* **2008,** *77* 092101. https://doi.org/10.1103/PhysRevB.77.092101
49. Chakrabarti, K.; Das, K.; Sarkar, B.; Ghosh, S.; De, S. K.; Sinha, G.; Lahtinen, J. *Appl. Phys. Lett.* **2012,** *101*, 042401. https://doi.org/10.1063/1.4738992
50. Naganuma, H.; Miura, J.; Okamura, S. *Appl. Phys. Lett.* **2008,** *93*, 052901. https://doi.org/10.1063/1.2965799
51. Wu, Y. J.; Chen, X. K.; Zhang, J.; Chen, X. J. *Appl. Phys.* **2012,** *111*, 053927. https://doi.org/10.1063/1.3693531
52. Liu, J.; Li, M.; Hu, Z.; Pei, L.; Wang, J.; Liu, X.; Zhao, X. *Appl. Phys. A* **2011,** *102*, 713. https://doi.org/10.1007/s00339-010-5965-9
53. Chen, L.; Ren, W.; Ye, Z. G.; Tian, A.; Wu, X.; Shi, P. *J. Appl. Phys.* **2012,** *112*, 052008. https://doi.org/10.1063/1.4746086
54. Zhang, J. X.; Li, Y. L.; Wang, Y.; Liu, Z. K.; Chen, L. Q.; Chu, Y. H.; Zavaliche, F.; Ramesh, R. *J. Appl. Phys.* **2007,** *101*, 114105. https://doi.org/10.1063/1.2743733
55. Annapu Reddy, V.; Pathak, N. P.; Nath, R. *Curr. Appl. Phys.* **2012,** *12*, 541. https://doi.org/10.1016/j.ssc.2013.07.032
56. Kharel, P.; Talebi, S.; Ramachandran, B.; Dixit, A.; Naik, V. M.; Sahana, M. B.; Sudakar, C.; Naik, R.; Rao, M. S. R.; Lawes, G. *J. Phys. Condens. Matter.* **2009,** *21*, 036001. https://doi.org/10.1088/0953-8984/21/3/036001
57. Raghavan, C. M.; Do, D.; Kim, J. W.; Kim, S. S. *J. Am. Ceram. Soc.* **2012,** *95*, 1933. https://doi.org/10.1111/j.1551-2916.2012.05081.x
58. Raghavan, C. M.; Kim, E. S.; Kim, J. W.; Kim, S. S. *Ceram. Int.* **2013,** *39*, 6057. https://doi.org/10.1016/j.ssc.2013.07.032
59. Annapu Reddy, V.; Varma, G. D.; Nath, R. *AIP Adv.* **2011,** *1*, 042140. https://doi.org/10.1063/1.3662093

60. Huang, J. Z.; Wang, Y.; Lin, Y.; Li, M.; Nan, C. W. *J. Appl. Phys.* **2009,** *106*, 063911. https://doi.org/10.1063/1.3225559
61. Wang, D. Y.; Chan, N. Y.; Zheng, R. K.; Kong, C.; Lin, D. M.; Dai, J. Y.; Chan, H. L. W.; Li, S. *J. Appl. Phys.* **2011,** *109*, 114105. https://doi.org/10.1063/1.3594745
62. Qi, X.; Dho, J.; Tomov, R.; Blamire, M. G.; Driscoll, J. L. M. *Appl. Phys. Lett.* **2005,** *86*, 062903. https://doi.org/10.1063/1.2965799
63. Tomashpol'skii, Y. Y.; Venevtsev, Y. N.; Zhdanov, G. S. *Kristallografiya* **1967,** *12*, 252. https://doi.org/10.1063/1.1866214
64. Wang, J.; Neaton, J. B.; Zheng, H.; Nagarajan, V.; Ogale, S. B.; Liu, B.; Viehland, D.; Vaithyanathan, V.; Schlom, D. G.; Waghmare, U. V.; Spaldin, N. A.; Rabe, K. M.; Wuttig, M.; Ramesh, R. *Science* **2003,** *299*, 1719. https://doi.org/10.4236/ape.2014.42011
65. Annapu Reddy, V.; Pathak, N. P.; Nath, R. *J. Alloys Compd.* **2012,** *543*, 206. https://doi.org/10.1016/j.jallcom.2012.07.098
66. Ren, S. B.; Lu, C. J.; Shen, H. M.; Wang, Y. N. *Phys. Rev. B* **2007,** *55*, 3485. https://doi.org/10.1063/1.5041893
67. Sekhar, K. C.; Nautiyal, A.; Nath, R. *J. Appl. Phys.* **2009,** *105*, 024109. https://doi.org/10.1063/1.3072626
68. Wangab, K. F.; Liuab, J. M.; Ren, Z. F. *Adv. Phys.* **2009,** *58*, 321. https://doi.org/10.1080/00018730902920554
69. Bai, F.; Wang, J.; Wuttig, M.; Li, J. F.; Wang, N.; Pyatakov, A. P.; Zvezdin, A. K.; Cross, L. E.; Viehland, D. *Appl. Phys. Lett.* **2005,** *86*, 032511. https://doi.org/10.1063/1.1851612
70. Hu, C. D. *Phys. Rev. B* **2008,** *77*, 174418. https://doi.org/10.1103/PhysRevB.77.174418
71. Toulemonde, P.; Bordet, P.; Bouvier, P.; Kreisel, J. *Phys. Rev. B* **2014,** *89*, 224107. https://doi.org/10.1103/PhysRevB.89.224107
72. Kimura, T.; Kawamoto, S.; Yamada, I.; Azuma, M.; Takano, M.; Tokura, Y. Phys. Rev. B **2003,** *67*, 180401(R). https://doi.org/10.1103/PhysRevB.67.180401
73. Belik, A. A.; Iikubo, S.; Yokosawa, T.; Kodama, K.; Igawa, N.; Shamoto, S.; Azuma, M.; Takano, M.; Kimoto, K.; Matsui, Y.; Muromachi, E. T. *J. Am. Chem. Soc.* **2007,** *129*, 971. https://doi.org/10.1021/ja0664032
74. Haumont, R.; Kreisel, J.; Bouvier, P.; Hippert, F. *Phys. Rev. B* **2006,** *73* 132101. https://doi.org/10.1103/PhysRevB.73.132101
75. Guennou, M.; Bouvier, P.; Toulemonde, P.; Darie, C.; Goujon, C.; Bordet, P.; Hanfland, M.; Kreisel, J. *Phys. Rev. Lett.* **2014,** *112*, 075501. https://doi.org/10.1103/PhysRevLett.112.075501
76. Igor Kornev, A.; Lisenkov, S.; Haumont, R.; Dkhil, B.; Bellaiche, L. *Phys. Rev. Lett.* **2007,** *99*, 227602. https://doi.org/10.1103/PhysRevLett.99.227602
77. Arnold, D. C.; Knight, K. S.; Morrison, F. D.; Lightfoot, P. *Phys Rev Lett.* **2009,** *102*, 027602. https://doi.org/10.1103/PhysRevLett.102.027602
78. Jaffari, G. H.; Samad, A.; Iqbal, A. M.; Hussain, S.; Mumtaz, A.; Awan, M. S.; Siddique, M.; Shah, S. I. *J. Alloys. Comp.* **2015,** *644*, 893. https://doi.org/10.1016/j.jallcom.2015.05.065
79. Rajput, S. S.; Katoch, R.; Sahoo, K. K.; Sharma, G. N.; Singh, S. K.; Gupta, R.; Garg, A. *J. Alloys Comp.* **2015,** *621*, 339. https://doi.org/10.1016/j.jallcom.2014.09.161
80. Yang, M. M.; Zhao, X. Q.; Wang, J.; Zhu, Q. X.; Zhang, J. X.; Li, X. M.; Luo, H. S.; Zheng, R. K. *Appl. Phys. Lett.* **2014,** *104*, 052902. https://doi.org/10.1063/1.4863825
81. Qi, X. D.; Dho, J.; Tomov, R.; Blamire, M. G.; MacManus-Driscoll, J. L. *Appl. Phys. Lett.* **2005,** *86*, 062**903,** https://doi.org/10.1063/1.2169905

82. Deng, H. L.; Zhang, M.; Hu, Z.; Xie, Q. F.; Zhong, Q.; Wei, J. Z.; H. Yan *J. Alloys. Comp.* **2014,** *582*, 273. https://doi.org/10.1016/j.jallcom.2013.07.187
83. Xue, F.; Liang, L. Y.; Gu, Y. J.; Takeuchi, I.; Kalinin, S. V.; Chen, L. Q. *Appl. Phys. Lett.* **2015,** *106*, 012903. https://doi.org/10.1007/s10854-015-3612-0
84. Karimi, S.; Reaney, I. M.; Han, Y.; Pokorny, J.; Sterianou, I. *J. Mater. Sci.* **2009,** *44*, 5102. https://doi.org/10.1007/s10853-009-3545-1
85. Dua, Y.; Chenga, Z. X.; Shahbazia, M.; Collingsb, E. W.; Doua, S. X.; Wang, X. L. *J. Alloys. Comp.* **2010,** *490*, 637. https://doi.org/10.1016/j.jallcom.2009.10.124
86. Deng, H. L.; Zhang, M.; Zhong, Q.; Wei, J. Z.; Yan, H. *Ceram. Int.* **2014,** *40*, 5869. https://doi.org/10.1016/j.jallcom.2016.04.029
87. Rout, J.; Choudhary, R. N. P. *Phys. Lett. A* **2016,** *380*, 288. https://doi.org/10.1016/j.physleta.2015.10.007
88. Pattanayaka, S.; Choudharya, R. N. P.; Piyush Dasa, R.; Shannigrahi, S. R. *Ceram. Int.* **2014,** *40*, 7983. https://doi.org/10.1016/j.jallcom.2016.04.029
89. Hermet, P.; Goffinet, M.; Kreisel, J.; PhGhose *Phys. Rev. B* **2007,** *75*, 220102. https://doi.org/10.1103/PhysRevB.75.220102
90. Makhdoom, A. R.; Akhtar, M. J.; Rafiq, M. A.; Hassan, M. M. *Ceram. Int.* **2012,** *38*, 3829. https://doi.org/10.1016/j.ceramint.2012.01.032
91. Saravana Kumar, K.; Venkateswaran, C.; Kannan, D.; Tiwari, B.; Ramachandra Rao, M. S. *J. Phys. D, Appl. Phys.* **2012,** *45*, 415302. http://dx.doi.org/10.1088/0022-3727/45/41/415302
92. Eerenstein, W.; F. D.Morrison, Dho, J.; Blamire, M. G.; Scott, J. F.; Mathur, N. D. *Science* **2005,** *307*, **1203,** https://doi.org/10.1126/science.307.5713.1203c
93. Yuan, G. L.; Or, S. W.; Chan, H. L. W. *J. Appl. Phys.* **2007,** *101*, 064101. https://doi.org/10.1063/1.2433709
94. Suresh, P.; Srinath, S. *J. Appl. Phys.* **2013,** *113*, 17D920. https://doi.org/10.1063/1.4863944
95. Nan, C. W.; Bichurin, M. I.; Dong, S.; Viehland, D.; Srinivasan, G. *J. Appl. Phys.* **2008,** *103*, 031101–031135. https://doi.org/10.1063/1.2836410
96. Folen, V. J.; Rado, G. T.; Stalder, E. W. *Phys. Rev. Lett.* **1961,** *6*, 607–608. https://doi.org/10.1103/PhysRevLett.6.607
97. Martin, L. W.; Crane, S. P.; Chu, Y. H.; Holcomb, M. B.; Gajek, M.; Huijben, M.; Yang, C. H.; Balke, N.; Ramesh, R. *J. Phys.* **2008,** *20*, 434220–434232. https://iopscience.iop.org/article/10.1088/0953-8984/20/43/434220/meta
98. Ryu, J.; Priya, S.; Uchino, K.; Kim, H. E. *J. Electroceram.* **2002,** *8*, 107–119. https://doi.org/10.1023/A:1020599728432
99. Zeng, M.; Wan, J. G.; Wang, Y.; Yu, H.; Liu, J. M.; Jiang, X. P.; Nan, C. W. . *J. Appl. Phys.* **2004,** *95*, 8069–8073. https://doi.org/10.1063/1.1739531
100. Ryu, J.; Priya, S.; Carazo, A. V.; Uchino, K.; Kim, H. . *J. Am. Ceram. Soc.* **2001,** *84*, 2905–2908. https://doi.org/10.1111/j.1151-2916.2001.tb01113.x
101. Bhimasankaram, T.; Suryanarayana, S. V.; Prasad, G. *Curr. Sci.* **1998,** *74*, 967–976. http://dx.doi.org/10.3390/ma13071520
102. Suchtelen, J. V. *Philips. Res. Rep.* **1972,** *27*, 28–37.
103. Yan, L.; Yang, Y.; Wang, Z.; Xing, Z.; Li, J.; Viehland, D. *J. Mater. Sci.* **2009,** *44*, 5080–5094. https://doi.org/10.1007/s10853-009-3679-1
104. Yang, H.; Wang, H.; He, L.; Shui, L.; Yao, X. *J. Appl. Phys.* **2010,** *108*, 074105–074110. https://doi.org/10.1063/1.3490782

105. Fawzi, A. S.; Sheikh, A. D.; Mathe, V. L. *J. Alloys Compd.* **2010,** *493*, 601–608. https://doi.org/10.1016/j.jallcom.2009.12.164
106. Kumar, M.; Yadav, K. L. *J. Phys. Chem. Solids* **2007,** *68*, 1791–1795. https://doi.org/10.1016/j.jpcs.2007.05.006
107. Devan, R. S.; Kanamadi, C. M.; Lokare, S. A.; Chougule, B. K. *Smart Mater Struct.* **2006,** *15*, 1877–1881. https:/doi.org/10.1088/0964-1726/15/6/043
108. Liu, X. M.; Fua, S. Y.; Huanga, C. *Mater. Sci. Eng. B* **2005,** *121*, 255–260. https://doi.org/10.1016/j.mseb.2005.04.009
109. Liu, X. M.; Gao, W. *Adv. Mater. Res.* **2011,** *197*, 456–459. https://doi.org/10.4028/www.scientific.net/AMR.197-198.456
110. Kumar, A.; Yadav, K. L.; Hemant, H.; Pandu, R.; Reddy, P. R. *Phys B* **2010,** *405*, 2362–2366. https://doi.org/10.1016/j.physb.2010.02.038
111. Adhlakha, N.; Yadav, K. L.; Singh, R. *Sci. Adv. Mater.* **2013,** *5*, 947–959. https://doi.org/10.1166/sam.2013.1543
112. Ajmal, M.; Maqsood, A. *Mater. Sci. Eng. B* **2007,** *139*, 164–170. https://doi.org/10.1016/j.mseb.2007.02.004
113. Bai, Y.; Xu, F.; Qiao, L.; Zhou, J.; Li, L. *J. Magn. Mater.* **2009,** *321*, 148–151. https://doi.org/10.1016/j.jmmm.2008.08.049
114. Kumar, M. M.; Srinivas, A.; Suryanarayana, S. V. *J. Appl. Phys.* **2000,** *87*, 855–862. https://doi.org/10.1063/1.371953
115. Mishra, R. K.; Pradhan, D. K.; Choudhary, R. N. P.; Banerjee, A. *J. Magn. Mater.* **2008,** *320*, 2602–2607. https://doi.org/10.1016/j.jmmm.2008.05.005
116. Devonshire A. F. *Philos. Mag.* **1949,** *40*, 1040–1063. https://doi.org/10.1080/14786444908561372
117. Palkar, V. R.; Kundaliya, D. C.; Malik, S. K.; Bhattacharya, S. *Phys. Rev. B* **2004,** *69*, 212102–212104. https://doi.org/10.1016/0038-1098(72)90280-3
118. Benguigui, L. *Solid State Commun.* **1972,** *11*, 825–828. https://doi.org/10.1016/0038-1098(72)90280-3
119. Uniyal, P.; Yadav, K. L. *J. Phys.* **2009,** *21*, 012205–012208. https://doi.org/10.1088/0953-8984/21/1/012205
120. Patankar, K. K.; Joshi, S. S.; Chougule, B. K. *Phys. Lett. A* **2005,** *346*, 337–341. https://doi.org/10.1016/j.physleta.2005.06.099
121. Bammannavar, B. K.; Naik, L. R. *Smart Mater. Struct.* **2009,** *18*, 085013–085021. https://iopscience.iop.org/article/10.1088/0964-1726/18/8/085013/meta
122. Patankar, K. K.; Patil, S. A.; Sivakumar, K. V.; Mahajan, R. P.; Kolekar, Y. D.; Kothale, M. B. *Mater. Chem. Phys.* **2000,** *65*, 97–102. https://doi.org/10.1016/S0254-0584(00)00216-9
123. Yua, Z.; Ang, C. *J. Appl. Phys.* **2002,** *91*, 794–797. https://doi.org/10.1063/1.1421033
124. Devana, R. S.; Ma, Y. R.; Chougule, B. K. *Mater. Chem. Phys.* **2009,** *115*, 263–268. https://doi.org/10.1016/j.matchemphys.2008.11.059

CHAPTER 4

Ferrites: Physics and Applications

JINCEMON CYRIAC[1], SUNNY MATHEW[1], TINA SEBASTIAN[2], SAJI AUGUSTINE[2], AMAL AJITH CHERIAN[1], L. R. SRIRAMACHANDRU[1], and NANDAKUMAR KALARIKKAL[3]

[1]*Department of Physics, Marian College Kuttikkanam (Autonomous), Kuttikkanam, Peerumade, Idukki, Kerala, India*

[2]*Department of Physics, Deva Matha College, Kuravilangad, Kottayam, Kerala, India*

[3]*International and Inter University Centre for Nanoscience and Nanotechnology, School of Pure and Applied Physics, Mahatma Gandhi University, Kottayam, Kerala, India*

ABSTRACT

Ferrites are highly resistant and permeable ferromagnetic oxide substances. This chapter provides information on ferrites, its types, such as normal and spinel, and the structural, dielectric, and magnetic properties possessed by them. There are numerous applications of ferrites in the modern world, some of which are discussed here, along with some futuristic applications of this material that are pertinent.

4.1 INTRODUCTION

For more than five decades, ferrites have been regarded as extremely significant electrical and magnetic materials due to their diverse physical properties. Lodestones (Fe_3O_4), the first magnetic material, were used as compass needles in the past.[1] However, the first attempt to prepare different

Physics and Mechanics of New Materials: Synthesis, Processing, and Emerging Applications. Devrim Balkose, Ana Cristina Faria Ribeiro, Nandakumar Kalarikkal, Ann Rose Abraham, A. K. Haghi (Eds.)

types of ferrites and industrialize ferrites did not occur until the turn of the twentieth century. Approximately 30 years after the commercialization of the ferromagnetic iron–silicon alloy, ferrites began to be sold on the open market. There are essentially six types of magnetic materials-diamagnetic, paramagnetic, ferromagnetic, antiferromagnetic, ferrimagnetic, and super-paramagnetic. This chapter will concentrate primarily on spinel ferrites, a subclass of ferrimagnetic materials. The ability to create ferrites in the form of nanoparticles has spawned a new, intriguing field of study with revolutionary applications in biotechnology and electronic technology. The term "spinel" in spinel ferrites is derived from the mineral magnesium aluminum oxide. William Henry Bragg was the first to recognize that their crystal structure is cubic. "Ferrites" are materials composed of ferric oxide, or Fe_2O_3, that resemble ceramics. Even though their saturation magnetization is less than half that of ferromagnetic alloys, they have several advantages over ferromagnetic alloys, including higher frequency application, lower cost, improved heat resistance, and superior corrosion resistance. By maximizing these advantages, the practical applications of ferrites have expanded.

4.2 TYPES OF FERRITES

Based on magnetic coercivity and resistance to demagnetization, ferrites can be classified as soft and hard ferrites. soft ferrites are the ones that lose their magnetic property after being magnetized, but hard ferrites can retain their magnetism even after being magnetized so they are also referred to as permanent magnets.[2]

4.2.1 SOFT FERRITES

A soft ferrite is a ceramic electromagnetic substance that has a dark grey or black hue and is extremely hard and brittle. The term "soft ferrites" refers to the magnetic properties of these materials, not to their physical qualities. Soft ferrites are incapable of retaining much magnetism, whereas hard ferrites are considered to have permanent magnetization. Soft ferrites are inverse spinel crystals composed of iron oxides and divalent metal oxides with a uniform cubic spinel structure.

Soft ferrites derive their magnetic properties from the interactions between the metal ions and oxygen ions in their crystal structure. Domains, which are magnetized regions responsible for a material's magnetic property, exist as a result of these interactions. In the absence of magnetic force, magnetic domains align in a random direction, resulting in a zero net flux. If there is a magnetizing force, the magnetic domains will align in the direction of the force, resulting in a significant total flux contribution. Due to their semiconducting nature, soft ferrites are capable of conducting electron flow through the material, which places them halfway between conductors and insulators.[3] The advantages of soft ferrites are their high resistivity, high magnetic permeability, and temperature resistance. These properties reduce eddy current loss over a broad frequency range.[4] Examples include Mn-Zn ferrites ($MnZnFe_2O_4$) and Ni-Zn ferrites ($NiZnFe_2O_4$).

4.2.2 HARD FERRITES

Hard ferrites maintain a large magnetization even after the applied magnetizing field has been removed, and residual magnetization is steady even if a demagnetizing field strength is applied. Hard ferrite is a ceramic material that is deep greyish and black in color, very fragile and rigid in nature.[5] Due to their low cost per unit of available energy, extensive availability of raw materials, and good chemical stability, hard ferrites dominate the market for permanent magnets.[6] They adopt the magnetoplumbite structure, which is characterized by the close packing of metal and oxygen ions along with Fe atoms in the interstitial site. This crystal structure can also be conceptualized in terms of hexagonal blocks containing metal ions and cubic blocks containing the spinel structure. Because they exhibit a greater coercive force than metallic magnetic materials, it is possible to create extremely thin magnets from hard ferrite. Hard magnetic ferrites are less structurally sensitive than soft ferrites, making them more susceptible to the effects of impurities during firing. As examples of hard ferrites, we have barium ferrite ($BaO{\cdot}6Fe_2O_3$) and strontium ferrite ($SrO{\cdot}6Fe_2O_3$).

4.2.3 BULK AND NANOFERRITES

The ferromagnetic oxide, which has a high resistance and permeability, is the primary constituent of ferrites. Although it has numerous applications as a bulk material, magnetism was first discovered at the nanoscale.[7] To produce more efficient devices, ferrites must be synthesized at the nanoscale level. In a size so small, these nanocrystals form a single domain condition to prevent domain wall resonance, allowing them to operate at higher frequencies.[8]

Nanoscale ferrites have numerous applications, including bioconjugated magnetic nanoparticles in the biomedical field, cancer diagnosis, hyperthermia, microwave absorbers, and numerous electronic applications. Due to the impressive properties of nanoferrites, researchers have recently lost interest in bulk ferrites. Ferrite materials such as $MgFe_2O_4$ and $ZnFe_2O_4$ are used for gas sensing purposes due to the microporosity of this material.

Thakur et al. examined the dielectric behavior of Mn-Zn ferrites, they discovered that the resistivity of nanoscale samples was increased in comparison to bulk samples.[9] Rath et al. developed superparamagnetic Mn-Zn microferrites using a hydrothermal technique.[10] They've demonstrated an increase in Curie temperature. From the study of nano and bulk forms of Li-Ni-Zn ferrite, Saafan et al. concluded that the saturation magnetization of nanosamples is lower than that of bulk samples.[11] Nanocomposites comprised of ferromagnetic particles implanted in noble metal matrices exhibit enormous magnetoresistance effects comparable to those observed in magnetic or nonmagnetic superlattices, according to recent research.

4.3 CLASSIFICATION OF FERRITES

Garnets, hexagonal ferrites, orthoferrites, and spinel ferrites are the four basic categories into which ferrites can be divided based on their structural characteristics. A type of ferrite known as garnet has "hard magnetic" properties with a chemical formula of R_3 (Fe_5O_{12}), where R is a rare-earth element. One typical example of a garnet is the yttrium garnet. The pbnm space group and orthorhombic crystal structure describe orthoferrites. These ferrites have the chemical formula $RFeO_3$, where R stands for a rare earth element. Orthoferrites are generally weak ferromagnetic compounds. $LaFeO_3$ and $DyFeO_3$ are a couple of examples of orthoferrites. The ferrites with the chemical formula hexagonal ferrites are those with the chemical formula M $Fe_{12}O_{19}$ (M might be barium, strontium, calcium, or lead).

Spinel ferrites are extremely stable materials with a variety of characteristics. Transformer cores (a device that converts power from one source to another), permanent magnets, memory chips, drug delivery systems, microwave devices, millimeter wave integrated circuitry, antenna rods, computer components, magnetic recorders, activators, and transducers are among their many applications.

The term "spinel" from spinel ferrites mainly represents the cubic structures which has a general formula of AB_2O_4, where B and A represent trivalent and divalent cations, respectively.[5] The full structural formula of spinel ferrites is AFe_2O_4, which has iron as a trivalent cation and is distinguished by its compact cubic form. Fe, Ni, Zn, Co, Mn, Cu, Cd, and the combination of these metals can be a divalent cation "A." The spinel ferrites are effective in metallurgical and other high-temperature industries because of their features such as strong thermodynamic stability, high electrical conductivity, and high corrosion resistance.[12,13] Due to their high magnetic losses and resistivity, spinel ferrites are widely utilized in the microwave absorber industry.[14] They also possess magnetic and electric properties such as large saturation magnetization, large squareness ratio, high magnetocrystalline anisotropy, small coercivity,[15] large electric resistivity,[16] large permeability,[17] low eddy current loss,[18] large curie temperature and mechanical hardness.[19]

Spinal ferrite (MFe_2O_4) is a cubic unit cell with 8 formula units and 56 ions.[5,20,21] The 56 ions are the sum of 32 oxygen anions, 8 M^{2+} cations, and 16 Fe^{3+} cations. By itself, the spinel unit cell has no net charge.

The spinel structure contains two different types of crystallographic sites as it has two different valance cations ($^{+}2$ and $^{+}3$). The "A" site is tetrahedral and the "B" site is octahedral. The "A" site is enclosed by four oxygen ions and the "B" site is enclosed by six oxygen ions. Overall, a spinel unit cell contains 16 B sites and 8 A sites. The tetrahedral ions are known as "network formers" whereas octahedral ions are known as "network modifiers.".[22] The distribution of divalent metal ions and trivalent ferric ions on the cationic sites characterizes two broad classes of spinel ferrites.[23]

1. **Normal spinels:** According to the following formula, these spinel ferrites contain 16 trivalent ferric cations (Fe^{3+}) at all 16 B sites and 8 divalent metal cations (M^{2+}) at all 8 A-sites. $[M^{2+}\downarrow]_A\ [Fe_3^{2+}\uparrow]_B\ O_4^{2-}$, where $\downarrow$ and $\uparrow$ stand for the opposing spin axes of the B and A sites. $ZnFe_2O_4$ and $CdFe_2O_4$ are examples of typical spinel ferrites, where Fe^{3+} ions act as network modifiers and M^{2+} ions as network formers, respectively.

2. **Inverse spinels:** There are 16 trivalent ferric cations (Fe^{3+}) in 8 B sites and 8 A sites, and 8 divalent metal cations (M^{2+}) in 8 of the 16 B sites in these spinels, as given in the following formula: $[Fe^{3+}\downarrow]_A[M^{2+}\uparrow Fe^{3+}\uparrow]_B O_4^{2-}$ where $\uparrow$and $\downarrow$are the opposite spin directions of A and B sites. Examples of spinel ferrites include Fe_2O_4, $CoFe_2O_4$, and $NiFe_2O_4$.

4.3.1 *NICKEL SPINEL FERRITES*

Nickel spinel ferrites are spinel ferrites that contain nickel (Ni), a divalent cation. Ni-ferrites have a resistivity of over 10^9 Ωcm, which makes them a good resistive material. They possess an inverse spinel structure. The eight tetrahedral voids of A sites are occupied by Fe^{3+} and the 16 octahedral voids of B sites are occupied by both Fe^{3+} and Ni^{2+} in an equal distribution. The value of the degree of inversion for pure nickel ferrite is one. Due to their high electrical resistivity, low coercivity, moderate saturation magnetization, and low hysteresis losses, nickel ferrites are classified as soft ferrites.

4.3.2 *PURE NICKEL FERRITES*

Since its discovery, pure nickel ferrite $NiFe_2O_4$ has been the focus of considerable interest. Researchers periodically modified the synthesis procedure for nickel ferrite $NiFe_2O_4$ and succeeded in developing its properties. Saito et al. and Takei et al. used a hydrothermal approach to create nickel ferrite in 1966.[24] The coprecipitation of hydroxide generated from a solution of constituent sulfates and the coprecipitation of two hydroxides (ferric hydroxide and nickel hydroxide) suspended in water at pH 11 were both used to synthesize nickel ferrite through a hydrothermal approach. In 1967, Poplawsky et al. used the arc image furnace technology to manufacture Ni ferrites and presented the results of their metallographic, chemical, structural, and electron probe studies.[25] Scientists' attention turned to "nickel deficiency ferrites" after carefully examining the properties of "nickel excess ferrites." The Maxwell Wien bridge was used by Rivas' research team to study nickel ferrites with a mixture of $Ni_xFe_{3-x-\Delta}O_4$ in temperature ranges from 80K to 500K, where the magnetic aftereffects were detected from self-induction of a coil with a nickel ferrite core.[26] Orehotsky et al. attempted to combine "nickel excess ferrites" and "nickel deficiency ferrites" in 1980.[27] Magnetic analysis of nanoferrite particles with the chemical formula $Ni_xFe_{3-x}O_4$ was done as a function of Ni(x). Magnetization at both ambient and Curie temperatures was calculated. Magnetization was found to follow a bell-shaped compositional dependency centered on $NiFe_2O_4$. It was reported that the Curie temperature (575°C) is independent of composition. Consequently, $NiFe_2O_4$ ferrite can be selected as having the most magnetically advantageous composition.

Several researchers attempted to improve the properties of ferrites by fabricating thin films. Such films were prepared by Lemaire et al. using the spray technique.[29] Whereas Banks et al. processed thin films of nickel ferrite of thickness 1000 Å using vacuum evaporation and high-temperature oxidation technology. The saturation magnetization of these thin films was discovered to be similar to that of bulk ferrites but with wider resonance linewidths. The ability to conduct optical studies of the material through transmission and the potential for determining magnetic exchange constants through the observation of spin wave resonances were the two potential benefits of deposition in thin-film form.[28] There is also ongoing research toward composites rather than pure nickel ferrites. Shamim et al. reported the composites of a ferrite material and a ferroelectric material (sodium potassium lithium niobate), in which an increase in the ferrite material increases the permittivity, coercivity, and losses.[30]

4.3.3 COBALT FERRITE

Due to its unique properties and extremely intriguing applications, cobalt ferrite has garnered considerable scientific interest. Numerous synthetic techniques, such as the sol–gel autocombustion method, ball milling method, coprecipitation method, microemulsion method, polyol approach, etc. have been utilized in its synthesis. The particle size of synthesized ferrite, which in turn determines its properties, is significantly affected by the synthetic processes employed and the sample's heat treatment. In addition, the properties of cobalt ferrite can be altered chemically in a number of ways, such as by substituting one or more metal ions from the lattice with others, altering the reaction conditions, combining the material with various inorganic matrices, etc. By substituting some of the cobalt or iron ions from the cobalt ferrite lattice with guest ions, it is possible to produce materials with completely different properties. Efforts are ongoing to incorporate new ions into the cobalt ferrite lattice as the number of studies in this field continues to expand. There is also a substantial amount of research on the effects of reaction conditions, such as annealing temperature, pH, the presence of additives, etc. In recent years, additional cobalt ferrite composites have been developed. Compared to cobalt ferrite, it is demonstrated that these composites possess superior properties. Therefore, it should not come as a surprise that cobalt ferrite will continue to attract a great deal of interest and has opened the door to a wide range of applications.

Pure nickel ferrite and cobalt ferrite have inverse spinel structures in their bulk counterparts.[31] The magnetic properties of Ni and Co ferrites are determined by the type and arrangement of metal cations at octahedral and tetrahedral sites. Doping an excellent magnetoresistive substance (cobalt) into a highly resistive substance (nickel) would produce a highly effective magnetoelectric material.[32] By partially doping Ni with Co, the electric and magnetic losses would also be increased, making the Ni-Co pair suitable for absorption applications.[33] Ni-Co ferrite can also be used to construct responsive and sensitive thin-film magnetic heads for high-density recording. Since it has lower coercivity and saturation magnetization than pure Ni ferrite, doping nickel ferrite with cobalt increases the likelihood of spinel phase formation. According to Rodrigues et al., the spinel ferrite phase (percent) in Ni ferrites increased by 72, 94, 90, and 87%, respectively, when the amount of Co was increased. It demonstrates that cobalt doping stabilizes Fe ions during network formation and nickel ions during system reorganization.[22]

Using L-threonine as a fuel, nanostructures of microwave-enhanced porous $CoFe_2O_4$ and TiO_2-doped $CoFe_2O_4$ were synthesized. The produced nanomaterials were characterized and examined to determine the photocatalytic degradation of Congo red dye upon exposure to visible light. According to the powder XRD pattern, the crystallite size of $CoFe_2O_4$ is 44 nm and the crystallite size of $CoFe_2O_4$ doped with TiO_2 is 25 nm. Compared to $CoFe_2O_4$, TiO_2-doped $CoFe_2O_4$ has significantly smaller crystalline sizes, which can be attributed to the addition of titania. In addition, the interfacial surface tension between cobalt ferrite and titania scatters the smaller particles with a significant degree of aggregation in the SEM images of titania-doped $CoFe_2O_4$. DRS spectra reveal that the measured bandgap of matching $CoFe_2O_4$ and TiO2-doped $CoFe_2O_4$ nanostructures is 3.01 and 2.88 eV, respectively. Microstructure and dimension of $CoFe_2O_4$ and TiO_2-doped $CoFe_2O_4$ spinel ferrites nanostructures were examined by means of high-resolution transmission electron microscopy (HR-TEM). Combining the photocatalytic properties of $CoFe_2O_4$ and $CoFe_2O_4$ doped with TiO_2 in a single particle is an advantage of this work.[34]

Nickel-cobalt ferrites have been a subject of study in the past decade. Various synthesis techniques, including SEM and XRD, were used to achieve morphological and structural properties to improve their quality. As a function of temperature, frequency, and composition, the dielectric and electric properties of synthesized ferrites are investigated.[35] Nickel-cobalt ferrites are distributed at a lower temperature and for a shorter period of time than Mg-Co ferrites. In the XRD characterization of pure spinel, the space group Fd_3m is observed. The SEM analysis revealed a consistent range of grain sizes with minimal porosity and an average size of approximately 0.5 nm.

Impedance analyses were conducted in the temperature range of 200–420 K and the frequency range of 10^2–10^7 Hz.[36] Ni-Co ferrite composites continues to attract the interest of scientists worldwide. Zhang et al. fabricated $Ni_xCo_{1-x}Fe_2O_4$/polyaniline composites and investigated their magnetic properties. Polyaniline has a wide range of applications in sensors, electrical condensers, electrode materials for secondary batteries, electromagnetic shielding, and anticorrosion due to its low cost, ease of production, and high conductivity. To create the composites, in situ polymerization and the sol-gel technique were utilized. Compared to pure ferrite, the researchers discovered that the composite had a lower saturation magnetization (37.36 emu g^{-1}) (57.57 emu g^{-1}). This was due to the composites' uneven nanoparticle distribution, smaller size, and reduced ferrite content (15%).[37]

As cobalt concentration increases, coercivity is said to increase due to its high anisotropy. Ati and Rodrigues observed a monotonic increase in coercivity for Co-doped Ni ferrites.[22,38] The relation $H_C = 0.96$ K/M_S, where M_S and H_C are coercivity and saturation magnetization, is utilized to analyze the variation of the anisotropy constant with cobalt ion concentration. At increasing dopant concentrations, the contribution of local configurations such as the Co^{2+} Co^{2+} pair becomes crucial, resulting in intricate dependence.[39] Co^{2+} ions provide a large positive input in ferrite anisotropy due to their quenchless orbital moment in the octahedral sites, allowing the contribution of strong spin-orbit interactions. Therefore, even with a small amount of Co doping, the anisotropy generated inhibits domain wall movement and amplifies ferromagnetic resonance. As a result, the frequency of ferromagnetic resonance rises.

4.3.4 ZINC FERRITE

Zinc ferrite belongs to the class of normal spinels and is thought to have a cation distribution of $Zn^{2+}(Fe^{3+})_2(O^{2-})_4$ while exhibiting no net magnetization. Using low-energy ion scattering, MÖssbauer spectroscopy, and magnetic measurements, samples of zinc ferrite were analyzed. In contrast to the assumption that zinc occupied only tetrahedral sites in a typical spinel, the results demonstrate that zinc occupies octahedral sites.

Ni-Zn spinel ferrites have several advantageous properties, such as high electrical resistivity, high magneto-crystalline anisotropy,[40] high Curie temperature, high permeability, low hysteresis loss,[41,42] and higher operating frequency. The 8-unit Ni ferrite unit cell has been observed to contain 4

Zn^{2+}, $4Fe^{3+}$ ions in tetrahedral sites and 4 Ni^{2+} and $12Fe^{3+}$ ions in octahedral sites. According to reports, the investigated composition $Ni_{0.5}Zn_{0.5}Fe_2O_4$ has a magnetic structure with a random distribution of nickel ions and an antiparallel distribution of iron ions between octahedral and tetrahedral sites.[43] Initial research was conducted by Kannon et al., who investigated how sintering temperature affected the structural and magnetic properties of $Ni_{0.5}Zn_{0.5}Fe_2O_4$ ferrite. To create the required composition, mechanical alloying was employed, followed by sintering at temperatures of 1000, 900, 800, and 700°C. As temperature increases, coercivity and lattice parameters decrease while saturation magnetization and particle size increase.

The optical properties and microstructure of Cr-doped $ZnFe_2O_4$ are highly dependent on chemical compositions, cation distribution, sintering temperature and time, the amount of additional cation, and preparation techniques. Cr-doped $ZnFe_2O_4$ nanoparticles were synthesized through a coprecipitation procedure. The XRD spectrum verifies the crystalline structure and particle size. The SEM image verifies the surface shape, porosity, and average particle size in the order of nanometers. The DC electrical conductivity is linear with respect to the Cr concentration. The temperature-dependent conductivity split around the Curie temperature exemplifies the behavior of a typical semiconductor.

Arboleda et al.[44] reported that Nickel-Zinc ferrites exhibit the spin Seebeck effect. Abu El-Fadl et al. synthesized Ni-Zn ferrites by means of the microwave combustion technique with glycine as the fuel.[45] The structure, site occupancy, and ferromagnetic activity of the prepared ferrites were studied. Crystallites range in size from 30 to 42 nanometers. As Ni content rises, saturation magnetization increases, whereas lattice constant and strain decrease. Recent reports indicate the use of nickel-zinc ferrites in composite architectures with a polypyrrole and multi-walled carbon nanotubes (MWCNT) matrix.[46] Babu et al. synthesized nickel-zinc ferrites using the sol-gel method, the sol–precipitate method, and the coprecipitation method.[47] According to his report and detailed information obtained from the XRD and FT-IR techniques of Ni-Zn ferrites, the sol–gel citrate method exhibits high initial permeability, high electrical resistivity, and high magnetization. Zinc doping replaces ferromagnetic Ni^{2+} ions with diamagnetic Zn^{2+} ions, resulting in a decrease in saturation magnetization.[10] Zinc doping lowers the resonance frequency of ferromagnetic magnetic materials and the resonance frequency of domain walls[48] due to a reduction in the exchange field acting on the cations present at B sites.

4.3.5 *MANGANESE FERRITE*

$MnFe_2O_4$ is one of the most promising and fascinating spinel ferrite nanoparticles, which combines the benefits of both iron and manganese-based materials. The Mn-Fe ferrite has superior biocompatibility to hematite, cobalt ferrite, magnetite, and nickel ferrite, making it a good choice for use in MRI and other biomedical applications.[49,50] According to Asghar et al.,[49] the thermal decomposition process yields more homogeneous Mn-Fe with smaller crystallite sizes (5 nm) than the coprecipitation method (9.5 nm) crystallite size. Wang et al. found that altering the amounts of Fe and Mn as precursors in oleic acid results in various shapes of manganese ferrite, including cubic, spherical, and even plate-like structures with TEM sizes of 16, 18, and 27 nm.[51] The development of manganese ferrite with various shapes and sizes has been shown to be generated by the adsorption of varying amounts of capping agent on surfaces with varying free surface energy. Using the coprecipitation synthesis process, Iranmanesh et al.[52] investigated the effect of pH on saturation magnetization. According to his findings, the saturation magnetization diminishes during the synthesis of small-sized manganese ferrite at a high pH.

4.4 STRUCTURAL PROPERTIES

As previously stated, nickel ferrites have an inverted spinel structure. Eight tetrahedral spaces and sixteen octahedral voids are equally occupied by $Fe^{3+}$and Ni^{2+} ions. The XRD patterns of nickel-zinc ferrites contain eight peaks: (111), (220), (311), (222), (400), (422), (511), and (440). The structural properties of nickel ferrites include lattice constant, crystallite size, interionic parameter, lattice strain, grain size, density, and porosity.

4.5 LATTICE CONSTANT

Due to the cubic structure of nickel ferrites, the three lattice constants have identical values ($a = b = c$). Several methods exist for calculating the lattice constant. The bulk of modern-day researchers determines the lattice parameter utilizing software applications such as Powder and Unicell Win. However, the lattice constant of ferrites can also be determined using a number of manual techniques:

1. Average lattice constant (a_{avg}):
The lattice constant is calculated using the formula,

$$a = d\sqrt{h^2 + k^2 + l^2} \tag{4.1}$$

where d is the interplanar spacing which is calculated using Bragg's law formula,

$$d_{hkl} = \frac{\lambda}{2\sin\theta_{hkl}} \tag{4.2}$$

h, k, and l are Miller indices of the crystal plane.[53] By averaging the values (a_{avg}) for each peak, the average lattice constant is determined.[5]

i. True lattice constant (a_{true}): The value of a is calculated by the Nelson–Riley method. In this method, the Nelson–Riley function also known as extrapolation function $F(\theta)$ is determined for each diffraction peak. The $F(\theta)$ is given by

$$F(\theta) = \frac{1}{2}\left[\frac{cos^2}{sin\theta} + \frac{cos^2\theta}{\theta}\right] \tag{4.3}$$

Then from the previous method (average lattice constant) the value of a is plotted against $F(\theta)$. The value of the true lattice constant is obtained by the straight lines passing through the values. This approach was applied to nickel-zinc ferrites by Babu et al.[47]

ii. Theoretical lattice constant (a_{th}): This theoretical method includes the calculation of mean ionic radii of A sites (r_A) and B sites (r_B) using the ideal cationic distribution.[54,55]

$$r_A = \sum_i a_i r_i \qquad rB = \frac{1}{2}\sum_i a_i r_i \tag{4.4}$$

where i is the element, α_i represents the concentration of element i, r_i is the ionic radius. lattice parameter is found using the equation,

$$a_{th} = \frac{8}{\sqrt[3]{3}}\left[r_a + R_0 + \sqrt{3}\left(r_b + r_0\right)\right] \tag{4.5}$$

where r_0 is the radius of oxygen ion 1.32 Å.[56] The reason for the continual difference among the experimental values and theoretical values of lattice constant is that this method considers the existing unit cell as a spinel structured ideal with the unit cell of all ions arranged in a rigid manner.[57] Yan Z. reported lower lattice constant values (8.310–8.347 Å) based on theoretical calculations than practically measured values (8.323–8.364 Å).[58]

4.6 GRAIN SIZE

The technique used to determine the structure and size of the grains is scanning electron microscopy. The Debye–Scherrer's formula used in the XRD pattern to calculate the crystallite size analysis is always smaller than the grain size determined using SEM analysis. This difference results from the grain size indicated by SEM analysis being that of aggregated particles, which are made out of many crystallites by soft reunion.[59] The average grain size of the ferrite is estimated using the line intercept method.[42] A straight line is drawn in the micrograph. It is counted how many grain boundaries cross the line. By multiplying the number of intersections by the actual line length, the average grain size is calculated.[59] Many scholars have used Cottrell's technique to calculate the typical grain size.[60]

It is reported that the cationic distribution affects the grain size.[61] Smaller-sized grains are produced by longer grinding periods. Because of their high surface free energy, fine powders are known to facilitate the process of diffusive acceleration and densification while solid-state sintering. For this reason, if the grinding interval is increased, the sintering temperature is dropped. The minimum loss and large Q-factor of the ferrites are caused by the huge particle size and minimum sintering temperature.[62]

4.7 CRYSTALLITE SIZE

By altering the sintering conditions and doping different ions, crystallite size can be regulated. The growth of crystallites steadily rises with an increase in sintering temperature. This causes crystallites to combine, which increases crystallite size.[63] Increased Ce-Zn substitution in nickel ferrites was associated with smaller crystallites, which was due to Ce's higher ionic radius necessitating more energy to complete the substitution and build the Ce-O bond. A decrease in crystallite size is observed because the energy source costs crystallization.[56] We can calculate the crystallite size by two methods (i) Scherrer's formula and (ii) Williamson Hall method.

i. Scherrer's formula: The Scherrer's method or full-width half maxima method is more efficient when the crystallite size is below 100 nm.[37] The Scherrer's equation is given by

$$D = \frac{k\lambda}{\beta \cos\theta} \qquad (4.6)$$

where k represents the Scherrer's constant or the shape factor, β is the full-width half maxima measured in radians is the wavelength of X-ray, θ is the Bragg's diffraction angle.

ii. Williamson Hall method: This method is used when the grain size is above 100 nm and the XRD data is taken to calculate the crystallite size.[64] Williamson Hall equation is given by

$$\beta cos\theta = \frac{K\lambda}{D} + 4\varepsilon sin\theta \tag{4.7}$$

where B represents the half-width full maxima, θ and K represent the peak position, and Scherrer's constant which is 0.89. ε is the lattice strain and λ represents the wavelength of the X-ray.

4.8 DENSITY AND POROSITY

When performing structural analysis, two densities are typically measured: X-ray density and the actual density. The theoretical density is also referred to as the X-ray density, because it involves knowledge of the crystal structure of the investigated molecule, which is inferred via X-ray diffraction analysis. X-ray density is calculated using the equation

$$D_X = \frac{\Sigma A}{NV} = \frac{8M}{Na^3} \tag{4.8}$$

where V is the volume of the unit cell, N is the Avogadro's number, M represents the molecular weight, and ΣA represents the total atomic weight of atoms in a unit cell. The molar masses of the doped cations are responsible for the variation in the density of X-ray and actual density that occurs with variations in the quantity of doped entity. The doped cations' molar masses were cited by Ali R et al. as the cause of the increase in densities with increasing substitution amount in their study on XRD characterization of Co-Zr doped Ni ferrites.[55] Porosity is a fundamental phase of ceramic materials produced through sintering as well as powder treatment. A rapid rate of sintering and small crystallite sizes considerably minimizes porosity.[65] The porosity (P) is calculated by

$$P = \frac{D_X - D_a}{D_X} * 100 \tag{4.9}$$

where D_a and D_x are actual and X-ray density.[65,66]

4.8.1 DIELECTRIC PROPERTIES OF FERRITES

Understanding the dielectric and electric characteristics of ferrites is vital because they govern how these materials are utilized to create electrical devices. Impedance analyzers are often used to test electric and dielectric characteristics. This gadget detects the complex impedance (*Z*) of the sample and converts it to the display's inductance (*L*), capacitance (*C*), and resistance (*R*) values. Batoo et al. have developed an impedance analyzer block diagram that is used to investigate the dielectric properties in the MHz frequency range.[67]

Wahba et al. used a low-frequency lock-in amplifier to measure the dielectric and electric characteristics (LIA, SR510). This method examines the phase angle (ϕ) between the resistance V_R and the current I as well as the voltage drop across V_R and a standard resistance *R*; *R*'s value was set to ensure that V_R does not surpass 1% of the signal generator's applied voltage's rms value.[61] To prevent moisture absorption on the surfaces, the pellets were housed in an evacuated silica tube. After the temperature had been stable for at least 10 min, the data was gathered. The dielectric constant (ε') was determined using the equation:

$$\varepsilon' = \frac{A_s t_s \sin \Phi}{\varepsilon^0 \omega_s} \tag{4.10}$$

where ε_0 is the permittivity of space, t_s and A_s are the thickness and the surface area of the sample, respectively.

A number of important parameters, including the synthesis technique, particle size, cation distribution, annealing temperature, stoichiometric composition, ratio of Fe^{3+}/Fe^{2+} ions, and oxygen vacancies, have a significant impact on the dielectric characteristics of nanoferrites.

4.8.2 DIELECTRIC CONSTANT

Since ferrites have a very high dielectric constant,[68,69] they are useful for developing microwave devices such as isolators, switches, etc. Numerous variables can affect the dielectric properties of ferrites, including production technique, sintering temperature, chemical composition, crystal structure, grain size, additives, applied field frequency, and humidity. The polarization process, or how quickly the polarizable units in a material position themselves to keep up with the oscillations of an alternating field, determines the dielectric constant of ferrites.[70] Dipolar, electronic, ionic, and interfacial polarizations are the four distinct kinds. Lower frequencies are dominated by dipolar and interfacial polarizations, while higher frequencies are dominated by electronic and ionic polarizations.[35]

The following factors affect the ferrite material's dielectric constant:

Sintering temperature: A low sintering temperature results in a low Fe^{2+} content as well as a lower density because the concentration of dipoles is lower. These mechanisms are in charge of the low values of the dielectric constant (the actual component of permittivity) for ferrites sintered at low temperatures.[71]

Particle size: Due to the presence of large surface boundaries in nanomaterials, their dielectric constant values are significantly lower than those of bulk materials (regions of high resistance). As a result, the interfacial polarization is reduced, which lowers the dielectric constant for nanoferrites.[72] With an increase in grain size, ferrite's dielectric constant typically rises. Large-sized grains have distinguishable grain and grain boundary differences, which improves polarization and, in turn, dielectric constant.[73]

Porosity: Atiq et al. found a correlation between the porosity variation and the dielectric constant variation.[5] Increased porosity causes poorer contact between adjoining grains and a rise in the surface area of the grain border region, which has a high resistive character. Increased porosity and grain boundaries result in more structural flaws, which lead to charge carrier accumulation at the interfaces. This reduces the carrier mobility, which in turn reduces the dipolar orientations, due to the applied oscillations in the AC electric field.[5]

Temperature: The dielectric constant rises as temperature rises, which may be related to an increase in molecular reorganization and orientation.[73]

4.8.3 DIELECTRIC LOSS

The dielectric loss of a dielectric medium may be used to determine the amount of energy it absorbs.[12] The dielectric loss occurs when the polarization lags behind the applied alternating field due to crystal lattice defects and flaws. As frequency increases, losses in a dielectric medium decrease.

4.9 MAGNETIC PROPERTIES

Magnetism in ferrites is caused by irregular 3D electrons dispersed unevenly at tetrahedral and octahedral locations.[23,22] In ferrimagnetism (a form of antiferromagnetism) the magnetic moments are aligned in opposition to one another but their magnitudes are not the same. As a result, the net magnetization of these materials is not zero. Yafet and Kittel[74] demonstrated that ferrimagnetism may occur when specific tetrahedral and octahedral cationic sites are divided into sublattices so that the vector formed by the sublattice magnetic moments is aligned in opposing directions.[75] In accordance with the Yafet Kittel model, at a given composition, the sublattice (A or B) begins to divide into two or four distinct portions, each of which is at an angle with the others known as the Yafet–Kittel angle.[76] The saturation magnetization (Ms), coercivity (H_c), and curie temperature are the magnetic characteristics of nickel ferrites.

4.10 SATURATION MAGNETISATION (M_S)

Saturation magnetization is determined by the cationic distribution of metal ions at the tetrahedral (A) and octahedral (B) sites. "Neel's two sublattice models of ferromagnetism" can be used to explain how spinel ferrites become magnetized. "The magnetic moments at the A and B cationic sites are said to be aligned antiparallel to one another, and their spins are said to have a collinear structure".[54] The difference between magnetization at B site and A site gives the total magnetization.

$$M = |M_B - M_A| \tag{4.11}$$

Between cationic sites A and B, there are three distinct exchange interactions that might take place: J_{AA} (inter A site), J_{AB} (between A and B site), and J_{BB} (inter B site). These are all antiferromagnetic interactions. The formula used to calculate the theoretical value of saturation magnetization is given by Ref. [77].

$$\mathrm{M}_s = \frac{8n\mu B}{10^3 a^3 \rho} \tag{4.12}$$

where μ_b is the size of the Bohr magnetron, a is the lattice parameter, and ρ is the density, n is the theoretical number of magnetic moments per formula unit. The following details were used to explain why M_S levels obtained experimentally were lower: A possible issue with the ion configuration at the octahedral and tetrahedral sites is that the spinel lattice's net magnetization may not support Neel's two-sublattice theory, and spins at B sites may not continue to be parallel in a canted structure as opposed to a collinear configuration.[78]

4.11 COERCIVITY (H_C)

Grain size has an inverse relationship with coercive force.[79] The volume concentration of grain boundaries increases as grain size decreases. Kannan et al. explained that coercive force depends on sintering temperature, the increase in sintering temperature increases the number of domain walls within the grains and it causes decreases the coercivity of the material.[80] Coercivity is affected by saturation magnetization, microstrains, magnetic particle shape, magnetocrystalline anisotropy, and magnetic particle size.

4.12 CURIE TEMPERATURE

The Curie temperature is an essential measure in the study of magnetic materials for determining the strength of the intersublattice interaction of ferrites. Above this point, the material loses its magnetic properties and the initial permeability rapidly decreases. Material changes from ferrimagnetic to paramagnetic at Curie temperature. Hopkinson peak refers to the first permeability peak that occurs just below the Curie temperature. Curie temperature is the temperature at which a rapid decline in permeability. Ferrites do have large anisotropy value at lower temperatures, which diminishes as the temperature rises and finally disappears at Curie temperature.

4.13 APPLICATIONS OF FERRITES

Ferrites are well-known for their exceptional qualities, which include high electric resistivity,[16] high saturation magnetization, low coercivity, and curie temperature, among others. Due to their high magnetic losses and high resistance,[81] spinel ferrites are frequently employed as microwave absorbers. These ferrites are utilized in the frequency range of 3–30 GHz.[23] Due to their thermodynamic stability, electrical conductivity, and corrosion resistance, these materials[82] are ideally suited for use in the metallurgical and other high temperature sector. In this part, we will briefly examine some applications of ferrites, including inductors, in bioscience, Solar cells, the process of water splitting, sensors, electromagnetic interference (EMI) suppressors, and other high frequency applications.

4.13.1 INDUCTORS

A metallic coil and a very soft ferrite are the two main essential material that required to design an inductor. Ferrites are typically used as inductive components in a variety of electronic circuits, such as filters, impedance matching networks, low-noise amplifiers, voltage-controlled oscillators, and amplifiers.

In addition, a process suitable for the fabrication of integrated circuits must be employed to manufacture the ferrite film to provide a high permeability at the operating frequency. Sputtering generates dense films; however, it can be difficult to accurately control the composition, and annealing processes can reach high temperatures. Pulsed laser deposition may generate high-quality films, but preparing the ferrite layer using a mix of sol-gel and spin-coating is simpler and less costly.[83] Ferrite thin-film inductors are available in several configurations, including single-level and dual-level metal connection topologies.

4.13.2 SOLAR CELL

Photovoltaic devices generate electricity by converting sunlight into direct current electricity through semiconducting materials that show the photovoltaic effect.[84] The majority of photovoltaic devices (solar cells) used in commerce are made of inorganic materials, particularly silicon. Photovoltaic systems have drawn a lot of interest in recent years as scientific and financial progress has increased.

Photovoltaic cells are often divided into first, second, and third-generation devices depending on the underlying technology. The three subcategories of photovoltaic cells are single- or multicrystalline, polycrystalline, and thin-film (amorphous) silicon solar cells. Major issues with silicon solar cells are the high temperature and purity requirements for manufacture, the high cost of the components, and the damaging effects on the environment.[85] Inorganic semiconductor materials developed in the 1970s are the foundation of the second-generation solar cell.[86] According to theoretical research, the absorption properties of crystalline silicon are fairly similar to the solar spectrum. The so-called "Shockley–Queisser limit" states that the greatest conversion efficiency of solar energy into electrical energy for absorber materials with band gaps E_g between 1.45 and 1 eV is 33%.

Third-generation solar cells are often used to describe systems that do not fall into the first or second group and/or those that try to surpass the 33.7% Shockley–Queisser limit.[87] The third-generation solar cell techniques focus on achieving high efficiency with widely available, harmless materials. The light source can be focused, tandem cells with numerous band gaps can be used, or photons can be converted such that several electron-hole pairs can be produced from a single photon.

Third-generation solar cells have lately been revitalized once more by ferroelectric and multiferroic photovoltaics. Because several ferroelectric and multiferroic materials, such as $Pb(Zr,Ti)O_3$, $BiFeO_3$, and Bi_2FeCrO_6, have recently been found to have very strong photoresponses. The photovoltaic effect in ferroelectric materials was discovered around 50 years ago, but it wasn't until the discovery of a high photovoltage (15 V) in $BiFeO_3$ thin films that the phenomenon gained widespread attention. Based on the characterization of bismuth ferrite, it is anticipated that the coupling of TiO_2 and $BiFeO_3$ will result in an efficient charge separation and transport system and a high solar light absorptivity. Without using any photovoltaic applications, Wu et al. studied the diode-like behavior of the bismuth ferrite/ZnO combination. It has been demonstrated that Bi deficiency acts as a *p*-type semiconductor by delivering hole carriers in bismuth ferrite.[88] The photovoltaic capabilities of ferroelectric materials such as the perovskites $BaTiO_3$ and $LiNbO_3$ have recently been studied.[89] $BiFeO_3$ has attracted a lot of attention recently due to its high-temperature ferroelectricity, magnetic ordering, extensive use in magnetoelectric random access memory, sensor and actuator, ferroelectric tunneling junction, nonvolatile ferroelectric-based random-access memory, and other systems.

4.14 HIGH FREQUENCY APPLICATIONS

The need for magnetic materials for high-frequency applications has increased as the frequencies and bandwidths needed by microwave technology, telecommunications, and radar systems have increased to 100 GHz. Ferrites are nonconducting oxides, which allow full electromagnetic field penetration, in contrast to metals, where the skin effect considerably limits the penetration of high-frequency fields.[23] Numerous ferrite applications depend on the fact that the spin rotation depends on the orientation of the external field to control the interaction with the microwave field. In one direction, the ferrite transmits the microwave field while substantially absorbing it in the other. This is the basis for nonreciprocal devices. Circulators, isolators, phase shifters, and antennas are common examples. The most widely used microwave ferrites at the moment are yttrium iron garnet[92,93] both in bulk and film form, spinel-type Ni-Zn, Mn-Mg-Zn, and lithium ferrites,[91] hexagonal-type barium ferrite.[90] Global demand for extremely low-loss and reasonably priced microwave devices made of ferrites has been further emphasized as a result of the explosive growth in microwave communication via mobile and satellite communications.

4.14.1 SUPERCAPACITORS

Although supercapacitors have very different technological characteristics, they function similarly to electrostatic and electrolytic capacitors. A supercapacitor has a capacitance that is 10 times greater than an electrolytic capacitor and an energy density, that is, 100 times greater than an electrostatic capacitor's.[94] Symmetric, asymmetric, and hybrid supercapacitors are the three basic kinds of commercial and experimental supercapacitors. In essence, symmetric supercapacitors are electric double-layer capacitors. The term "symmetric" in this context indicates that the materials, mass, and design of the electrodes are same.[87] The two major kinds of symmetric capacitors are double-layer electric capacitors and pseudocapacitors.[95] A supercapacitor that uses two distinct electrode materials is known as an asymmetric supercapacitor (ASC). One electrode relies mostly on electric double-layer (non-Faradic or electrostatic) absorption/desorption, whereas the other is dependent on redox (Faradic) reactions with or without non-Faradic reactions.[96] The fundamental principle behind hybrid capacitors is to use the optimal geometry and materials for each component to successfully improve technical features and even reduce the price of conventional supercapacitors. Deng et al. have described the production method and electrochemical properties of $CoFe_2O_4$ nanorods and nanorings. Ferrites, such as nickel ferrite, cobalt ferrite, manganese ferrite, and $CoFe_2O_4$, have also been the focus of extensive research. By generating unique hybrids based on $CoFe_2O_4$, performance can be enhanced.[97] Sreedhar et al. created $CuFe_2O_4$ nanoparticles as a magnetically reversible and reusable catalyst for the synthesis of 1*H*-tetrazoles. The one-step impregnation approach developed by Yen et al. utilizes mesoporous silica.[98,99]

4.14.2 POWER APPLICATIONS

Ferrites are utilized in numerous power-related items, including computers, televisions, video systems, and small and medium-sized machinery. Recent research has investigated how to make ferrite cores more effective by decreasing eddy current and increasing resistivity. In addition to employing nonconductive additives that frequently stick to grain boundaries, Mn-Zn and Ni-Zn are mixed using the citrate precursor method to produce $Mn_xNi_{0.5-x}Zn_{0.5}Fe_2O_4$.[100] To achieve minimum losses at high operating temperatures of up to 140°C, Mn-Zn ferrites have been modified and included into vehicle power devices.[101] Losses in ferrites are mostly caused by hysteresis loss at low frequency, eddy current loss, and relaxation resonance loss at high frequency.

4.14.3 SENSORS (GAS AND HUMIDITY)

Components of a chemical sensor or gas sensor include a transducer and an active layer, which transform the chemical input into another type of electrical signal, such as a change in frequency, current, or voltage.[102] It is a device for detecting hazardous gases in the environment. The ferrite-based gas sensors are mechanically reliable, have powerful detecting capabilities, and are reasonably priced. During a chemical reaction, when there is an electron transfer process between gas molecules and a ferrite surface, the electric resistivity and conductivity of a ferrite surface can change. Multiple factors affect the performance of a sensor, including sensitivity, selectivity, detection limit, reaction time, and recovery time. Selectivity is the capacity of a sensor to respond selectively to a single analyte or a collection of analytes. The detection limit is the lowest analyte concentration detectable by the sensor under a given set of conditions. Response time is the length of time required for a sensor to respond to a step change in concentration. Recovery time is the time required for a signal to return to its initial value following a step concentration value.

Researchers have investigated the characteristics and responses of ferrites as sensors. According to Liu et al., $CdFe_2O_4$ has a high sensitivity to vaporized ethanol. $CdFe_2O_4$ demonstrates a high sensitivity to vaporized ethanol. Reddy et al.[103] examined $NiFe_2O_4$ as a Cl_2 sensor for the atmosphere. Chen et al.[104] report that $MgFe_2O_4$ and $CdFe_2O_4$ are sensitive and specific to C_2H_2 and liquefied petroleum gas. As observed by Kapse et al.,[105] the addition of palladium to Ni-Zn ferrite enhances ethanol sensitivity and response times. Kadu et al.[106] revealed that the addition of palladium to Mn-Zn ferrite enhanced sensitivity, response time, and decreased working temperature toward ethanol from 300 to 230°C.

Numerous industries, including the paper and sugar industries, the chemical and pharmaceutical industries, the automobile industry, climatology, heating, ventilation, and air conditioning (HVAC) systems, agriculture, etc. require accurate humidity monitoring. The ability to collect and analyze data on environmental humidity in real time is essential for meteorological applications since it enables the forecasting of precipitation, dew, and fog. Materials composed of nanoparticle-sized polycrystalline spinel ferrite can detect dampness. As humidity sensors, several researchers have examined spinel ferrites. Lipare et al.[107] investigated the impact of Li-Cl substitution on Cu-Zn spinel ferrites. Wang et al. The reaction time of composite materials comprised of polymer quaternary acrylic resin or lanthanum ferrite has improved,[108], as determined by a study of their humidity-dependent electrical properties. It is considered that the presence of a protonic conduction mechanism explains how materials react to humidity. In their research of the humidity response of Li-Mg ferrites, Kotnala et al.[109] discovered that lithium makes magnesium ferrite more sensitive to humidity.

4.14.4 PROGRESS IN FERRITES

Due to their outstanding magnetic properties, remarkable heat and chemical resistance, low cost, and other attributes, ferrites have garnered a great deal of interest for a very long time.[110] Ferrite-based technologies that are "mature" include permanent magnets, magnetic recording, and microwave devices. Since more than fifty years ago, ferrites have been studied and utilized.

Electromagnetic pollution and EMI resulting from the rapid growth of commercial and military radars and other electronic devices[111–113] are a significant issue. Electromagnetic interference, which is simply electromagnetic disturbance from other sources, degrades the performance of a device. Soft ferrites,[114] ferromagnetic metals,[115] ferromagnetic metal/hexaferrite composites,[116] encapsulated magnetic particles,[117] and carbon nanotube composites[118] are among the materials that can be used to generate EMI suppressors. In recent years, however, the need for these materials has increased due to shrinking trends, improved integration density, and higher clock frequencies, especially in communication, computer, and information technologies. Composites of carbon nanotubes and polystyrene foam have proven[119] significant EMI shielding efficiency via a method involving the reflection of electromagnetic radiation. Hexagonal ferrites offer an intriguing alternative to cubic ferrites as EMI suppressor components due to their higher resonance frequencies, relatively high microwave permeabilities, and high electrical resistance.

Ni-Cu-Zn ferrite is utilized in numerous products, including multilayer ferrite chip components, noise filters, and rotary transformers.[120] Mn-Zn ferrite is mostly utilized in flyback transformers, television, and carrier telephones. In recent years, however, Mn-Zn has been utilized mostly in switching transformers. In response to the rising demand for electromagnetic compatibility (EMC) countermeasures, ferrite wave absorbers have garnered substantial interest from all around the world.[121,122] Wave absorber composites research has recently advanced using carbon-based ferrite composites, polymer-based ferrite composites, ceramic-based ferrite composites, and biomass-based ferrite composites.

Numerous biotechnological applications that rely on magnetic micro- and nanoparticles that are both biogenic and synthetic[123] have been created. Radionuclides have been targeted on specific tissues using magnetic nanoparticles. Utilizing thermal energy derived from the hysteresis loss of ferrites, hyperthermia, or the heating of particular tissues or organs is carried out. As tumor tissues become warmer, they become more amenable to radiotherapy and chemotherapy.

In recent years, there has been considerable interest in hydrogen as a potential future energy source. Therefore, ferrites are a potentially promising material in the creation of hydrogen to replace TiO_2 as a photocatalyst. The primary benefits of transition metal ferrites are their low cost, excellent catalytic behavior, resistance to corrosion, and, most importantly, their huge bandgaps in the visible light spectrum.[124]

4.15 CONCLUSION

In this chapter, we examine the properties of several ferrites. Initially, we talked about the different types of ferrites and what sets them apart. After that, we talked about how spinel ferrites can be classified as either ordinary or inverse spinel based on their stability. When ferrites are doped with certain elements, they take on new properties and become more useful in numerous applications. Here, we have zeroed in on nickel, cobalt, zinc, and manganese in their ferrite forms. The structural and dielectric properties of ferrites, arguably their most vital characteristics, are also covered. In this article, we explore the current uses and potential future developments of ferrites, highlighting the fascinating nature of this material.

KEYWORDS

- **spinel ferrite**
- **nickel spinel ferrite**
- **zinc ferrite**
- **cobalt ferrite**

REFERENCES

1. Smit, J.; Wijn, H. P. J. Ferrites; Philips Technical Library, Eindhoven, **1959**.
2. Hajarpour, S.; Raouf, A. H.; Gheisari, K. Structural Evolution and Magnetic Properties of Nanocrystalline Magnesium Zinc Soft Ferrites Synthesized by Glycineenitrate Combustion Process. J. Magn. Magn. Mater. 2013, 363, 21–25. https://doi.org/10.1016/j.jmmm.2014.03.027.
3. Suarez, A.; Victoria, J.; Alcarria, A.; Torres, J.; Martinez, P. A.; Martos, J.; Soret, J.; Garcia-Olcina, R.; Muetsch, S. Characterization of Different Cable Ferrite Materials to Reduce the Electro-Magnetic Noise in the 2–150 kHz Frequency Range. Materials 2018, 11, 174. https://doi.org/10.3390/ma11020174.
4. Naseri, M. G.; Halimah, M. K.; Dehzangi, A.; Kamalianfar, A.; Saion, E. B.; Majlis, B. Y. A Comprehensive Overview on the Structure and Comparison of Magnetic Properties of Nanocrystalline Synthesized by a Thermal Treatment Method. J. Phys. Chem. Solid 2014, 75, 315–327. https://doi.org/10.1016/j.jpcs.2013.11.004.
5. Mathew, D. S.; Juang, R. S. An Overview of the Structure and Magnetism of Spinel Ferrite Nanoparticles and Their Synthesis in Microemulsions. Chem. Eng. J. 2007, 129, 51–65. https://doi.org/10.1016/j.cej.2006.11.001.
6. Srivastava, R.; Yadav, B. C. Ferrite Materials: Introduction, Synthesis Techniques, and Applications as Sensors. Int. J. Green Nanotechnol. 2012, 4, 141–154. https://doi.org/10.1080/19430892.2012.676918.
7. Beringer, R. H. Electron Spin Magnetic Moment in Atomic Hydrogen. Phys. Rev. 1954, 95, 1474.. https://doi.org/10.1103/PhysRev.95.1474.
8. Kumar, M. K.; Rao, M. S. R. R. Structural, Morphological and Hydrogen Sensing Studies on Pulsed Laser Deposited Nanostructured Palladium Thin Films. S. J. Phys. D Appl. Phys. 2006, 39, 2791.
9. Thakur, A.; Mathur, Singh, P.; M. Study of Dielectric Behaviour of Mn-Zn Nano Ferrites. J. Phys. Chem. Solids 2007, 68, 378. https://doi.org/10.1016/j.jpcs.2006.11.028.
10. Rath, C.; N. C Mishra, Anand, S.; Das, R. P.; Sahu, K. K.; Upadhyay, C.; Verma, H. C. Appearance of Superparamagnetism on Heating Nanosize Mn0.65Zn0.35Fe2O4. Appl. Phys. Lett. 2000, 76, 475. https://doi.org/10.1063/1.125792.
11. Saafan, S. A.; Assar, S. T.; Moharram, B. M.; El Nimr, M. K. Comparison Study of Some Structural and Magnetic Properties of Nano-Structured and Bulk Li-Ni-Zn

Ferrite Samples. J. Magn. Magn. Mater. 2010, 322 (6), 628. https://doi.org/10.1016/j.jmmm.2009.10.027.
12. Olsen, E.; Thonstad, J. Nickel Ferrite as Inert Anodes in Aluminium Electrolysis: Part I Material Fabrication and Preliminary Testing. J. Appl. Electrochem. 1999, 29, 293. https://doi.org/10.1023/A:1003460220418.
13. Berchmans, L. J.; Selvan, R. K.; Augustin, C. O. Evaluation of Mg2+—Substituted NiFe2O4 as Green Anode Material. https://doi.org/10.1016/j.matlet.2003.12.008.
14. Xie, J. L.; Han, M.; Chen, L.; Kuang, R.; Deng, L. Microwave-Absorbing Properties of Ni-Co-Zn Spinel Ferrites. J. Magn. Magn. Mater. 2007, 314, 37–42. https://doi.org/10.1016/j.jmmm.2007.02.124.
15. Heiba, Z. K.; Mohamed, M. B.; Hamdeh, H. H.; Ahmed, M. A. Structural Analysis and Cations Distribution of Nanocrystalline Ni1-xZnxFe1.7Ga.3O4. Alloy. Compd. 2015, 618, 755–760. https://doi.org/10.1016/j.jallcom.2014.08.241.
16. Liu, Q.; Lv, L.; Zhou, J. P.; Chen, X. M.; Bian, X. B.; Lu, P. Influence of Nickel-Zinc Ratio on Microstructure, Magnetic and Dielectric Properties of Ni1-xZnxFe2O4 Ferrites. J. Ceram. Proc. Res. 2012, 13, 110–116.
17. Mukharjee, S.; Pradip, S.; Mishra, A. K.; Das, D. Appl. Phys. A 2013, 114, 389–393. https://doi.org/10.1016/S0304-8853(02)01544-5.
18. Ghasemi, A.; Mousavinia, M. Structural and Magnetic Evolution of Substituted NiZnFe2O4 Particles Synthesized by Conventional Sol-Gel Method. Ceram. Int. 2014, 40, 2825–2834. https://doi.org/10.1016/j.ceramint.2013.10.031.
19. Wu, X.; Chen, W.; Wu, W.; Li, H.; Lin, C. J. Electron. Mater. 2017, 46, 199–207.
20. Kumar, V. K.; Reddy, S. A.; Ravinder, D. High Frequency Dielectric Behaviour of Erbium Substituted Ni-Zn Ferrites. J. Magn. Magn. Mater. 2003, 263, 121–126. https://doi.org/10.1016/S0304-8853(02)01544-5.
21. Sayed, F.; El Haj Hassan, F. Hollow Magnetic Nanoparticles Experimental and Numerical Study. In: Half-Metallic Ferromagnetism in New Chalcopyrite Materials, 2016. DOI:10.13140/RG.2.2.33777.94565
22. Rodrigues, A. P. G.; Gomes, D. K. S.; J. H. Araújo, Mela, D. M. A.; Oliveira, N. A. S.; Braga, R. M. Nanoferrites of Nickel Doped with Cobalt Influence of Co2+ on the Structural and Magnetic Properties. Magn. Magn. Mater. 2015, 374, 748–754. https://doi.org/10.1016/j.jmmm.2014.09.045.
23. Pardavi-Horvath, M. Microwave Application of Soft Ferrites. J. Magn. Magn. Mater. 2000, 215–216, 171–183. https://doi.org/10.1016/S0304-8853(00)00106-2.
24. Saito, S.; Takei, T. J. Jap Soc. Powder Powder Metall. 1966, 13 (2), 60–66.
25. Poplawsky, R.; Wright, M. R. Nickel Iron Ferrite Crystals Using Arc Image Furnace Techniques. J. Crystals Growth 1967, 1 (3), 139–148. https://doi.org/10.1016/0022-0248(67)90024-3.
26. Rivas, J.; Iniguez, J.; Ayala, J. Magnetic After Effects in Nickel Ferrites. Appl. Phys. 1979, 19 (1), 71–74. https://doi.org/10.1007/BF00900541.
27. Orehotsky, J.; Davidson, C. R. Magnetic Properties of Fine Particle NixFe3-xO4 Spinels. Magn. Lett. 1980, 1 (5–6), 117–122.
28. Banks, E.; Riederman, N. H.; Schleuning, H. W.; Silber, L. M. Preparation and Properties of Thin Ferrite Films. J. Appl. Phys. 1961, 32 (3), S44–S45. https://doi.org/10.1063/1.2000491.
29. Lemaire, H. P.; Croft, W. J. Ferrite Thin Films. J. Appl. Phys. 1961, 32 (3), S46–S47. https://doi.org/10.1063/1.2000492.

30. Shamim, M. K.; Sharma, S.; Choudhary, R. J. J. Magn. Magn. Mater. 2019, 469, 1–7. https://doi.org/10.1016/j.jmmm.2018.08.029.
31. Osaka, T.; Electrodeposition of highly functional thin films for magnetic recording devices of the next century, Electrochim. Acta 45 (2000) 3311–3321. https://doi.org/10.1016/S0013-4686(00)00407-2.
32. Kamble, R. C.; Shaikh, P. A.; Kamble, S. S.; Kolekar, Y. D. Effect of Cobalt Substitution on Structural, Magnetic and Electric Properties of Nickel Ferrite. J. Alloy Compd. 2009, 478, 599–603.
33. Shaikh, P. A.; Kambale, R. C.; Rao, A. V.; Kolekar, Y. D. J. Magn. Magn. Mater. 2010, 322 (6), 718–726. https://doi.org/10.1016/j.jallcom.2008.11.101.
34. Maria Magdalane, C.; Maria Assuntha Priyadharsini, G.; Kaviyarasu, K.; Irudaya Jothi, A.; Gnanamani Simiyon, G. Synthesis and Characterization of TiO2 Doped Cobalt Ferrite Nanoparticles via Microwave Method: Investigation of Photocatalytic Performance of Congo Red Degradation Dye. Surface Interphases, ScienceDirect 2021. https://doi.org/10.1016/j.surfin.2021.101296.
35. Mathe, V. L.; Kamble, R. B. Anomalies in Electrical and Dielectric Properties of Nanocrystal-Line Ni–Co Spinel Ferrite. Mater. Res. Bull. 2008, 43 (8–9), 2160–2165. https://doi.org/10.1016/j.materresbull.2007.09.001.
36. Hcini, S.; Selmi, A.; Rahmouni, H.; Omri, A.; Bouazizi, M. L. Structural, Dielectric and Complex Impedance Properties of Ti0.6Co0.4Fe2O4 (Ti, Ni, Mg) Ferrite Nanoparticles Prepared by Sol Gel Method. Ceram. Int. 2017, 43, 2529–2536. https://doi.org/10.1016/j.ceramint.2016.11.055.
37. Zhang, C. S.; Yang, L. Preparation and Magnetic Properties of the Conductive Co1-xNixFe2O4/Polyaniline Microsphere Composites. J. Magn. Magn. Mater. 2012, 324, 1469–1472. https://doi.org/10.1016/j.jmmm.2011.09.036.
38. Ati, A. A.; Othaman, Z.; Samavati, A. Influence of Cobalt on Structural and Magnetic Properties of Nickel Ferrite Nanoparticles. J. Mol. Struct. 2013, 1052, 177–182. https://doi.org/10.1016/j.molstruc.2013.08.040.
39. Wohlfarth, E. P. Ferromagnetic Materials, Vol. 3; North-Holland: New York, 1982; p 248.
40. Stergiou, C. Microstructure and Electromagnetic Properties of Ni-Zn-Co Ferrite up to 20 GHz. Adv. Mater. Sci. Eng. 2016, 1934783. DOI: 10.1155/2016/1934783.
41. Ünal, B.; Baykal, A. Effect of Zn Substitution on Electrical Properties of Nanocrystalline Cobalt Ferrite. J. Supercond. Nov. Magn. 2014, 27, 469–479. https://doi.org/10.1007/s10948-013-2285-2.
42. Ghodake, J. S.; Kamble, R. C.; Shinde, T. J.; Maskar, P. K.; Suryavanshi, S. S. Magnetic and Microwave Absorbing Properties of Co2+ Substituted Nickel–Zinc Ferrites with the Emphasis on Initial Permeability Studies. J. Magn. Magn. Mater. 2016, 401, 938–942. https://doi.org/10.1016/j.jmmm.2015.11.009.
43. Wilson, V. C.; Kasper, J. S. Phys. Rev. 1954, 95 (6), 1408–1411.
44. Arboleda, J. D.; Olmos, O. A.; Aguiree, M. H.; Ramos, R.; Anadon, A.; Ibarra, M. R. Spin See-Beck Effect in a Weak Ferromagnet. Appl. Phys. Lett. 2016, 108, 232401. https://doi.org/10.1063/1.4953229.
45. Abu El-Fadl, A.; Hassan, A. M.; Mahmound, M. H.; Tatarchuk, T.; Yaremiy, I. P.; Gismelssed, A. M.; Ahmad, M. A. Synthesis and Magnetic Properties of Spinel Zn1-xNixFe2O4 ($0.0 \leq x \leq 1.0$) Nanoparticles Synthesized by Microwave Combustion

Method. J. Magn. Magn. Mater. 2019, 471, 192–199. https://doi.org/10.1016/j.jmmm.2018.09.074.

46. Jafarian, M.; Afghahi, S. S. S.; Atassi, Y.; Salehi, M. Insights on the Design of a Novel Multi-Component Microwave Absorber Based on SrFe10 Al2 O19 and Ni0.5 Z 0.5 Fe2O4. J. Magn. Magn. Mater. 2019, 471, 30–38. https://doi.org/10.1016/j.jmmm.2018.09.047.
47. Babu, B. R.; Prasad, M. S. R.; Ramesh, K. V. Role of Synthesis on Physical Properties of Ni0.5Z0.5Fe2O4 Nanoferrite: A Comparative Study. J. Supercond Nov. Magn. 2017, 30 (6), 1609–1617. https://doi.org/10.1007/s10948-016-3964-6.
48. Stergiou, C. Magnetic, Dielectric and Microwave Absorption Properties of Rare Earth Doped Ni–Co and Ni–Co–Zn Spinel Ferrites. J. Magn. Magn. Mater. 2016, 426, 629–635. https://doi.org/10.1016/j.jmmm.2016.11.001.
49. Asghar, K.; Qasim, M.; Das, D. Preparation and Characterization of Mesoporous Magnetic MnFe2O4SiO2 Nanocomposite for Drug Delivery Application. Mater. Today. Proc. 2020, 26, 87–93. https://doi.org/10.1016/ j.matpr.2019.05.380.
50. Baig, M. M.; Yousuf, M. A.; Agboola, P. O.; Khan, M. A.; Shakir, I.; Warsi, M. F. Optimization of Different Wet Chemical Routes and Phase Evolution Studies of MnFe2O4 Nanoparticles. Ceram. Int. 2019, 45, 12682–12690. https://doi.org/10.1016/j.ceramint.2019.03.114.
51. Wang, Z.; Liu, J.; Li, T.; Liu, J.; Wang, B. Controlled Synthesis of MnFe2O4 Nanoparticles and Gd Complex-Based Nanocomposites as Tuneable and Enhanced T1/T2-Weighed MRI Contrast Agents. J. Mater. Chem. B 2014, 2, 4748–4753. https://doi.org/10.1039/C4TB00342J.
52. Hergt, R.; Dutz, S.; Muller, R.; Zeisberger, M. Magnetic Particle Hyperthermia: Nanoparticle Magnetism and Materials Development for Cancer Therapy. J. Phys. Condens. Matter. 2006, 18, s2919–s2934. https://doi.org/10.1088/ 0953-8984/18/38/S26.
53. Patil, N. D.; Velhal, N. B.; Puri, V. R. Electromagnetic Properties of Co Substituted Ni–Cd Ferrite Thick Films on Inorganic Substrate by Auto Combustion Technique. J. Mater Sci.: Mater. Electron. 2017, 28, 1793–1798. https://doi.org/10.1007/s10854-016-5727-3.
54. Ünal, B.; Ünver, I. S.; Güngüneş, H.; Topal, U.; Baykal, A.; Sozeri, H. Microwave, Dielectric and Magnetic Properties of Mg-Ti Substituted Ni-Zn Ferrite Nanoparticles. Ceram. Int. 2016, 42, 17317–17331. https://doi.org/10.1016/j.ceramint.2016.08.028.
55. Ali, M. A.; Uddin, M. M.; Khan, M. N. I.; Chowdhury, F. U. Z.; Haque, S. M. Structural, Morphological and Electrical Properties of Sn-Substituted Ni-Zn Ferrites Synthesized by Double Sintering Technique. J. Magn. Magn. Mater. 2017, 424, 148–154. https://doi.org/10.1016/j.jmmm.2016.10.027.
56. Yan, Z.; Luo, J. Effects of Ce Single Bond Zn Co-substitution on Structure, Magnetic and Microwave Absorption Properties of Nickel Ferrite Nanoparticles. J. Alloys Compd. 2017, 695, 1185–1195. https://doi.org/10.1016/j.jallcom.2016.08.333.
57. Zaki, H. M.; Al-Heniti, S. H.; Elmosalami, T. A. Structural, Magnetic and Dielectric Studies of Copper Substituted Nano-Crystalline Spinel Magnesium Zinc Ferrite. J. Alloys Compd. 2015, 633, 104–114. https://doi.org/10.1016/j.jallcom.2015.01.304.
58. Zhuang, J.; Chen, X. P.; Chi, Y. H.; Liu, Y. M. Gongneng Cailiao. J. Funct. Mater. 2006, 37 (1), 43–46.

59. Vijaya Babu, K.; Chandra, M. R.; Kumar, G. V. S.; Jagadeesh, K. Effect of Cobalt Substitution on Structural, Electrical and Magnetic Properties of NiFe2O4. Process. App. Ceram. 2017, 11 (1), 60–66. DOI: 10.2298/PAC1701060V.
60. Cottrell, A. An Introduction to Metallurgy; Edward Arnold: London, 1967. https://doi.org/10.1201/9780429293917.
61. Wahba, A. M.; Mohamed, M. B. Structural and Magnetic Characterization and Cation Distribution of Nanocrystalline CoxFe3-xO4 Ferrites. J. Magn. Magn. Mater. 2015, 378, 246–252. https://doi.org/10.1016/j.jmmm.2014.10.164.
62. Wu, X.; Yan, S.; Liu, W.; Feng, Z.; Chen, Y.; Harris, V. G. Influence of Particle Size on the Magnetic Spectrum of Ni-Cu-Zn Ferrites for Electromagnetic Shielding Applications. J. Magn. Magn. Mater. 2016, 401, 1093–1096. https://doi.org/10.1016/j.jmmm.2015.10.129.
63. Toksha, B. G.; Shirsath, S. E.; Patange, S. M.; Jadhav, K. M. Structural Investigations and Magnetic Properties of Cobalt Ferrite Nanoparticles Prepared by Sol–Gel Auto Combustion Method. Solid State Commun. 2008, 147, 479. https://doi.org/10.1016/j.ssc.2008.06.040.
64. George, M.; John, A. M.; Nair, S. S.; Joy, P. A.; Anantharam, M. R. Finite Size Effects on the Structural and Magnetic Properties of Sol–Gel Synthesized NiFe2O4 Powders. J. Magn. Magn. Mater. 2006, 302, 190–195. https://doi.org/10.1016/j.jmmm.2005.08.029.
65. Chavan, P.; Naik, L. R.; Belavi, P. B.; Chavan, G.; Ramesha, C. K.; Kotnala, R. K. Studies on Electrical and Magnetic Properties of Mg-Substituted Nickel Ferrites. J. Electron Mater. 2017, 46, 188–198. https://doi.org/10.1007/s11664-016-4886-6.
66. Khan, K. Microwave Absorption Properties of Radar Absorbing Nanosized Cobalt Ferrites for High Frequency Applications. J. Super Cond. Nov. Magn. 2014, 27, 453–461. https://doi.org/10.1007/s10948-013-2283-4.
67. Batoo, K. M.; Kumar, S.; Lee, C. G. Alimuddin. Curr. Appl. Phys. 2009, 9, 1072.
68. Ravinder, D.; Reddy, P. V. B.; Shalini, P. Frequency and Composition Dependence of Dielectric Behavior of Copper Substituted Lithium Ferrites. J. Mater. Sci. Lett. 2003, 22, 1599. https://doi.org/10.1023/A:1026388509062.
69. Ravinder, D.; Rao, S. S.; Shalini, P. Room Temperature Electric Properties of Cadmium-Substituted Nickel Ferrites. Mater. Lett. 2003, 57, 4040. https://doi.org/10.1016/S0167-577X(03)00089-2.
70. Xiao, H. M.; Liu, X. M.; Fu, S. Synthesis, Magnetic and Microwave Absorbing Properties of Core-Shell Structured MnFe2O4 /TiO2 Nanocomposites. Comp. Sci. Tech. 2006, 66, 2003. https://doi.org/10.1016/j.compscitech.2006.01.001.
71. Stergiou, C.; Litsardakis, G. Structural and Magnetic Properties of Yttrium and Lanthanum-Doped Ni-Co and Ni-Co-Zn Spinel Ferrites. AIP Conf. Proc. 2014, 1627, 117–122. DOI: 10.1063/1.4901668.
72. Jalaiah, K.; Babu, K. V. Structural, Magnetic and Electrical Properties of Nickel Doped Mn-Zn Spinel Ferrite Synthesized by Sol-Gel Method. J. Magn. Magn. Mater. 2017, 423, 275–280. https://doi.org/10.1016/j.jmmm.2016.09.114.
73. Amir, M.; B. Ünal, Geleri, M.; H. Güngüneş, Shrisath, S. E.; Baykal, A. Electrical Properties and Hyperfine Interactions of Boron Doped Fe3O4 Nanoparticle. Superlattices Microstruct. 2015, 88, 450–466. https://doi.org/10.1016/j.spmi.2015.10.005.
74. Yafet, Y.; Kittel, C. Antiferromagnetic Arrangements in Ferrites. Phys. Rev. 1952, 87, 290. https://doi.org/10.1103/PhysRev.87.290.

75. Lee, S. H.; Yoon, S. J.; Lee, G. J.; Kim, H. S.; Yo, C. H.; Ahn, K.; Lee, D. H.; Kim, K. H. Electrical and Magnetic Properties of NiCrxFe2-xO4 Spinel. Mater. Chem. Phys. 1999, 61, 147–152. https://doi.org/10.1016/S0254-0584(99)00136-4.
76. Sinfrônio, F. S. M.; Santana, P. Y. C.; Coelho, S. F. N.; Silva, F. C.; Menezes, A. S. D.; Sharma, S. K. Magnetic and Structural Properties of Cobalt- and Zinc-Substituted Nickel Ferrite Synthesised by Microwave-Assisted Hydrothermal Method. J. Electron. Mater. 2017, 46 (2), 1145–1154. https://doi.org/10.1007/s11664-016-5081-5.
77. Valenzuela, R. Magnetic Ceramics; Cambridge University Press: Cambridge, 2005.
78. Safari, A.; Gheisari, K. H.; Farbod, M. Characterization of Ni Ferrites Powders Prepared by Plasma Arc Discharge Process. J. Magn. Magn. Mater. 2017, 421, 44–51. https://doi.org/10.1016/j.jmmm.2016.07.024.
79. Van der Zaag, P. J.; Van der Valk, P. J.; Rekveidt, M. Th. A Domain Size Effect in the Magnetic Hysteresis of Ni-Zn-Ferrites. Appl. Phys. Lett. 1996, 69, 2927. https://doi.org/10.1063/1.117326.
80. Kannan, Y. B.; Saravanan, R.; Srinivasan, N.; Ismail, I. Sintering Effect on Structural, Magnetic and Optical Properties of Ni0.5Zn0.5Fe2O4 Ferrite Nano Particles. J. Magn. Magn. Mater. 2017, 423, 217–225. https://doi.org/10.1016/j.jmmm.2016.09.038.
81. Xie, J. L.; Han, M.; Chen, L.; Kuang, R.; Deng, L. Microwave-Absorbing Properties of Ni-Co-Zn Spinel Ferrites. J. Magn. Magn. Mater. 2007, 314, 37–42. https://doi.org/10.1016/j.jmmm.2007.02.124.
82. Berchmans, L. J.; Selvan, R. K.; Augustin, C. O.; Evaluation of Mg 2+-Substituted NiFe2O4 a Green Anode Material. Mater. Lett. 2004, 58, 1928. https://doi.org/10.1016/j.matlet.2003.12.008.
83. Yang, C.; Liu, F.; Ren, T. et al. Fully Integrated Ferrite-Based Inductors for RF ICs. Sens. Actuators A 2006, 130–131, 365–370.
84. Surek, T. Crystal Growth and Materials Research in Photovoltaics: Progress and Challenges. J. Cryst. Growth. 2005, 275, 292–304. https://doi.org/10.1016/j.jcrysgro.2004.10.093.
85. Loferski, J. J. The First Forty Years: A Brief History of the Modern Photovoltaic Age. Prog. Photovoltaics Res. Appl. 1993, 1, 67–78. https://doi.org/10.1002/pip.4670010109.
86. Kouhnavard, M.; Ikeda, S.; Ludin, N. A.; Khairudin, N. B. A.; Ghaffari, B. V.; Mat-Teridi, M. A.; Ibrahim, M. A.; Sepeai, S.; Sopian, K. A Review of Semiconductor Materials as Sensitizers for Quantum Dot-Sensitized Solar Cells. Renew. Sustain. Energy Rev. 2014, 397–407. https://doi.org/10.1016/j.rser.2014.05.023.
87. Ponseca, C. S.; Chabera, P.; Uhlig, J.; Persson, P.; Sundstrom, V. Ultrafast Electron Dynamics in Solar Energy Conversion. Chem. Rev. 2017, 117, 10940–11024. https://doi.org/10.1021/acs.chemrev.6b00807.
88. Wu, J.; Fan, Z.; Xiao, D.; Zhu, J.; Wang, J. Multiferroic Bismuth Ferrite-Based Materials for Multifunctional Applications: Ceramic Bulks, Thin Films and Nanostructures. Prog. Mater. Sci. 2016, 84, 335–402. https://doi.org/10.1016/j.pmatsci.2016.09.001.
89. Zhang, Y.; Sun, J.; Perdew, J.; Xifan, W. Comparative First-Principles Studies of Prototypical Ferroelectric Materials by LDA, GGA, and SCAN meta-GGA, Phys. Rev. B 2017, 96. https://doi.org/ 10.1103/PhysRevB.96.035143.
90. Ings, J. B.; Simmins, J. J.; May, J. L.; Tolino, D. Fabrication of Thin Bulk Ceramics for Microwave Circulator Applications, 1995.

91. Wei, F.; Baoshun, L.; Jizhong, Y.; Xi, L.; Muyu, Z. Synthesis and Microwave Absorptive Properties of Lithium Ferrite Nanocrystals. J. Magn. Soc. Jpn. 1998, 22 (Suppl. S1), 366–368.
92. Prakash, O.; Paraolia, R. S. Preparation of Low-Loss Garnets. In Proceedings of 3rd International Conference on Ferrites (Kyoto, Japan. Sept.–Oct. 1980); Watanabe, H., Iida, S., Sugimoto, M., Eds.; Centre for Academic Publications: Tokyo, Japan, 1981; pp 815–817.
93. Kim, M.; Jun, D. S.; Park, J. R.; Lee, S. S. 'Fabrication of a Resonance Absorption-Type Microstrip Line Isolator Using Polycrystalline YIG. J. Magn. Soc. Jpn. 1998, 22 (Suppl. S1) 363–365.
94. Panda, P. K.; Grigoriev, A.; Mishra, Y. K.; Ahuja, R. Progress in Supercapacitors: Roles of Two Dimensional Nanotubular Materials. Nanoscale Adv. 2020, 2, 70–108. https://doi.org/10.1039/ C9NA00307J.
95. Sharma, P.; Bhatti, T. A Review on Electrochemical Double-Layer Capacitors. Energy Convers. Manag. 2010, 51, 2901–2912. https://doi.org/10.1016/j.enconman.2010.06.031.
96. Wang, Y.; Song, Y.; Xia, Y. Electrochemical Capacitors: Mechanism, Materials, Systems, Characterization and Applications. Chem. Soc. Rev. 2016, 45, 5925–5950. https://doi.org/10.1039/ C5CS00580A.
97. Deng, D. H.; Pang, H.; Du, J. M.; Deng, J. W.; Li, S. J.; Chen, J.; Zhang, J. S. Fabrication of Cobalt Ferrite Nanostructures and Comparison of Their Electrochemical Properties. Cryst. Res. Technol. 2012, 47, 1032–1038. https://doi.org/10.1002/crat.201200161.
98. Sreedhar, B.; Kumar, A. S.; Yada, D. CuFe2O4 Nanoparticles: A Magnetically Recoverable and Reusable Catalyst for the Synthesis of 5-Substituted 1H-Tetrazoles. Tetrahedron Lett. 2011, 52, 3565–3569. https://doi.org/10.1016/j.tetlet.2011.04.094.
99. Yen, H.; Seo, Y.; R. Guillet-Nicolas, Kaliaguine, S.; Kleitz, F. One-Step-Impregnation Hard Templating Synthesis of High-Surface-Area Nanostructured Mixed Metal Oxides (NiFe2O4, CuFe2O4 and Cu/ CeO2). Chem. Commun. 2011, 47, 10473–10475. https://doi.org/10.1039/C1CC13867G.
100. Verma, A.; Alam, M. I.; Chatterjee, R.; Goel, T. C.; Mendiratta, R. G. Development of a New Soft Ferrite Core for Power Applications. J. Magn. Magn. Mater. 2006, 300 (2), 500–505. https://doi.org/10.1016/j.jmmm.2005.05.040.
101. Zaspalis, V.; Tsakaloudi, V.; Papazoglou, E.; Kolenbrander, M.; Guenther, R.; and Valk, P. V. D. Development of a New Mn-Zn Ferrite Soft Magnetic Material for High Temperature Power Applications. J. Electro-Ceram. 2004, 13 (1–3), 585–591. https://doi.org/10.1007/s10832-004-5162-3.
102. Fenner, R.; Zdankiewicz, E. Micro Machined Water Vapour Sensors: A Review of Sensing Technologies. IEEE Sens. J. 2001, 1, 309–317. https://doi.org/10.1109/7361.983470.
103. Reddy, C. V. G.; S.Manorama, V.; Rao, V. J. Semiconducting Gas Sensor for Chlorine Based on Inverse Spinel Nickel Ferrite. Sens. Actuators B Chem. 1999, 55, 90–95. https://doi.org/10.1016/S0925-4005(99)00112-4.
104. Chen, N. S.; Yang, X. J.; Liu, E. S.; Huang, J. L. Reducing Gas-Sensing Properties of Ferrite Compounds MFe2O4 (M¼Cu, Zn, Cd and Mg). Sens. Actuators B Chem. 2000, 66, 178–180. https://doi.org/10.1016/S0925-4005(00)00368-3.
105. Kapse, V. D.; Ghosh, S. A.; Raghuwanshi, F. C.; Kapse, S. D.; Khandekar, U. S.; Nanocrystalline Ni0.6 Zn0.4Fe2O4: A Novel Semiconducting Material for Ethanol Detection. Talanta, 2009, 78, 19–25. https://doi.org/10.1016/j.talanta.2008.10.031.

106. Kadu, A. V.; Jagtap, S. V.; Chaudhari, G. N. Studies on the Preparation and Ethanol Gas Sensing Properties of Spinel Zn0.6Mn0.4Fe2O4 Nanomaterials. Curr. Appl. Phys. 2009, 9, 1246–1251. https://doi.org/10.1016/j.cap.2009.02.001.
107. Lipare, A. Y.; Vasambekar, P. N.; Vaingankar, A. S. Dielectric Behaviour and A. C. Resistivity Study of Humidity Sensing Ferrites. Mater. Chem. Phys. 2003, 81, 108–115. https://doi.org/10.1016/S0254-0584(03)00165-2.
108. Wang, J.; Wu, F. Q.; Shi, K. H.; Wang, X. H.; Sun, P. P. Humidity Sensitivity of Composite Material of Lanthanum Ferrite/Polymer Quaternary Acrylic Resin. Sens. Actuators B Chem. 2004, 99, 586–591. https://doi.org/10.1016/j.snb.2004.01.012.
109. Kotnala, R. K.; Shah, J.; Singh, B.; kishan, H.; Singh, S.; Dhawan, S. K.; Sengupta, A.; Humidity Response of Li-Substituted Magnesium Ferrite. Sens. Actuators B Chem. 2008, 129, 909–914. https://doi.org/ 10.1016/j.snb.2007.10.002.
110. Kaur, P.; Bahel, S.; Narang, S. B. Broad-Band Microwave Absorption of Sr0.85 La0.15 (MnZr)xFe12–2xO19 Hexagonal Ferrite in 18–40 GHz Frequency Range. J. Magn. Magn. Mater. 2018, 460, 489. https://doi.org/10.1016/j.jmmm.2018.01.042.
111. Wu, H. et al. Enhanced Low-Frequency Microwave Absorbing Property of SCFs TiO2 Composite. PowderTechnol. 2018, 333, 153. https://doi.org/10.1016/j.powtec.2018.04.015.
112. Wu, H.; Wu, G.; Wang, L. Peculiar Porous a-Fe2O3, c-Fe2O3 and Fe3O4 Nanospheres: Facile Synthesis and Electromagnetic Properties. Powder Technol. 2015, 269, 443. DOI: 10. 1016/j.powtec.2014.09.045.
113. Jia, Z. R. et al. Effects of Filler Loading and Surface Modification on Electrical and Thermal Properties of Epoxy/Montmorillonite Composite. J. IOP Sci. Chin. Phys. B 2018.
114. Li, Z. W.; Guoqing, L.; Chen, L.; Yuping, W.; Ong, C. K. Co2+ Ti4+ Substituted Z-Type Barium Ferrite with Enhanced Imaginary Permeability and Resonance Frequency. J. Appl. Phys. 99 (6). https://doi.org/10.1063/1.2180431.
115. Feng, Y. B.; Qiu, T.; Shen, C. Y.; Li, X. Y. Electromagnetic and Absorption Properties of Carbonyl Iron/Rubber Radar Absorbing Materials. IEEE Trans. Magn. 2006, 42 (3), 363–368. DOI: 10.1109/TMAG.2005.862763.
116. Li, B. W.; Shen, Y.; Yue, Z. X.; Nan, C. W. Enhanced Microwave Absorption in Nickel/Hexagonal-Ferrite/Polymer Composites. Appl. Phys. Lett. 2006, 89 (13), Article ID 132504. https://doi.org/10.1063/1.2357565.
117. Che, R. C.; Zhi, C. Y.; Liang, C. Y.; Zhou, X. G. Fabrication and Microwave Absorption of Carbon Nanotubes CoFe2O4 Spinel Nanocomposite. Appl. Phys. Lett. 2006, 88 (3), 1–3. https://doi.org/10.1063/1.2165276.
118. Xiang, C.; Pan, Y.; Liu, X.; Sun, X.; Shi, X.; Guo, J. Microwave Attenuation of Multiwalled Carbon Nanotube-Fused Silica Composites. Appl. Phys. Lett. 2005, 87 (12), 1–3. https://doi.org/10.1063/1.2051806.
119. Yang, Y.; Gupta, M. C.; Dudley, K. L.; Lawrence, R. W. Novel Carbon Nanotube—Polystyrene Foam Composites for Electromagnetic Interference Shielding. Nano Lett. 2005, 5 (11), 2131–2134. https://doi.org/10.1021/nl051375.
120. Nakamura, T.; Okano, Y. Low-Temperature Sintered Ni-Zn-Cu Ferrite. J. Phys. IV (Paris) 1997, (Suppl. III), 91–92. DOI: 10.1051/jp4:1997126r.
121. Naito, Y.; Mizumoto, T. Ferrite Electromagnetic Wave Absorbers with, Wide-Band Characteristics, pp 1320–1325.

122. Ueno, H.; Yasuyoshi, T.; Yoshikado, S. Fabrication of the CompositeFerrite Electromagnetic Wave Absorber. J. Magn. Soc. Jpn. 1998, 22 (Suppl. S1), 369–371".
123. Safaˇrˇ´ık, I.; Safaˇrˇ´ıkova, M. Magnetic Nanoparticles and Biosciences. Monatshefte fur Chemie 2002, 133 (6), pp. 737– 759. DOI: 10.1007/978-3-7091-6740-3_1.
124. Hernandez-Alonso, M.; Fresno, F.; Suarez, S.; Coronado, J. Development of Alternative Photocatalysts to TiO2: Challenges and Opportunities. Energy Environ. Sci. 2009, 2, 1231–1257. https://doi.org/ 10.1039/B907933E.

CHAPTER 5

A Rudimentary Analysis on Room Temperature Multiferroic: Bismuth Ferrite

VIDHYA SIVAN[1,2], KAVYA RAVINDRAN[1,2], S. L. NAVYA[2], and PIUS AUGUSTINE[1,2,3]

[1]*Material Research Laboratory, Sacred Heart College (Autonomous), Thevara Kochi, Kerala, India*

[2]*Department of Physics, Sacred Heart College (Autonomous), Thevara Kochi, Kerala, India*

[3]*Material Research Centre, Indian Institute of Science, Bangalore, Karnataka, India*

ABSTRACT

Bismuth ferrite (BFO) is a versatile multiferroic material that can be used at ambient temperature which offers a comprehensive package of multifunctionality among the many multiferroic compounds. The coexistence of many order parameters in BFO multiferroics and their novel physical properties finds new scope in research, and utility in a variety of applications in electronics and engineering. This chapter discusses the background and timeline of BFO along with advances in its different forms as ceramics, thin films, and nanostructures followed by a review of various synthesis techniques, a summary of the material's general properties, synthesis-related challenges, and possible device applications. Being an eco-friendly material that exhibits excellent functional properties, BFO is an ideal candidate for electronics and applications.

Physics and Mechanics of New Materials: Synthesis, Processing, and Emerging Applications. Devrim Balkose, Ana Cristina Faria Ribeiro, Nandakumar Kalarikkal, Ann Rose Abraham, A. K. Haghi (Eds.)

5.1 INTRODUCTION

Multiferroics is a class of materials that exhibits at least two of the ferroic orders viz., ferroelectricity, ferromagnetism, ferroelasticity, or ferrotoroidicity. Generally, the coexistence of two switchable states shown by a multiferroic system makes the material suitable for various multifunctional device applications. Bismuth ferrite ($BiFe_2O_3$), widely known as BFO, is considered as one of the well-known room temperature multiferroic materials that show antiferromagnetism and ferroelectric properties at room temperature. Various nanostructured forms of BFO include 0D nanoparticles, 1D nanowires, nanorods, and 2D thin film form. Several reviews have been reported on the magnetoelectric property of this material.[1–3] In 1970, it was discovered that BFO single crystal could be spontaneously polarized to 3.5 $\mu C/cm^2$ in the α0 0 1ρ direction of the psuedocubic unit cell. However, the issue of leakage current noticed with ferroelectric response was a hurdle to be overcome before being absorbed as a potential ferroelectric system for applications. Despite significant attempts having done by multiple research groups to settle the problem, it took more than two to three decades to reach a conclusive solution. Prof. Ramesh's group had set the groundwork for a surge of research explorations on BFO[4] when they reported a high enhancement (about 15-fold) in residual polarization over the bulk.[3] This multiferroic material finds application in various fields including spintronics, data storage, photovoltaic, electronics, and memory-based applications.[2] Other major hurdle was the possible formation of secondary phases during the processing of BFO and the existence of spiral spin modulation, which can completely cancel out the net spontaneous magnetization in the coalesced material.[3]

Numerous studies conducted since the initial discovery of BFO (BFO) by Royen et al. in 1957 have shown that this material may also take multiple crystal shapes and have distinct physical characteristics. It has been reported that BFO system is expected to show maximum piezoelectric coefficient the morphotropic phase boundary (MPB) region of the phase diagram.[5] At room temperature, it exhibits multiferroic ferroelectric characteristics and has an ABO_3-type perovskite structure.[6] It possesses high Curie temperature T_c ~ 1103 K and Neel temperatures of T_N ~ 643 K.

In 2003, epitaxial thin films of BFO with ferroelectric polarization ~60 $\mu C/cm^2$[7] became a significant leap in the thin-film study of this material. Since then, several studies have been conducted on various forms of BFO, including those found in bulk ceramics, thin films, and nanostructures. Thin film form of BFO has demonstrated high remnant polarization of the order 55–200 μC/cm^2 which depend on various thin-film yardsticks like epitaxial growth of the film, nature of substrate, strain in the lattice, tuning composition, and influence of buffer layer (like SRO, LSMO etc).[8] In addition to the previously mentioned ferroelectric feature, the high Curie temperature demonstrated by BFO is useful for the development of prototype applications for non-volatile memories and electronics devices. Moreover, nanostructured BFO demonstrate high magnetization and superparamagnetism with decreasing size in comparison to other forms.

5.2 STRUCTURAL ANALYSIS OF BIFEO$_3$

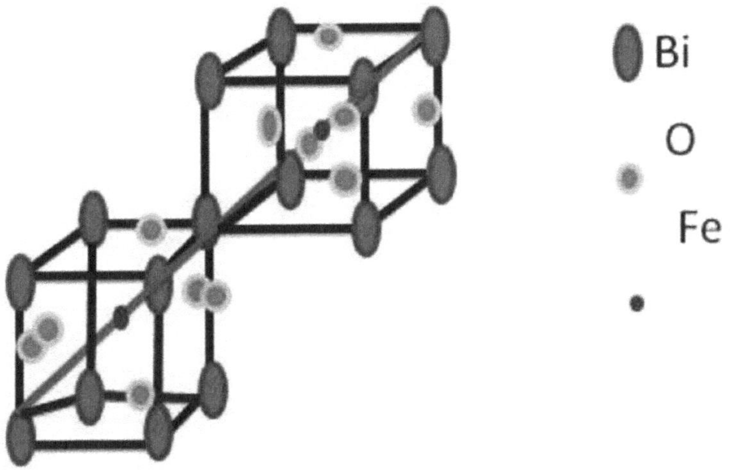

FIGURE 5.1 Schematic showing the ABO$_3$ crystal structure of BiFeO$_3$.

Bismuth ferrite (BiFeO$_3$) possesses ABO$_3$ perovskite structure with distorted rhombohedral shape in R3c#161 space group. The unit cell may be psuedocubic, hexagonal or rhombohedral along $[111]_{pc} \parallel [111]_{rh} \parallel [001]_{hex}$ with lattice parameters a = b = c = 5.63 Å and α = β = γ = 59.4°. The unit cells [111] psuedocubic/[001] hexagonal[6] diagonals are aligned with the ferroelectric polarization and spontaneous polarization along $[111]_{pc}$ direction due to high Curie temperature.

In ABO$_3$ perovskite structure of BiFeO$_3$, Bi and Fe occupy A- and B-–sites, respectively. The Fe^{3+} ions of the material displays up-spin with antiferroic ordering and six nearest neighbors exhibit down spin around it. The 6s lone pair electrons in Bi and partially filled d-orbital's of Fe are speculated to be responsible for magnetic ordering and the ferroelectric characteristics. The existence of Dzyaloshinskii–Moriya (DM) interaction between the nearest neighbors and Fe^{3+} ions with up-spins, results in an antiferromagnetic long-cycloid spin structure of 62 nm wavelength. The ferroelectric spin cycloid nullify magnetoelectric (ME) coupling due to polarization coupling with magnetization gradient, averages the net canted magnetism.[7] Additionally, the presence of spin cycloid cancels out the ME coupling between polarization and magnetization and hence only higher order ME effects is anticipated up to many orders of applied field.

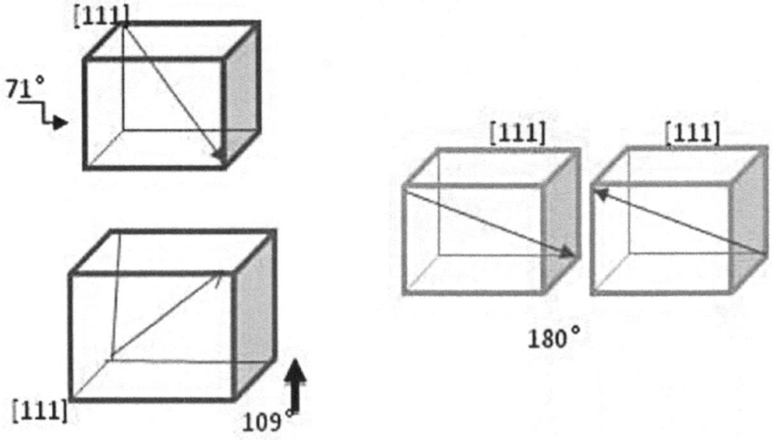

FIGURE 5.2 Different polarization switching exhibited by BFO is depicted.

The cubic diagonals at 180°, 109°, and 71° as shown in Figure 5.2 correspond to the eight distinct ferroelectric polarizations with regard to negative and positive orientations. BFO exhibits polarization switching under the application of external field. By the first principle calculation of the epitaxial grown BFO thin film, it has been presented that there exists an easy magnetization plane orientation for magnetic moment in BFO and it is always normal to the ferroelectric polarization.

Generally, magnetic property in a system depends on the type of spin orientation available with the system.[9] In BFO, each Fe^{3+} spin aligns anti-parallel to nearest Fe^{3+} leading to short-range local magnetic ordering, and possess G-type antiferroic nature. But local magnetic coupling polarization makes the spin not to align antiparallel which results in a weak spin canting.[2,6] The spins are not entirely antiparallel because of a minor canting moment brought by local magnetic coupling to the polarization. Canting will then be overlaid by a wide-range superstructure made up of an antiferromagnetically ordered sublattice with disproportionate spin. Hence, the spin canting at oxide interfaces can improve the magnetization in antiferromagnetic materials. This results in controllable large magnetization in multiferroic BFO which can deliver electric field control of magnetism and opens out its application in spintronics devices[10] which opens a broader utility of the system for electronic and memory applications.

5.3 PHYSICAL FORMS OF BISMUTH FERRITE

5.3.1 BFO CERAMICS

Ceramic bulk form of BFO was synthesized for the first time in 1957 by processing precursor's bismuth oxide and iron oxide via solid-state synthesis method. Apart from solid-state synthesis, there are various other methods to realize single-phase BFO, which are also discussed in the forthcoming sections. Since the bismuth (Bi^{3+}) is volatile in room temperature and may create a number of oxygen vacancies during processing, secondary phase may be formed along with BFO during sintering, which may deteriorate the quality of the multiferroic ceramic. The leakage current associated with conversion of Fe^{3+} ion to Fe^{2+} ion and oxygen vacancies that arise during high-temperature sintering would hamper the strength of the ceramic during poling. Therefore, in order to improve the functional aspects of the bulk BFO ceramics, various chemical modulations are practiced by the researchers to stabilize the system for multiferroic prospects. Easy way out is substitution at A-site and B-site ions in this ABO_3 compound by suitable dopants, which can significantly enhance the ferroelectric and magnetic characteristics of BFO.[11] Substitution in B-site is primarily responsible for ferroelectric polarization, whereas doping at the A-site can alter the phase structure.[12] Substitution in either of sites of ABO_3 structure will have a considerable impact on improving phase structure and other physical attributes of the ceramic. Also, magnetoelectric properties exhibited by bulk ceramic can be modified by chemical modifications at either or both of the perovskite sites.

5.3.2 BFO THIN FILM FORM

Epitaxial grown BFO thin films on $SrRuO_3$/STO substrate has displayed a high remnant polarization (Pr) ~60 $\mu C/cm^2$ which is comparable to that of theoretically predicted value for the system. Various novel techniques have been tried to enhance the functional properties of BFO in thin film form which includes strain engineering of the domains to increase polarization. Also, improved properties could be realized on BFO thin film by optimizing various thin-film growth techniques and ideal substrates with off-cuts and/or or lattice misfit with $BiFeO_3$[13] Thus, a number of studies on the development of high-quality thin-film samples produced by appropriate synthesis techniques and fine-tuning procedures were published, offering a range of applications including non-volatile memories and spintronics devices.

5.3.3 BFO NANOSTRUCTURES

Scaling downsize can bring anomalous properties in materials. This enhancement in material properties due to reduction in size piqued the curiosity in researchers to bring out novel applications such as memory and tunnel junctions. Surface area-to-volume is the major parameter that plays a major role in material properties. The nanostructures have wide surface area; surface area-to-volume ratio allows the study of surface states in carrier transport and optical properties. In addition to this, nanostructured forms exhibit magnetism much stronger than thin film and bulk ceramics.[14]

The BFO nanostructure finds application in various avenues such as solar cells, spintronics, data storage, microelectronics, and others. The nanostructured form of BFO exhibits weak ferromagnetism. In BFO antiferromagnetic coupling occurs between nearest neighboring planes while ferromagnetic coupling takes place along (111) plane, associated with the magnetic ordering. Moreover, in nanostructure form, antiferromagnetism and ferroelectricity are coupled together.[15]

5.4 SYNTHESIS TECHNIQUES

Various synthesis techniques tried for realizing single-phase BFO ceramics are briefly analyzed below.

5.4.1 SOLID-STATE REACTION ROUTE

The conventional solid-state approach entails combining and grinding the right amount of metal oxides or carbonates, followed by calcination at the stabilized temperature. Calcination allows interdiffusion of cations. Since ion diffusion in solid state takes place at slow rate of kinetics. The rate of reaction can be enhanced by giving a thorough grinding of the mixed precursors followed by high-temperature calcination, which allows interdiffusion of ions to for BFO ceramic powder. Even though this method is easy and economically feasible, it has some limitations, such as unfavored sintering performance, non-uniform shape and size of particle, and multiphase character.

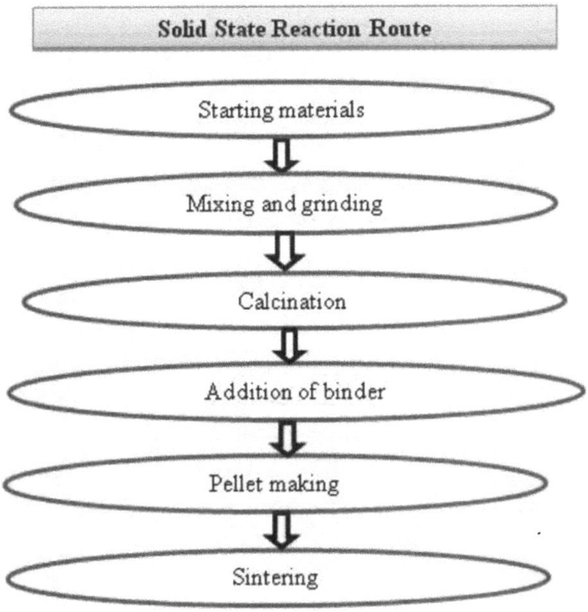

FIGURE 5.3 Schematic of the solid-state synthesis route to realize BFO ceramic.

5.4.2 *RAPID LIQUID SINTERING*

Secondary phase formation is a major drawback in solid-state route synthesis of BFO sample. An easy method to circumvent this problem is to improve the kinetics of reaction which is possible in rapid liquid phase sintering. Since it takes place by higher kinetics, the calcination stage is skipped to eliminate the unwanted secondary phases and thus phase pure sample can be prepared. Processing involves mixing precursors such as high-purity bismuth oxide and ferric oxide in an agate mortar, the fine-ground powder sample will be pressed into pellets of appropriate density. As this method does not involve calcination, the powder sample is kept for 7–8 min at 880°C, which is regarded as the maximum heating rate and shortest sintering length. The melting point of Bi_2O_3 is 817°C, thus the appearance of liquid phase during the synthesis procedure is critical. The liquid phase speeds up the process, makes sintering easier, and inhibits the formation of a secondary phase. Mechanical activation is another method of synthesis with the advantage of fast reactivity of the sample material. As a result, the calcination can be completed at extremely low temperatures using this approach. In contrast to the solid-state technique, where the BFO phase occurs at 825°C, this method can produce BFO ceramic with secondary phases reasonably suppressed by calcination at 700°C.

5.4.3 SOL–GEL METHOD

It is a wet chemical process that involves the dissolution of molecular precursors in water or alcohol, stirred in low temperature to create gel via the process of hydrolysis. The method involves preparation of multicomponent solution by mixing appropriate precursors like bismuth nitrate pentahydrate mixed at room temperature with 2-methoxyethanol and acetic acid. Once transparent solution is obtained, it is mixed with iron nitrate non-anhydrate followed by constant stirring to obtain blackish red clear solution. Solution is then dried to obtain xerogel powder. The xerogel is then grinded and the powder so obtained is annealed at appropriate temperature to obtain the required phase of BFO.[16]

5.4.4 CO-PRECIPITATION METHOD

In co-precipitation method, solids with diverse ionic species can be separated from the solution phase. Ideally, all cations precipitate quantitatively and simultaneously, resulting in a totally mixed-metal precursor. Since most of the solutions contain more than one metal ion, this scenario occurs rather rarely. Furthermore, homogenous co-precipitation of metal ions at the atomic level is difficult to achieve. However, this limitation can be overcome by using cation-insoluble precipitating agents.

5.4.5 HYDROTHERMAL SYNTHESIS

Inorganic material can be synthesized via hydrothermal method. In the synthesis of hydrated materials, reactivity rises under huge pressure and reasonable temperatures. The solvent will be heat treated in an autoclave or in a sealed chamber to a pressure higher than the ambient pressure. As a result the boiling point of the solvent may rise. The synthesis technique is referred to as hydrothermal if water is used as solvent.[17]

5.5 ATTRIBUTES OF BFO

5.5.1 FERROELECTRIC AND PIEZOELECTRIC PROPERTIES

Ferroelectric and piezoelectric properties of the ferroelectric materials are crucial in various applications like actuators and sensors. In paraelectric and ferroelectric materials, polarization can be varied by applied electric field, but in piezoelectric materials, polarization is induced by varying temperature or pressure. In BFO, electric dipole moment induces spontaneous polarization due to relative variations of Bi^{3+} ions in oxygen octahedron. The volatile nature of bismuth oxide and fluctuating Fe during processing results in secondary phases due to leakage current issues, which make it difficult to obtain good hysteresis loop. The first hysteresis loop of BFO ceramic bulk was analyzed in the 1970s, with a polarization value of 6.1 μC cm^{-2} along [111] direction. Later on ferroelectric, dielectric, and piezoelectric properties were enhanced by substitution of rare earth metals in either or both sites of ABO_3 perovskite and by appropriately manipulating the phase structure with ions; effective enhancement in ferroelectric and piezoelectric performance could be achieved.

In thin film form, the ferroelectric and piezoelectric properties of $BiFeO_3$ are influenced by epitaxial strain effects in addition to the other variables mentioned above. With increase in compressive strain, there is a transition in phases from rhombohedral to tetragonal. As a result, the induced polarization can be altered. A compressive strain of about 4–5% induces R-like/T-like transition, which has a considerable impact on modulating the piezoelectric response. Furthermore, piezoelectric properties can also be related to the mobility of the phase border between various phases.[18–20]

5.5.2 MMAGNETIC PROPERTIES

BFO shows a G-type antiferromagnetic order due to Fe^{3+} spins in rhombohedral distorted perovskite. Magnetic order of BFO depends on periodicity of G-type antiferromagnetic spins of 62–64 nm along [111] direction. The size and shape also determine its magnetic characteristics of BFO. Below 62 nm, the periodic length of BFO spin cycloid and magnetic sensitivity grows dramatically. Canted G-type antiferromagnetic materials have weak magnetism because of their asymmetrical spin cycloid structure.

5.5.3 MAGNETOELECTRIC COUPLING

A distinguishing characteristic of BFO-based nanomaterials is the magnetoelectric coupling (ME coupling), that is, coexistence and coupling of magnetic and ferroelectric order parameters. This coupling facilitates to tune and switch electrical polarization with applied magnetic field. Single-phase multiferroic materials show weak magnetic coupling due to independent magnetic and ferroelectric order. BFO shows ME coupling at room temperature. Zhao et al. demonstrated substantial coupling of two orders. For the implementation of large magnetization and for device applications, anisotropy at the interface of ferromagnetic and antiferromagnetic heterostructures were used so that composite magnetoelectric materials could bring ME coupling between various phases at room temperature or above.[21]

5.6 FUNCTIONAL ASPECTS OF $BIFEO_3$

Ferroelectric materials are capable of switching between two equilibrium orientations of the spontaneous polarization vector in the absence of an external field. The temperature at which materials switch from a high-temperature non-ferroelectric phase to a low temperature ferroelectric phase is the Curie point. BFO, which has been described in various articles over the past few decades, is considered to be the perfect material for switching between two ferroic regimes. For different memory-based applications, it is regarded as the optimum ferroelectric material.

An important functional aspect of BFO is its ability to make transitions at different temperatures. At 643 K, corresponding to the Neel temperature, the first phase transition occurs. It is a second-order phase transition accompanied by lattice expansion from antiferromagnetic to paramagnetic state. Second shift from ferroelectric to paraelectric state happens at a high temperature of 1098 K. The third transition with strain is the metal–insulator transition, which happens at 1203 K and is known as spontaneous strain. This strain can cause irregularities in the material's dielectric, elastic, thermal, and other properties.

Frequency-dependent dielectric response is another interesting feature of $BiFeO_3$. Dielectrics are materials with a high electrical resistivity. The frequency ranges and dielectric loss tangent will be crucial to consider such materials for various applications.[22] $BiFeO_3$ exhibit dielectric response in the low frequency range of the electric field (mHZ–MHz). The intrinsic dielectric permittivity exhibited by BFO in the GHz range is very small ($\varepsilon_r \sim 30$), where as in the low frequency range (MHz), it is ~10,000 (in room temperature), in ceramics and single crystal form of BFO. These frequency fluctuations influence electrical conductivity and hence directly affect real and imaginary part of dielectric permittivity. Identifying the dominant mechanism is still under concern, but still evokes the fact that dielectric response of BFO is very much case sensitive and difficult to control.[5]

Another interesting functional aspect of BFO is the occurrence of domains at predetermined sites due to structural defect. The domain wall motion occurs under the application of applied electric field. The domain wall movement is influenced by the presence of local defects, which pins the domain wall motion. Many of the oxide ferroelectrics such as $BaTiO_3$, PZT, etc. also exhibit this behavior. The presence of defects would harden the material and facilitate the formation of acceptor-oxygen-vacancy dipoles. Cycling electric field method is an easy way out for de-pinning the domain walls.[23]

In $BiFeO_3$, the electromechanical properties can be enhanced by partial substitution of isovalent rare earth ions such Gd^{3+}, Eu^{3+}, La^{3+}, Nd^{3+}, and Sm^{3+}, which have a considerable impact on multifunctional properties. Also the piezoelectric properties of material can be improved by inducing lattice distortion chemically. Among the many isovalent rare earth elements, Sm^{3+} is reported to be effective for enhancing the Bi_2O_3 localization. Chen et al. reported that Sm-modified BFO can exhibit ferroelectric polarization $P_r \sim 15$ μCm–[2] at 180 kV cm–[1] and $d_{33} \sim 42$ pCN–[1] after polling at 170 kVcm–[1] DC at room temperature.

The domain wall movement is hindered by the presence of charge defects and will subsequently reduce piezoelectric properties. Moreover, bismuth oxide (Bi_2O_3) is volatile in nature and can even create defects during sintering. The accumulation of defect dipole would also favor pinning of domain walls.[23–25] Another issue that would hamper the quality of the system is the presence oxygen vacancies, which can be resolved effectively by site engineering. Addition of lanthanides is found to be effective in electrical and magnetic property enhancement. There are reports which established the advantage of doping lanthanides at the A-site of ABO_3 perovskite structures to reduce the leakage current, and to improve the long-range ferroic order and remnant magnetization.[26–28]

Magnetic field-induced polarization is another important functional aspect which defines the polarization induced by the magnetic field which is proportional to the square of the magnetic field over several Tesla. The ME coupling to polarization can create improper antiparallel spin alignment that results in canting moment. While the polarization and propagation vectors together define the magnetic easy plane. Its sublattice magnetization is roughly 0.43 from bifringence and 0.37 from Mossbauer hyperfine splitting as a function of temperature, with magnetic Neel temperature at 370°C.[29]

5.7 DEVICE APPLICATIONS OF BFO

Being a room temperature eco-friendly multiferroic exhibiting large remnant polarization along polar direction [111], $BiFeO_3$ can even replace PZT in multiple applications. BFO is identified as an ideal alternative to PZT in ferroelectric memory devices. Ferroelectric tunnel junctions (FTJs) for neuromorphic computing have been realized recently with brain like computing has attracted companies such as Intel, IBM. The ability to switch between two ferroic orders makes BFO a good candidate for potential applications, such as non-volatile resistive memory with fast switching and large ON/OFF ratios. Memristive behavior of this material system is also a current topic of research that works as artificial synapse and very recently implemented for neuromorphic computing applications.[30,31] The field of memory also investigates photovoltaic effects of multiferroic BFO. Wang et al. found V_{oc} and I_{sc} found reversed in 10 ns indicating high writing speed for BFO-based memory devices. Also, photovoltaic effect-based ferroelectric memory was used to demonstrate prototype 16-cell memory.[32]

The inherent multiferroic property of BFO-based nanomaterials can be explored in two-dimensional electron gas (2DEG). Modulation of two-dimensional anisotropic electron transport in BFO can be explained by ferroelectric polarization control. The interfacial 2-DEG generated by ferroelectric materials is expected to exhibit strong anisotropic conductivity in 109° domain. The alternate *p* and *n*-type conducting channels that emerge owing to domain structure create a potential barrier at the junction (interface of thin films), resulting in an insulating interface perpendicular to this direction, producing either electron or hole along 109° domain stripes. The ME coupling explored BFO make it a good contender for spintronics devices. Giant magnetoresistive devices of BFO are developed due to spin valve exchange of BFO-based material in electrically written spintronics devices.

BFO-based materials have been identified as suitable choice for solar cell applications due to their reduced band gap, good carrier transport, and light absorption properties. One of the important areas of concern for promoting green energy devices based on BFO is self-powered photodetectors.

Acousto-optic and photonic devices are another key application of BFO-based material. For this, the epitaxial strain induced optical response in thin film form of BFO was used, and the strain can affect the optical index. As a result, the system exhibits high elasto-optic coefficient. The shift in the optical band gap refers to Piezochromism. This electrically controlled Piezochromism is reversible and persistent, extending the range of applications for BFO-based materials in photonics and acousto-optics applications.[27]

5.8 CONCLUSION

This chapter presented a brief overview of "Bismuth Ferrite," one of the most promising room temperature multiferroic. Structural and physical forms along with challenges in synthesis of single-phase BFO is outlined. Rudimentary analysis on the state of the art functional features of BFO are present for establish its utility in variety of applications along with research prospects.

ACKNOWLEDGMENT

Pius is thankful to Science and Engineering Research Board (SERB), DST, Govt. of India for the research grant, Teacher Associate for Research Excellence (TARE- TAR/2020/000241), to carry out research in association with Prof. Karuna Kar Nanda, Material Research Centre, Indian Institute of Science Bangalore. Also, the financial support to Material Research Laboratory from SH College (SH MRP/01/2021) is acknowledged.

KEYWORDS

- **$BiFeO_3$**
- **thin films**
- **synthesis**
- **applications**

REFERENCES

1. Catalan, B. G.; Scott, J. F. Physics and Applications of Bismuth Ferrite. *Adv. Mater.* **2009,** 2463–2485.
2. Gumiel, C.; Jardiel, T.; Bernardo, M. S.; Villanueva, P. G.; Urdiroz, U.; Cebollada, F.; et al. Combination of Structural and Microstructural Effects in the Multiferroic Response of Nd and Ti Co-doped BiFeO3 Bulk Ceramics. *Ceram. Int.* **2019,** *45,* 5276–5283.
3. Wang, N.; Luo, X.; Han, L.; Zhang, Z.; Zhang, R.; Olin, H.; et al. Structure, Performance, and Application of $BiFeO_3$ Nanomaterials. *Nano-Micro Lett.* **2020,** *12.*
4. Wang, J.; Neaton, J. B.; Zheng, H.; Nagarajan, V.; Ogale, S. B.; Liu, B.; et al. Epitaxial $BiFeO_3$ Multiferroic Thin Film Heterostructures. *Science* **2003,** *299,* 1719–1722.
5. Daniels, J. E.; Damjanovic, D. $BiFeO_3$ Ceramics: Processing, Electrical, and Electromechanical Properties. **2014,** *19,* 1–19.
6. Wu, J. Advances in Lead-Free Piezoelectric Materials. *Adv. Lead-Free Piezoelectric Mater.* **2018.**
7. Article R. Multiferroics: Progress and Prospects in Thin Films. *3,* 21–29.
8. Uchida, H.; Ueno, R.; Funakubo, H.; Koda, S. Crystal Structure and Ferroelectric Properties of Rare-earth Substituted $BiFeO_3$ Thin Films Crystal Structure and Ferroelectric Properties of Rare-earth Substituted $BiFeO_3$ Thin Films. **2006,** 014106.
9. Kotnala, R. K.; Shah, J. Ferrite Materials : Nano to Spintronics Regime [Online]. BS:HMM. Elsevier; 2015. http://dx.doi.org/10.1016/B978-0-444-63528-0.00004-8
10. Carranza-Celis, D.; Cardona-Rodríguez, A.; Narváez, J.; Moscoso-Londono, O.; Muraca, D.; Knobel, M.; et al. Control of Multiferroic Properties in $BiFeO_3$ Nanoparticles. *Sci. Rep.* **2019,** *9,* 1–9.
11. Hussain, A.; Xu, X.; Yuan, G. The Development of $BiFeO_3$-based Ceramics. **2014,** *59,* 5161–5169.
12. Kim, H. P.; Kang, W. S.; Hong, C. H.; Lee, G. J.; Choi, G.; Ryu, J.; et al. Piezoelectrics. *Adv. Ceram. Energy Convers. Storage.* **2019.**
13. Wang, N.; Luo, X.; Han, L.; Zhang, Z.; Zhang, R.; Olin, H. Structure, Performance, and Application of Nanomaterials. **2020.**
14. Multiferroic Bismuth Ferrite-based Materials for Multifunctional Applications: Ceramic Bulks, Thin Films and Nanostructures Jiagang Wu.
15. Yuan, G. L.; Or, S. W. Multiferroicity in Polarized Single-Phase Bi0.875Sm $0.125FeO_3$ Ceramics. *J. Appl. Phys.* **2006,** *100,* 0–5.
16. Afzal, A. M.; Manzoor, A.; Iqbal, M. Z.; Javed, Y.; Deshmukh, A. H.; Khan, R.; et al. Environmental Effect on Structural, Magnetic, and Dielectric Properties of BFO Nanostructure and Its Solar Cell Applications. *J. Mater. Sci. Mater. Electron.* **2021,** *32,* 3313–3323. https://doi.org/10.1007/s10854-020-05079-8
17. Zhang, X. Y.; Dai, J. Y.; Lai, C. W. Synthesis and Characterization of Highly Ordered $BiFeO_3$ Multiferroic Nanowire Arrays. *Prog. Solid State Chem.* **2005,** *33,* 147–151.
18. Teague, J. R.; Gerson, R.; James, W. J. Dielectric Hysteresis in Single Crystal $BiFeO_3$. *Solid State Commun.* **1970,** *8,* 1073–1074.
19. Wu, S. M.; Cybart, S. A.; Yu, P.; Rossell, M. D.; Zhang, J. X.; Ramesh, R.; et al. Reversible Electric Control of Exchange Bias in a Multiferroic Field-Effect Device. *Nat. Mater.* **2010,** *9,* 756–761.

20. Daumont, C.; Ren, W.; Infante, I. C.; Lisenkov, S.; Allibe, J.; Carrétéro, C.; et al. Strain Dependence of Polarization and Piezoelectric Response in Epitaxial $BiFeO_3$ Thin Films. *J. Phys. Condens. Matter.* **2012,** *24,* 3–9.
21. Zavaliche, F.; Zhao, T.; Zheng, H.; Straub, F.; Cruz, M. P.; Yang, P. L.; et al. Electrically Assisted Magnetic Recording in Multiferroic Nanostructures. *Nano Lett.* **2007,** *7,* 1586–1590.
22. Peled, D.; Pratt, V.; Holzmann, G. Report Documentation Page. **1997,** *298,* 405–405.
23. Preethi, A. J.; Ragam, M. Effect of Doping in Multiferroic BFO: A Review. *J. Adv. Dielectr.* **2021,** *11,* 1–17.
24. Yuan, G. L.; Yang, Y.; Or, S. W. Aging-induced Double Ferroelectric Hysteresis Loops in BiFeO3 Multiferroic Ceramic. *Appl. Phys. Lett.* **2007,** *91,* 2005–2008.
25. Gil-González, E.; Perejón, A.; Sánchez-Jiménez, P. E.; Criado, J. M.; Pérez-Maqueda LA. Thermoanalytical Characterization Techniques for Multiferroic Materials. *Handb. Therm. Anal. Calorim.* **2018.**
26. Wu, J.; Wang, J. Bilayered BiFe0.95Mn0.05O_3/Bi 0.90La0.10FeO3 Thin Films with Low Ferroelectric Coercivity and Large Remanent Polarization. *J. Am. Ceram. Soc.* **2010,** *93,* 2113–2116.
27. Vengalis, B.; Devenson, J.; Oginskis, A. K.; Lisauskas, V.; Anisimovas, F.; Butkute, R.; et al. Optical and Electrical Properties of Nd-Doped $BiFeO_3$ Thin Films and Heterostructures. *Phys. Status Solidi Curr. Top. Solid State Phys.* **2009,** *6,* 2746–2749.
28. Gupta, S.; Tomar, M.; James, A. R.; Gupta, V. Ce-Doped Bismuth Ferrite Thin Films with Improved Electrical and Functional Properties. *J. Mater. Sci.* **2014,** *49,* 5355–5364.
29. Scott, J. F.; Singh, M. K.; Katiyar, R. S. Critical Phenomena at the 140 and 200 K Magnetic Phase Transitions in $BiFeO_3$. *J. Phys. Condens. Matter.* **2008,** *20.*
30. Catalan, G.; Scott, J. F. Physics and Applications of Bismuth Ferrite. *Adv. Mater.* **2009,** *21,* 2463–2485.
31. Boyn, S.; Girod, S.; Garcia, V.; Fusil, S.; Xavier, S.; Deranlot, C.; et al. High-Performance Ferroelectric Memory Based on Fully Patterned Tunnel Junctions. *Appl. Phys. Lett.* **2014,** *104,* 14–16.
32. Guo, R.; You, L.; Zhou, Y.; Lim, Z. S.; Zou, X.; Chen, L.; et al. Non-Volatile Memory Based on the Ferroelectric. *Nat Commun.* **2013,** *4,* 2–6.

SECTION 2

Energy Materials: Fundamental Physics and Latest Advances in Relevant Technology

CHAPTER 6

Electrochemical Energy Storage in 2D Nanomaterials: Present Status and Future Challenges

GOPAL SANYAL[1], RAVI KUMAR TRIVEDI[2], and
BRAHMANANDA CHAKRABORTY[2,3]

[1]*Mechanical Metallurgy Division, Bhabha Atomic Research Centre, Trombay, Mumbai, India*

[2]*High Pressure and Synchroton Radiation Physics Division, Bhabha Atomic Research Centre, Trombay, Mumbai, India*

[3]*Homi Bhabha National Institute, Mumbai, India*

ABSTRACT

Supercapacitors are extensively acknowledged as a preferred energy stowage expedient in the form of electricity. These have arisen as a decent substitute to old-style electrical condensers and fuel cells for their advanced energy compactness and power concertation in comparison with batteries. Nevertheless, they do possess a few limitations such as poorer energy density and sequence lifetime than batteries. To surmount these difficulties, scientists are focusing on fabrication of 2D nanostructures and their adjustment by diverse tactics, such as nobbling, oxygen vacancies' introduction, and crossbreeding with carbon allotropes for heightened electrochemical attributes. In this chapter, we have offered the aforementioned themes with the help of newly published reports. Furthermore, we have presented theoretical intuitions after density functional theory (DFT) for electrochemical response of the probe substances since recently reported literatures. This appraisal tersely gifts the progression in the supercapacitor

Physics and Mechanics of New Materials: Synthesis, Processing, and Emerging Applications. Devrim Balkose, Ana Cristina Faria Ribeiro, Nandakumar Kalarikkal, Ann Rose Abraham, A. K. Haghi (Eds.)

energy stowage arena and the unalike methods tangled in the construction of supercapacitor electrode ingredients, which will be quite convenient to the scholars at work in the turf of energy stowing. Additionally, the contests and forthcoming standpoints of this thrilling research pitch are thoroughly elaborated.

6.1 INTRODUCTION

The present universal estimated electric production capacity is about 20 terawatt hours.[1] Figure 6.1 is to illustrate the contribution of electricity from different sources.[2] Even with intense preservation and expansion of new competent skills, the global electricity mandate is foretold to twofold by the mid of the current century and tripartite by the century's culmination. Electricity is the overriding formula of energy castoff, and the need for electricity is swelling at a quicker stride than total energy consumption considering all forms. Simultaneously, natural gas and oil mining is forecast to top over the following limited periods. Coal has been the leading basis of electricity production;[3] copious coal assets may uphold current consumption echelons extended compared with oil and gas. Though, every kWh of electricity produced by scorching coal generates a mean 1 kg lifecycle CO_2 emission, a conservatory gas that is extensively well-thought-out as the main sponsor to worldwide warming.[4,5] In USA on its own, coal power generation units discharge annually 1.5 billion tons of CO_2, and releases since unindustrialized nations are hastening. To decrease greenhouse gas emanations, many countries are espousing emission protocols (i.e., cap-and-trade or alternatives) and carbon "trading," which aids industries with a minor "carbon footprint" and needs those creating higher releases to buy carbon "allowances." The ecological anxieties over the usage of fossil fuels and their reserve restraints, joint with energy refuge worries, have urged countless attention in producing electricity from renewable foundations. Wind and solar energy are among the maximum plentiful and possibly voluntarily available.[4,6,7] Interestingly, the amount of radiation energy that the Earth obtains in 1 h is sufficient to meet the worldwide energy needs for a year. Seizing a trivial fraction of latent wind energy could also subsidize meaningfully to fulfilling global electrical energy necessities. Whereas developments in expertise are yet required to produce renewable energy frugally, solar and wind power technologies have advanced rapidly. Globally, since the date wind power was installed, the total electricity has reached about 1870 GW in

2021.[8] The World Energy Council estimates that the new wind capacity worldwide would total up to 8000 GW by 2030 for achieving a Net Zero Emission Scenario. The yield from photovoltaic (PV) element settings is presently rising at 40% per year globally.[6] The US targets 160 GW solar control by 2050. Nevertheless, solar and wind power are not relentless and steadfast bases of power. The capricious characteristics of these unconventional foundations cause noteworthy contests for the electric lattice machinists since other power plants (typically fossil-fueled units) need to recompense for the inconsistency. For instance, as revealed in Figure 6.2a, wind power outlines in Tehachapi, California, diverge over minutes, hours, and days while cresting at night when mandate is low. All through the day, wind power may be few GW at some instants and just a few megawatts (MW) and even zero at others. Likewise, as shown in Figure 6.2b, solar power is produced only through the daylight and fluctuates as soon as clouds pass by.[9] An additional fear is the circumstance that the unconventional resources are confined and are frequently far from load centers. In the USA, wind bases are intense in the Midwest areas, and solar foundations in southwest counties. To flat out the irregularity of unconventional energy creation, cheap electrical energy storage (EES) will turn out to be essential. EES has been well-thought-out as a main enabler of the smart grid or future grid, which is anticipated to assimilate an important quantity of unconventional energy capitals while supplying fuel (i.e., electricity) to electrical and hybrid automobiles,[10] even though the price of executing EES is of excessive concern.[11]

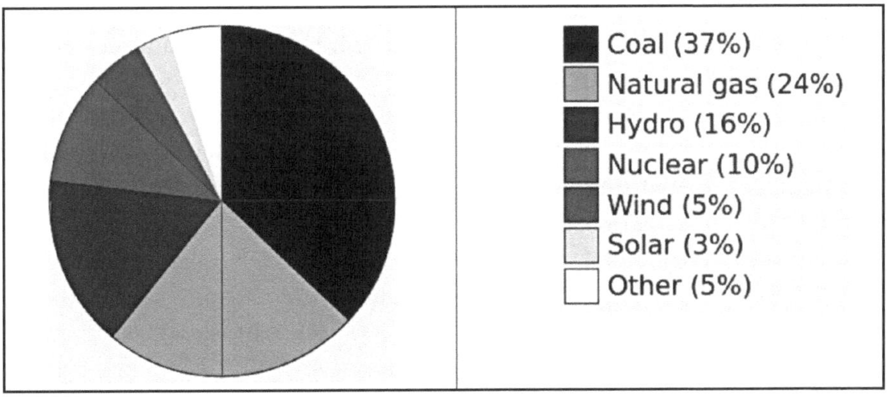

FIGURE 6.1 2019 world electricity generation by source (total generation was 27 petawatt-hours).

Source: Reprinted with permission from Ref. [2]

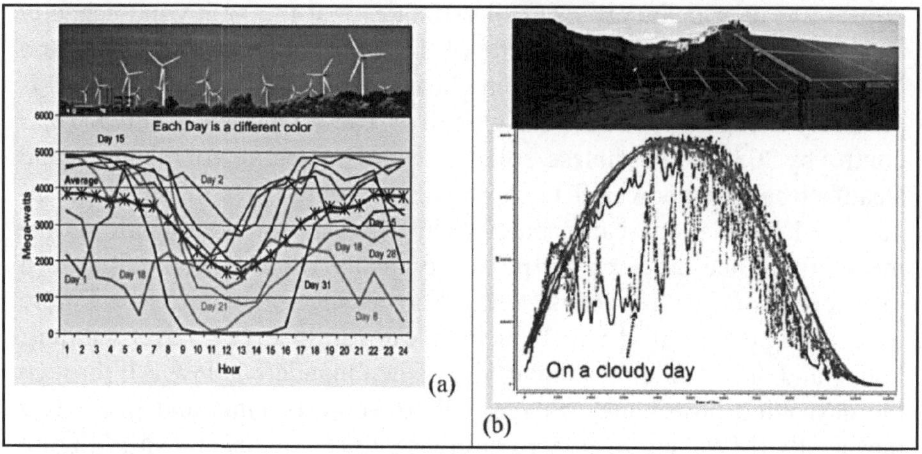

FIGURE 6.2 (a) Day-to-day outlines of wind power predictable by 7× output in April 2005 for the year 2011 in Tehachapi, California (Courtesy of ISO California). (b) 5 MW PV power over an extent of 6 days in Spain (Courtesy of AES).

Source: Reprinted with permission from Ref. [9]

Nanomaterials bid significantly enhanced ionic conveyance and electronic conduction equated with conformist battery and supercapacitor resources. They allow the fulfilment of all intercalation places accessible in the subdivision capacity too, causing large specific capacities and rapid ion dispersal. These characteristics turn nanomaterial-based probes competent to endure high currents, presenting a gifted resolution for great energy and superior power stowage. There are yet numerous contests allied with their usage in energy stowing skill, and with the exclusion of multiwall carbon nanotube extracts and carbon coverings on silicon subdivisions in lithium-ion battery (LIB) electrodes, the practice of nanomaterials in marketable strategies is quite restricted. After many years of progress, a collection of nanomaterials with multipurpose chemical conformations and forms remains, extending from carbides, oxides and chalcogenides, to carbon and elements founding mixtures with lithium. This chapter contains several subdivision morphologies, for example, zero-dimensional (0D) quantum dots and nanoparticles; 1D nanotubes, nanowires, and nanobelts; 2D nanosheets and nanoflakes; and 3D permeable nanowebs. Collective with lithium and yonder lithium ions, these chemically varied nanoscale subunits are obtainable for generating energy storing answers such as wearable and organizational energy stowage expertise, that are not attainable with traditional resources.[12]

6.1.1 NEED FOR ENERGY STORAGE

Speculation in energy stowage changes with network, generating long-standing advantage and dependability for eons to arise. Energy storing is an important pivot for the complete grid, supplementing capitals from renewable to conventional base-load nuclear and fossil fuels to mandate lateral resources and scheme competence properties. It can perform as a creation, spread or delivery asset—occasionally in a solitary asset. Eventually, stowing is an empowering knowhow. It can save customers' money, advance dependability and suppleness, assimilate production bases, and aid to lessen ecological impressions.[13]

Energy stowing might save functioning expenses in running the grid, in addition to conserve currency for electricity customers who fix energy stowage in homes and industries. Energy storing can decrease the cost to deliver incidence directive and whirling standby facilities, along with counterbalance the prices to patrons by stowing cheap energy and consuming it later, throughout topmost stages at sophisticated electricity charges. By means of energy storage all over transitory outages, trades can evade expensive disturbances and linger usual processes. Inhabitants can protect themselves from gone food and medicines, and the troublesomeness of not partaking electricity. And there is a selection for both commerce and housing clients to contribute in mandate retort agendas when obtainable.[14]

Energy storage may offer holdup power all through commotions. The identical idea that smears to gridlock power for a discrete expedient (e.g., a smoulder fright that wads into a home but also possess battery backup), may be clambered up to a whole structure or even the grid at huge. Stowage delivers suppleness for the grid, to safeguard continuous power to clients, wherever and whenever they require it. This suppleness is important to both trustworthiness and pliability. As the price of outages lingers to upsurge, the worth of improved consistency and enhancements in bounciness also surges.[15]

Energy storing may flat out the transport of adjustable or sporadic resources, for example, solar and wind, by stowing extra energy once the airstream is gusting and the sun is brilliant, and sending it when the reverse is trendy. But storing might also back the effectual conveyance of electricity for obstinate, baseload possessions. When mandate fluctuates rapidly, and litheness is essential, energy storage can insert or excerpt electricity as wanted to precisely bout load—whenever, and wherever it is required. Energy stowing is a qualifying knowhow. While the sun is not outstanding or the breeze is not wafting, energy stowage can be there. When demand swings and baseload resources cannot respond fast adequately, energy storing may be beneficial.[16]

In meekest footings, energy stowage permits electricity to be saved for later, where and when it is most required. This forms competences and abilities for the electric grid—including the capability to decrease greenhouse gas (GHG) discharges. By familiarizing more litheness into the grid, energy storing can help assimilate more wind, solar and disseminated energy possessions. It may also advance the competence of the grid—swelling the capacity factor of current resources—and counterpoise the necessity for the construction of new hazardous peak power stations. As energy source combination gets cleaner with low- and no-carbon possessions, energy stowage helps that stock mix grow more straightforwardly and steadfastly.[17]

6.1.2 ENERGY STORAGE DEVICES: BATTERY AND SUPERCAPACITOR

Battery storing knowledge is vital to fast-moving the substitution of hydrocarbon deposit with renewable energy. These schemes will perform a progressively key role amid clean energy provisions and retorting to electricity mandates.[18] Battery stowage, or battery energy storage systems (BESS) are strategies that permit energy since renewables, such as wind and solar, to be stockpiled and then unconfined when consumers require power furthermost. Lithium–ion batteries, which are utilized in cellular phones and electric vehicles, are now the leading storage skill for great gage set-ups to assist electricity networks guarantee a consistent stock of renewable energy. The technology has a crucial chunk to play in safeguarding establishments and may be power-driven by clean energy, even when the sun is not brilliant or the wind is still. For instance, the UK possesses the major fitted bulk of offshore wind in the world, yet the skill to tap this energy and decisively organize it may upsurge the worth of this green energy by raising production and possibly dipping expenses. Everyday, engineers at National Net and electricity networks worldwide must link supply and mandate. Handling these summits and mangers turns out to be more stimulating when the bull's eye is to attain net zero-carbon emission by abolishing fossil fuel plants that have conventionally been cast-off as a snarl to deliver a dependable, baseload source of energy.[19] The UK administration approximates skills like battery storing systems—backing up the amalgamation of further low-carbon power, heat and transportation knowledge—could economize the UK energy structure up to £40 billion (or $48 billion) by the year 2050, at last plummeting commons energy expenses. Battery energy storing schemes are significantly more progressive than domestic small batteries. A battery stowage arrangement can be replenished by electricity produced from renewable energy such as solar and wind power. Intelligent battery software utilizes set of rules to organize energy creation and computerized regulator schemes are applied to choose when to preserve the energy to make available reserves or discharge it to network. Energy is unconfined since the battery stowing structure all through the times of crowning mandate, retaining the costs low and electricity smooth. This writing is elaborate with extensive battery stowage arrangements, but household energy stowage systems work on the same philosophies. Stowage of renewable energy needs cheap technologies that survive long—discharging and charging millions of times—are innocuous and may stock enough energy economically to compete mandate. LIBs were technologically advanced by a British researcher in the 1970s and were foremost used for business by Sony in 1991 for the corporation's handheld video recorder.[20] Although they are now the most frugally feasible energy stowage keys, there are a sum of other skills for battery storing presently being industrialized. These comprise: (a) Compressed air energy storage: with these schemes, usually positioned in big cavities, excess power is utilized to compress air and then hoard it. While energy is required, the trampled air is unconstrained and moves through an air turbine to produce electricity. (b) Mechanical gravity energy storage: One instance of this kind of scheme is when energy is utilized to boost concrete chunks up a barbican. When the energy is required, the concrete slabs are dropped back down, producing electricity using the wrench of gravity. (c) Flow batteries: In these batteries, which are fundamentally rechargeable fuel cells, chemical energy is supplied by two chemical constituents liquified in fluids confined within the scheme and parted by a sheath.

Supercapacitors are a kind of energy-storing devices that are better than both batteries and usual capacitors in their performance. They partake a superior capacity for energy storing to conventional capacitors and may carry it at a bigger power yield in disparity to batteries. These physiognomies, combined with their longer stability and huge cyclability, turn supercapacitors an outstanding energy stowage expedients.[21] These are presently installed in an assortment of submissions, either in aggregation with other energy-stowing strategies (habitually batteries) or as autonomous energy bases. For their great conduction and surface area, permeable carbons are being incorporated in the electrodes of saleable supercapacitors. A supercapacitor's main characteristic is to amass energy through the binge of exciting ions in the electrolyte on the electrode shells. The original kind of supercapacitor, the electrical dual coat capacitor, ropes the rescindable electrostatic build-up of ions on the superficial of a porous electrode. This grouping comprises carbon complexes with a great surface area.[22] Secondly, the pseudocapacitor class includes the revocable redox response with electrolyte ions, which ions respond to surface functional moieties. Pseudocapacitive energy storing is completed up of some oxide complexes of transition metals (TM) such as ruthenium and manganese, conducting polymers (CPs) and unlike-atom-nobbled carbon complexes.[23] The third type, a hybrid capacitor, is fundamentally a combination of a faradaic battery-kind electrode and a non-faradaic electrical dual-layer capacitor-type electrode.[24] The faradaic battery electrode is composed of phosphides, sulfides, and TM oxides, among other resources. Furthermore, battery-type electrodes are often muddled with pseudocapacitive electrodes, in spite of the circumstance that their charge stowage machineries are basically diverse. Pseudocapacitive charge stowage is attained utilizing an extremely rescindable surface redox response that does not need phase transition. On the other hand, battery-type electrodes stock charge through an alterable faradaic procedure where the charged and discharged electrodes experience a phase conversion. Supercapacitors are utilized as impartial energy bases or in combination with other strategies. The succeeding are some of the plans that are utilizing supercapacitors: (a) Transferable strategies: Supercapacitors are used as an energy basis in portable screwdrivers and camera flashes, as they need just spurts of energy and quick and incessant replenishment. (b) Recollection holdups: Supercapacitors by a fast response time offer a provisional resolution to a fleeting break of the memory‘s power cause. (c) Decoupling of energy and power necessities: Supercapacitors are utilized to encounter energy needs whereas a dissimilar scheme delivers the main foundation of energy. Decoupling of this kind is castoff in cross and electric vehicles. (d) Revival strategies: Supercapacitors can recuperate energy unconfined by machineries that accomplish monotonous and fixed activities. They are seen in a diversity of tenders, counting cranes and lifts, as well as in the slowing schemes of electric or mixture cars, such as trains, buses, and trash or distribution vans. (e) Electricity circulation and stowage: Supercapacitors are proficient of recompensing for small-spell variations in voltage indications in the circulation line and for linking the break between making and burning of electricity. Moreover, supercapacitors are applied to resume power schemes that have botched or to deliver energy till the actual source is restored. Supercapacitors are also utilized as energy stowage devices in renewable production units, most particularly wind energy, for their low preservation necessities.

6.2 MECHANISM FOR SUPERCAPACITOR

Decrypting the charge storing machinery of conformist supercapacitors (SCs) may be an important tread in the direction of the progress of great energy compactness SCs with protracted reusability, which can counter energy catastrophe to a prodigious level. Even though ex situ depiction methods have aided to fix the charge-storing knowhows of SCs, big uncharted gray areas with unidentified troupes are still existent, which should not be ignored. Through the previous decade, in situ investigative description gears, such as "in situ" XRD, XAS, XPS, IR/FTIR, NMR, AFM, SEM, TEM, and EQCM practices have entirely emerged to the vanguard to shed luminance on charge stowage machinery of SCs.[25] The merger of in situ methods and theoretical recreations can competently explicate the ion subtleties and charge transmission in SC electrode schemes. The following is to discuss most noteworthy SC mechanisms in detail.

6.2.1 ELECTRIC DOUBLE-LAYER SUPERCAPACITOR (EDLC) MECHANISM

The Coulombic charge storing machinery in EDLCs is contingent on the preoccupation and release of ions on the probe/medium crossing point deprived of a charge transmission response. Big surface part and electronically steering electrode resources are used for the manufacture of EDLC-based SCs. Figure 6.3 exemplifies an abstract drawing of EDLCs.[26] In the 19th century, the primary prototype for EDLC was presented by Helmholtz. This is called the Helmholtz model (Fig. 6.3a), which reiterates that an atomic aloofness splits up two films of conflicting charge at the probe/medium border. The Helmholtz prototype is alike to that of traditional dielectric capacitors. Far ahead, the Helmholtz EDLC classic was more improved by Gouy and Chapman, that is recognized as the Gouy–Chapman theory (Fig. 6.3b). This theory conditions that ions are considered as point charges, and the population of ions in the medium mixture close the probe exterior trails a Boltzmann distribution. However, the Gouy–Chapman hypothesis fails to explain the extremely charged double layers in the EDLC model. Therefore, to offset the limitations of the Gouy–Chapman model, Stern joined the Helmholtz conjecture and Gouy–Chapman hypothesis, which differentiates two areas of ion delivery, that is, the internal region termed as the Stern layer and the diffuse layer which is outer (Fig. 6.3c). In the Stern film, the ions are powerfully sucked by the probes, which contain a multilayer assembly, counting the inner Helmholtz plane (IHP) and outer Helmholtz plane (OHP). A verbose layer is situated following the outer Helmholtz plane (OHP).

Henceforth, the entire double-layer capacitance (C_{dl}) of the probe may be defined by means of the subsequent equation:

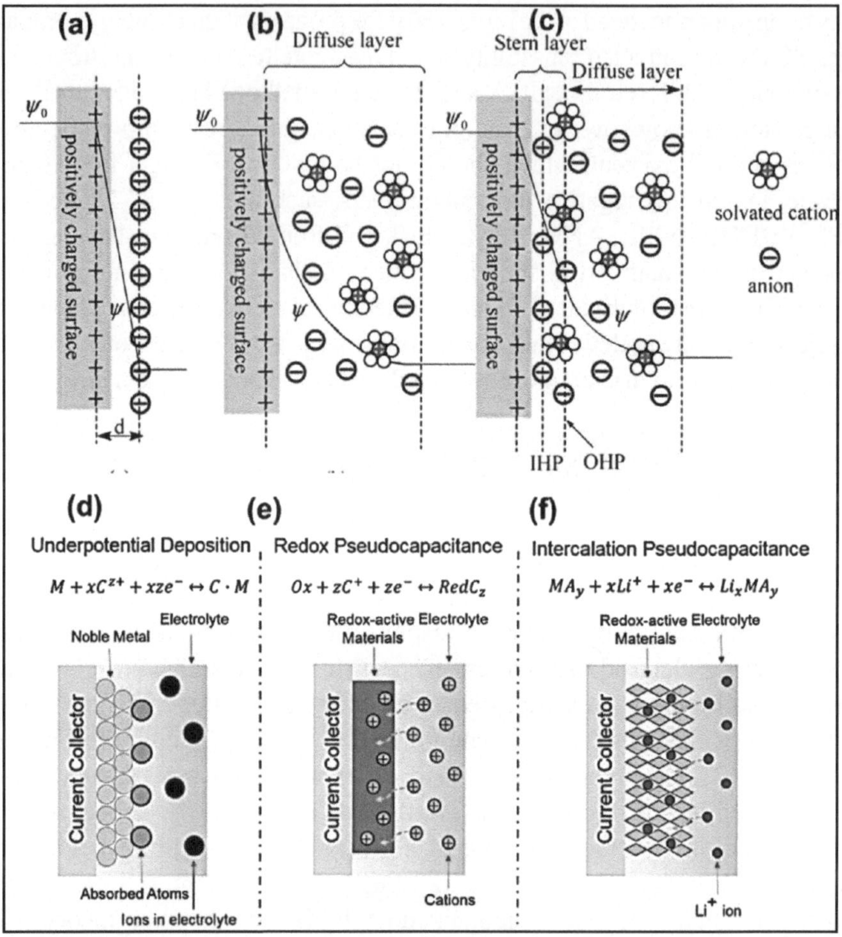

FIGURE 6.3 Models of the electrical double layer: (a) Helmholtz model, (b) Gouy–Chapman model, and (c) Stern model; Schematic diagram of various types of pseudocapacitance mechanisms: (d) underpotential deposition, (e) redox pseudocapacitance, and (f) intercalation pseudocapacitance.

Source: Reprinted with permission from Ref. [26]

$$\frac{1}{C_{dl}} = \frac{1}{C_{Stern}} + \frac{1}{C_{diff}} \tag{6.1}$$

where C_{Stern} signifies the double-layer capacitance of the Stern coat and C_{diff} demonstrates the double-layer capacitance of the diffuse film. Usually, carbon-based ingredients, for example, graphene, AC, and CNT are extensively utilized for EDLCs. AC is typically utilized as a probe resource for supercapacitors for its high superficial area, decent electrical conduction, and reasonable price. The carbonization procedure can build AC from numerous kinds of carbonaceous supplies such as coal, wood, and nutshell. Graphene is a discrete sort of graphite, entailing of a one-atom-thick leaf made of sp^2-attached carbon atoms in a polyaromatic honeycomb crystal lattice that is utilized in ultra-performance energy stowage systems for its outstanding physicochemical attributes, for example, high superficial area, decent electrical conduction, great thermal and chemical steadiness, and excellent litheness. Characteristically, CNTs can be shaped from rolled-up graphene sheets and reliant on the fabrication process, can be categorized as SWCNTs and MWCNTs. The arrival of CNTs has unlocked new potentials in energy storing devices, indicating to higher capacitance values due to their knotted webs with an exposed vital canal.[27]

6.2.2 PSEUDOCAPACITORS' CAPACITANCE

Comparing EDLCs, PCs stock charge in the probe constituents via a faradaic course, involving exterior or near-surface rapid rescindable redox responses.[28] As per the reviewed theory established by Conway about the functioning machinery of PCs, the charge stowage machinery of PCs can be separated as the following: (a) Underpotential deposition (UD), (b) Surface redox pseudocapacitance (SRP), and (c) Intercalation pseudocapacitance (IP). The UD course comprises the faradaic fascination or detachment of metal ions on diverse kinds of noble metals' faces to build an adsorbed monolayer overhead their redox potential (Fig. 6.3d).[26] SRP happens on the superficial or close to the exterior of ingredients for the associated faradaic charge-transfer course (Fig. 6.3e).[26] Likewise, IP designates the method of intercalation of ions into the channels or coatings of a redox-lively electrode substance, that is related with the faradaic charge-transfer course deprived of a crystallographic phase transition (Fig. 6.3f).[26] These three charge stowage machineries happen in PCs as a result of diverse corporal procedures and with numerous kinds of electrode substances. The knowhow of charge storing in PC electrodes rests on the succeeding equation:

$$E \sim E - \frac{RT}{nf} ln \frac{X}{1-X} \tag{6.2}$$

Characteristically, metal oxides, for example, RuO_2, MnO_2, IrO_2, NiO, SnO_2, Co_2O_3, MoO_x, and V_2O_5, metal chalcogenides, metal hydroxides, and conducting polymer (CP)-founded probe substances are incorporated as pseudocapacitors. Typically, MnO_2 and RuO_2 are researched as pseudocapacitive probes. CPs, counting polyaniline, polypyrrole, and polythiophene, are employed as pseudocapacitive probes for their high suppleness and conduction.[29] The charge–discharge profiles and cyclic voltammetry (CV) of several kinds of charge stowing machineries of energy stowage resources are shown in Figure 6.4a.[26] Type-A signifies the CV and charge–discharge profile of the capacitive probe supplies, manifesting quadrangular cyclic voltammograms and a rectilinear voltage retort all through persistent current charging/discharging. Correspondingly, type-B pseudocapacitive is faradaic in sense, while type-C faradaic is not the classical capacitive in conduct, precisely it is sort of battery (Fig. 6.4).[26]

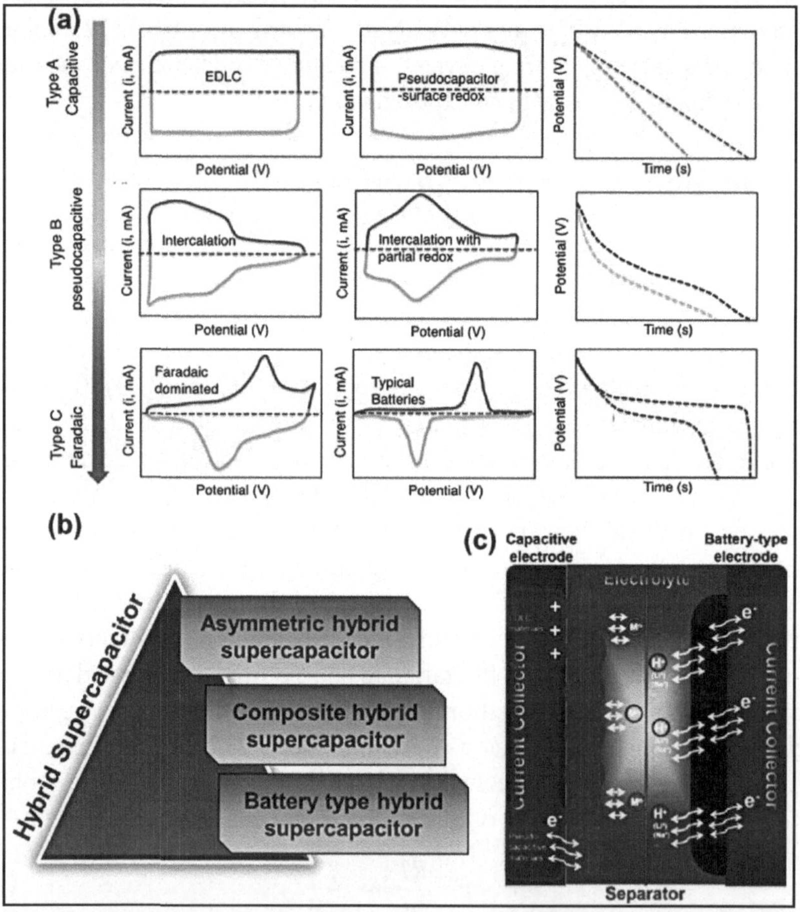

FIGURE 6.4 (a) Cyclic voltammograms and corresponding charge–discharge profiles for various types of energy storage materials. (b) Classification of hybrid supercapacitors according to their design or working mechanism. (c) Schematic diagram of a battery–supercapacitor hybrid system.

Source: Reprinted with permission from Ref. [26]

6.2.3 HYBRID SUPERCAPACITOR

Hybrid supercapacitors (HS) are embraced of polarizable probes (EDLC-based) and non-polarizable probes (PCs) to stock charges. This kind of SC parades the energy concentration of a battery-analogous electrode and power compactness of a capacitor-equivalent probe in the identical expedient with the amalgamation of together faradaic and non-faradaic courses. Crossbreeding can be attained at either the maneuver or substance level. As per the interbreeding progression of supercapacitors, HSs may be alienated into three classes as the following (Fig. 6.4b and c)[26] (a) Asymmetric, (b) Composite, and (c) Battery-type. Asymmetric HSs are premeditated utilizing carbon-derivative materials as anode and metal oxide or sulfide cathode to attain high energy and power concentration concurrently. The belongings of both carbon-founded ingredients and chalcogenides and transition metal oxides are united to create complex HS. Therefore, composite HS demonstrate better electrochemical presentations, counting high energy density, specific capacitance, and power density, contrasted with pristine EDLC-based or PCs-based supercapacitors for the high electric conduction, vast surface area, and excellent mechanical strength of composite hybrid substances. The battery-type hybrid supercapacitor joins two diverse electrodes, where the battery-type cross electrodes undertake faradaic redox responses during the charging and discharging courses.[30]

6.2.4 SYMMETRIC AND ASYMMETRIC SUPERCAPACITOR

Symmetric supercapacitors (SSCs) partake mutually positive and negative electrodes that are fabricated with precisely the same resources. These materials comprise metal oxides, CPs, transition metal dichalcogenides (MX_2), transition metal trichalcogenides (MX_3), MXenes and carbonaceous substances. The charge storing machinery of these ingredients is not very complicated while compared with assymetric supercapacitors (ASCs). The charge transference and charge stowage machinery are principally EDLC type or pseudocapacitive type, administered by the electrode substance. Sporadically, complex materials (combination of EDLC and pseudocapacitive) are utilized as the probes, the charge stowage knowhow of which is either prevailed by one of the resources or it may be a cross type. Lately, several researchers have progressed on SSC strategies to attain great capacitance along with decent cycling steadiness employing dual and treble metal sulfides and oxides and their mixtures. Isacfranklin et al. organized hydrogen-deprived carbon nanorod probes for a huge energy compactness (22 Wh/kg @ 1 A/g) and outstanding cycling constancy (96.65%)-based SSC. The charge stowing machinery is totally EDLC kind, that displayed a capacity of 198 F/g in a dual-electrode scheme.[31] Mane et al. of late incorporated MnO_2/RGO electrodes to illuminate 211 LEDs through a timespan of 3 min and their supple balanced device exhibited a huge energy and power concentration of 57 Wh/kg and 1.6 kW/kg, correspondingly. The charge stowage machinery demonstrated pseudocapacitive conduct, as palpably represented from the electrochemical act.[32] A great energy concentration of 81 Wh/kg at 714 W/kg was unveiled by an IrO_2@Mn_3O_4 symmetric expedient with rectangular-shaped CV arcs, which was created by Beknalkar et al. The synergetic result of cooperating oxides underwrote to the huge specific capacity and dazzling cycling constancy.[33] Lately, numerous MXenes and metal carbides have been explored as SS expedients, which outstripped majority of the fine discovered substances with respect to their energy stowage skill.[34]

If the electrodes' pair in a supercapacitor are not alike in all conceivable means, for example, owning the equivalent properties, mass share, functional clusters, charge stowage machineries, and lively redox places, then they create an asymmetric expedient. Furthermore, mixture strategies are ASSC, given that their two introduced probes (one capacitive sort and another, battery faradaic kind) have diverse charge stowage schemes. The noteworthy main benefit of ASSC is that it offers a widespread potential functional opening, leading to enormous energy compactness and power concentration that is not fairly practicable in the circumstance of SSC strategies. A redox-lively medium may be acquainted with the symmetric scheme as a standby for an inactive electrolyte to amass an ASSC expedient if adequate redox classes occur inherently. Recently, through the former years, a surfeit of ASS plans has brought in the pursuit for conventional energy bases. Wu et al. made-up an ASSC device by means of cobalt–nickel double hydroxide as anode and cobalt metal–organic framework (MOF) as cathode, realizing an energy mass of 42.5 Wh/kg with a steadiness of 81.5% and a widespread working potential.[35] Ghose et al. brought together exfoliated MXene and $MoS_{(3-x)}$ as cathodes and anodes, correspondingly, to deliver the dispersal associatory pseudocapacitance apparatus at a potential space of 1.6 V with great energy concentration and power compactness.[36] The comprehensive device strategy and machinery of ASS were richly assessed by two diverse research groups.[37,38]

6.3 SUPERCAPACITOR BASED ON METAL OXIDE AND ITS HYBRIDS

Midst the several nanomaterials utilized for SCs, metal-oxides have congregated noteworthy consideration because of their corporeal and natural attributes. In contrast with metal chalcogenides, metal oxides are laidback to concoct (and therefore enable big-volume manufacture), chemically steady, ecologically welcoming, and well-matched with widespread diversities of electrolytes, which are vital to achieve sturdy and cheap SC strategies. Additionally, metal oxides display higher theoretic specific capacitance, better energy concentration, and improved electrochemical presentation contrasted with graphene and other 2D resources by reason of the joint consequence of big probe superficial area and manifold oxidation stages.[39,40] Till present day, a wide range of metal oxides, counting single metal oxides, for example, TiO_2, RuO_2, Co_3O_4, MnO_2, V_2O_5, NiO, Cr_2O_3, and MoO_3, and dual metal oxides for instance $NiCo_2O_4$, $CuCo_2O_4$, $FeCo_2O_4$, $ZnCo_2O_4$, etc., have been efficaciously prepared through what‘s more straight exfoliation since their conforming bulk complements or self-assembly after salt forerunner-composing mixtures.[41] For illustration, manganese oxide (MnO_2), and ruthenium oxide (RuO_2) have been expansively researched as electrode resources for SCs for their rapid, rescindable redox responses.[42,43]

Typically, the supercapacitive performance of metal oxides hinges on some critical aspects, for example, (a) crystallinity, (b) morphology, (c) specific superficial area, (d) electronic conduction, and (e) mass concentration of the electroactive substance.[44,45] Amorphous metal oxides in general demonstrate superior electrochemical presentations while contrasted with their high-crystalline complements.[46,47] SCs, concerning 2D assemblies, for example nano-lamellar, nanosheet, and nano-film constructions, display improved corporeal and natural attributes like mechanical steadiness, suppleness, transparency, and chemical firmness contrasted with other nanostructures (OD and 1D).[48] An electrode with a high specific superficial expanse can cause heightened electrochemical performance (with respect to both specific capacity and energy compactness) for obtainability of further electroactive places, enabling manifold redox rejoinders.[49] Maximum of the metal oxide complexes (counting 0D, 1D, and 2D resources) show meagre electronic conduction, which limits the electron and ion transmission in the electrochemical charge–discharge course. At tall charge–discharge paces, the hindrance in the direction of the electronic movement will be suggestively high for an electroactive substance (metal oxides), that reduces the specific capacity, energy concentration, power compactness, and rate competence.[46] Huge mass concentration of electroactive ingredients (metal oxides) leads to lessened electrical conduction and delimited admittance to the electrolyte ions, resulting in deprived electrochemical performance. In essence, the mass loading factor is the wrong way round proportionate to the worth of specific capacity, which implies that an upsurge in the mass concentration would reduce the specific electrical capacity.[50,51]

Even though metal oxides own fascinating electrochemical performances founded on the faradic redox response, they might not be executed in their unspoiled arrangement as SC electrodes for the succeeding restrictions: (a) the electrical conduction of majority of metal oxides is quite poor, resulting in an upsurge in the pane hindrance and electrode's charge transport confrontation, and that leads to a huge IR falling at high current concentrations in the charge–discharge sequences. This causes a deterioration in the rate competence and power compactness of SC expedients; (b) all through the cycling of metal oxide founded SCs, strain is settled in the metal oxide exteriors, generating organizational dilapidation of the electrode substance, resulting in meagre cycle lifetime, and (c) it is awkward to modify few attributes of metal oxides, for example, to upsurge the specific surface area and the stoma extent dispersal. These are the main disadvantages of metal oxide-based SCs.[52] To circumvent these shortcomings of metal oxide-based SCs and improve their electrochemical presentation, emerging complexes and/or cross probe constituents is indispensable by compounding metal oxides with carbonaceous substances or conductive polymers.[53]

6.3.1 METAL OXIDE–GRAPHENE CROSSES

Usually, graphene amalgams of metal oxides partake countless budding in SC tenders for the succeeding whys and wherefores: (a) the metal oxide/graphene cross-structure may lead to great conduction for the inoculation of electrons since graphene coat hooked on the metal oxides, swelling the hole density in graphene,[54] (b) metal oxide–graphene mixes possess a huge superficial area that improves the collaboration among the electroactive substance and the medium, leading to better electrochemical act,[55] and (c) in metal oxide–graphene fusions, metal oxide nanosheets are habitually squeezed in amid graphene nanosheets to evade the rearrangement of graphene coats. Zhang et al. organized a 2D MnO_2/graphene cross via the electrostatic co-precipitation process.[56] The morphologic investigation (Fig. 6.5a and b) of the 2D MnO_2/

graphene mixture exhibits that the MnO_2 nanosheets are distributed on the superficial of the graphene substratum as a result of the electrostatic communication. This 2D MnO_2/graphene hybrid displays improved electrochemical presentation contrasted with the primeval MnO_2 and graphene for the synergistic influence of the outstanding electrical transference in the graphene and the pseudocapacitive attributes of the MnO_2 nanosheets. Figure 6.5(c) characterizes the CV silhouette of the 2D MnO_2/graphene cross probe and it attained a specific capacity of around 188 F/g at a mass regularized current of 0.25 A/g. In the same way, Ge et al. organized a NiO–graphene mix through a simplistic hydrothermal course.[57] Figure 6.5(d and e) characterize the SEM images of plain NiO nanocoats and the NiO–graphene amalgam. The electrochemical depiction of graphene, NiO nanocoat and the NiO–graphene cross-probes was done in a three-electrode arrangement in 2 M KOH watery medium. The NiO–graphene mixture probe attained a precise capacity of nearby 348 F/g, that is more than the magnitudes got since both the simple NiO and graphene. Figure 6.5(f) shows the cycling constancy presentation for the NiO–graphene cross-probe over 500 sequences at a current concentration of 1.5 A/g. For constructing a supple flat SC, Peng et al. utilized the fusion nanostructures of quasi-2D ultraslim MnO_2–graphene nanosheets. The two-dimensional SC expedient showed a unit capacity of 267 F/g at a weight standardized current of 0.2 A/g with exceptional driving steadiness (capacitance preservation of 92% even afterward 7000 charge–discharge rounds).[58] Besides, Tan et al. equipped a 2D Co_3O_4 nanoflake–graphene mixture on a Ni foam substratum by CVD method trailed by hydrothermal handling and attained a high capacity of 1.75 F/cm^2 at a current compactness of 1 mA/cm^2.[59] Lately, Shrestha et al. manufactured $MnCo_2O_4$/nitrogen nobbled graphene 2D/2D amalgam nanostructures by a one-pot polyol process and made electrochemical analysis by manufacturing a lop-sided expedient.[60] Figure 6.5(g) exemplifies the SEM photograph of the $MnCo_2O_4$/nitrogen-nobbled graphene 2D/2D cross-structure, which exposed that the communication of the 2D $MnCo_2O_4$ NFs with the planer nitrogen-decorated graphene (with a crumpled construction) evades the rearrangement of nitrogen-nobbled graphene. Electrochemical presentation of the bare $MnCo_2O_4$ and $MnCo_2O_4$/nitrogen-incapacitated graphene 2D/2D cross-probe was assessed in a trio-electrode scheme with 2 M KOH as the water-borne medium and the relative CV outline at an examination pace of 10 mV/s is revealed in Figure 6.5(h). The CV contour discloses that the $MnCo_2O_4$/nitrogen-defected graphene cross-probe displays improved electrochemical act than the pristine $MnCo_2O_4$ probe for harmonious collaboration of spinel planer $MnCo_2O_4$ with the exterior of layered nitrogen nobbled graphene. The $MnCo_2O_4$/nitrogen-decorated graphene cross-founded asymmetric maneuver realized a supreme energy concentration of nearly 48.5 Wh/kg and a power compactness of nearby 16,000 W/kg. Figure 6.5(i) shows the cycling solidity presentation of the lopsided stratagem over 10,000 sequences and the expedient recollects no less than 85.9% of its initial specific capacity upon 10,000 charging–discharging sequences

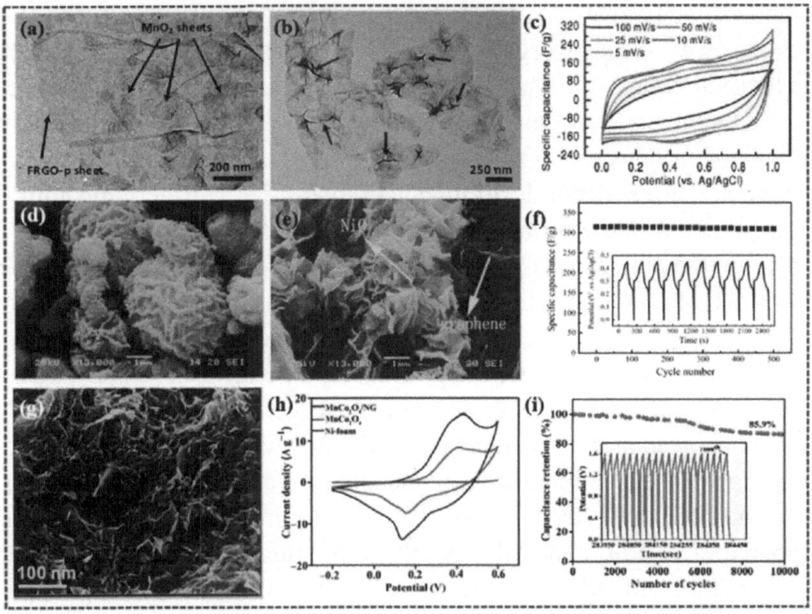

FIGURE 6.5 (a and b) TEM image of the 2D MnO_2/graphene cross at diverse amplifications, and (c) CV outline of the MnO_2/graphene mixture at several scan paces. SEM photograph of (d) plain NiO nanosheets and (e) the NiO/graphene mix, and (f) cycling steadiness silhouette of the NiO/graphene mixture probe. (g) SEM photo of the $MnCo_2O_4$/nitrogen-doped graphene 2D/2D amalgam nanostructure, (h) the CV contour of plain nickel foam, MnO_2O_4 and $MnCo_2O_4$/nitrogen-doped graphene 2D/2D cross electrodes, and (i) the cycling constancy curve of the $MnCo_2O_4$/nitrogen-doped graphene 2D/2D hybrid-based asymmetric expedient.

Source: Reproduced with permission from [53].

6.3.2 METAL OXIDE-CNT CROSSES

Carbon nanotubes (CNTs) are another extensively discussed ingredients as substrata or development templates backing up metal oxides for SC tender for their tall electrical conduction, and decent thermal and structural firmness.[61] Zhou et al. organized many transition-metal-oxide/CNT amalgams by wet chemical technique trailed by annealing as revealed in Figure 6.6(a) for both SC and lithium–ion battery usage.[62] The SEM image in Figure 6.6(b) displays the nanosheet constructions of the $NiCoO_2$ on the CNT mainstay. The electrochemical depiction of the $NiCo_2O_4$/CNT cross-probe was done in a three-electrode arrangement with 2 M KOH medium. The $NiCo_2O_4$/CNT mixture electrode attained a great specific capacity of around 1360 F/g at a weight-regulated current flow of 10 A/g with outstanding cycling firmness (3000 sequences). Equally, Mondal et al. created cross-MWCNT/$NiCo_2O_4$ nanosheets through a microwave-associated technique shadowed by succeeding calcination in air.[63] Figure 6.6(c and d) demonstrates the SEM and TEM photos of the MWCNT/$NiCo_2O_4$ cross, implying that the $NiCo_2O_4$ nanosheets have grownup round the exterior of the MWCNTs in the mixed MWCNT/$NiCo_2O_4$ nanosheets. For SC investigation, the electrochemical portrayal was done in 2 M KOH watery medium. The CV outline of the MWCNT/ $NiCo_2O_4$ nanosheet probe is displayed in Figure 6.6(e) and the MWCNT/$NiCo_2O_4$ cross-electrode brought a supreme precise capacity of around 1395 F/g at a nominal current of 1 A/g. The bi-electrode electrochemical examination for MWCNT/$NiCo_2O_4$ nanosheets was steered in a coin cell arrangement.

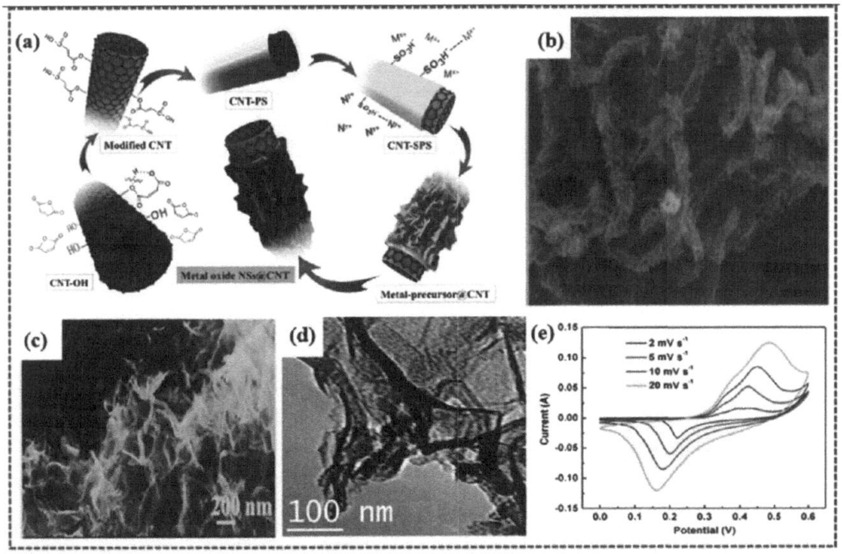

FIGURE 6.6 Representative illustration for preparation approach adapted for synthesis of the transition metal oxide/CNT hybrid, and (b) SEM image of the $NiCoO_2$/CNT cross (c and d) SEM and TEM photo of the MWCNT/$NiCo_2O_4$ mix correspondingly, and (e) CV outline of the MWCNT/$NiCo_2O_4$ hybrid probe.

Source: Reprinted with permission from Ref. [53]

6.3.3 METAL OXIDE-ACTIVATED CARBON CROSSES

Activated carbon (AC) is one more widespread carbonous sustenance ingredient for the manufacture of metal oxide/AC mixes for SC usage. AC owns modest electronic conduction, and huge surface area that conveys great actual superficial expanse for the development and backing of the metal-oxide, in metal oxide/AC crosses.[64] Zhang et al. organized a 2D-coated carbon–metal oxide complex probe for SC tender.[65] Figure 6.7(a) demonstrates the representation illustration for the creation of the 2D-thin carbon–metal oxide mixes by a three-stage procedure. A cobalt coating is first deposited electrolytically on Ni foam, and thereafter, yet, once more electrodeposition is made to accumulate incrusted MOFs. The SEM image of the thin carbon–metal oxide compound is revealed in Figure 6.7(b), which showed the creation of the amalgam's 2D-coated construction and hexagon form. The electrochemical depiction of the thin carbon–metal oxide complex was done in 4 M KOH at probe rates amid 10 and 50 mV/s as exposed in Figure 6.7(c). The coated carbon–metal oxide complex achieved peak spatial capacity of nearly 2098 mF/cm^2 at current flow compactness of 1 mA/cm^2.

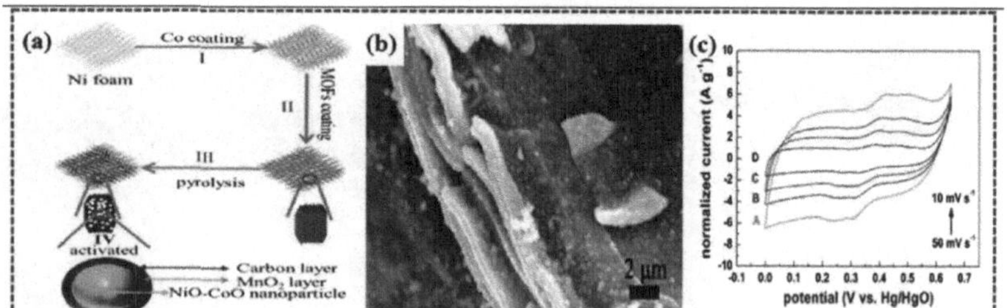

FIGURE 6.7 (a) Picture relating the creation of the 2D-layered carbon–metal oxide cross. (b) SEM photo of the 2D-layered carbon–metal oxide mix noted at numerous enlargements. (c) CV outline of the 2D-layered carbon–metal oxide cross at several probe rates. (d) The schematic representation of synthesis of 2D ternary cross of Co_3O_4, PANI, and graphene. (e) The CV outline of the ternary hybrid quantified in 6 M KOH watery medium.

Source: Reprinted with permission from Ref. [53]

6.3.4 METAL OXIDE–POLYMER CROSSES

CPs have expanded noteworthy courtesy as backing media to implant metal oxides for SC usage for their advanced conduction than metal oxides and the comfort of construction procedures concerning an extensive array of polymeric complexes. To construct metal oxide/polymer cross, the greatest utilized polymers are poly(3,4-ethylenedioxythiophene) (PEDOT), polypyrrole (PPy), polyaniline (PANI), and several offshoots of polythiophene.[66] Li et al. established a competent stratagem for the regulated development of 2D triplex hybrids of Co_3O_4, graphene, and PANI for SC usage.[67] Figure 6.7(d) schematically shows the total creation process for 2D ternary mixes of Co_3O_4, graphene, and PANI. Graphene nanocoats were primarily chemically activated associating PANI by an onsite polymerization procedure which shaped the graphene-founded PANI. Then, graphene–PANI nanocoats were assorted with the conforming metal salts as the forerunners of metal oxide/hydroxide trailed by hydrothermal handling to generate a treble mixture. Figure 6.7(e) displays the CV outline of the ternary hybrid probe in the latent space of 0.0–0.4 V at several examination paces. The synthesized treble cross attained an all-out capacity of nearby 1000 F/g at a probe pace of 5 mV/s.

6.3.5 METAL OXIDE–METAL OXIDE CROSSES

The metal oxide/metal oxide cross-probe along with a multifaceted construction can bid rapid electron and ion transmission alleyways and plentiful electroactive superficial places for electrochemical response, resulting in greater specific capacity and energy compactness than the sole constituent electrode.[68] Liu et al. developed Co_3O_4 nanowire@2D MnO_2 superthin nanosheet mixture core/shell arrangements by hydrothermal technique for SC usage.[69] The graphic illustration for the creation of the electrode is revealed in Figure 6.8(a). The nanostructure was shaped with a bi-step hydrothermal process shadowed by an annealing course. The SEM photographs are shown in Figure 6.8(b), depicting that the ultrathin sheets are interrelated to each other, creating an extremely permeable surface structure. The cross arrangement displays the peak specific capacity of around 480 F/g at a flow concentration of 2.67 A/g with outstanding capacitance retaining (2.7% holding even upon 5000 sequences). Kuang et al. confirmed the manufacture [as exposed in Figure 6.8(c)] of 2D $CuCo_2O_4$ nanosheet@MnO_2 nanoflake core/shell arrangements on Ni foam and used it for ASC compartments.[68] The hybrid arrangements are grownup on Ni foam substrata using mixed nucleation procedures associating heat and water. Figure 6.8(d) shows the SEM images of 2D $CuCo_2O_4$ nanosheet@MnO_2 nanoflake core/shell ranges organized at a diverse reaction period of 24 h. An ASC cell was built using $CuCo_2O_4$@-MnO_2 hybrids as the anode and triggered graphene as the cathode with the extractor and 1 M Na_2SO_4 electrolyte. The formed ASC expedient (2D $CuCo_2O_4$ nanosheets@MnO_2 nanoflakes//AG) distributed an energy concentration of around 43.3 Wh/Kg in the voltage span of 2 V. Anon, Xu et al. arranged a $CuCo_2O_4$@2D NiO nanosheet cross for SC usage.[70] Figure 6.8(e) schematically displays the picture for the trio-stage construction procedure of the $CuCo_2O_4$@ 2D metal oxide nanosheet cross probe on a Ni foam substratum. Figure 6.8(f) demonstrates the SEM images of the $CuCo_2O_4$@ 2D NiO nanocoat cross, which exposed that NiO sheets (mean depth of 50–80 nm) consistently shelter the whole surfaces of the $CuCo_2O_4$. The electrochemical trial of the hybrid electrode was done in treble-electrode manner using 1 M KOH medium which displayed a supreme areal capacity of nearby 2.3 F cm$-^2$ at a current compactness of 2 mA cm$-^2$ with outstanding cycling constancy. Supplementarily, an ASC was manufactured using the $CuCo_2O_4$@ 2D NiO nanolayer as the anode and AC as the cathode and attained a thoroughgoing energy compactness of 38.9 W h/kg in a voltage space of 1.5 V.

6.3.6 SYNERGIC PARAPHERNALIA IN THE MIX EDIFICE

The harmonious impacts in cross assemblies (metal-oxide joint with graphene, or CNT, or AC or CP) can be momentarily emphasized as follows: (a) graphene may play a dynamic role in metal oxide/graphene cross constructions, as graphene delivers a 2D conducting pattern to indorse an interrelated conducting permeable web that would improve the electrical conduction and charge conveyance trails of metal oxides; (b) graphene found a decent face-to-face communication and electrical connection amid itself and metal oxides for the obtainability of oxygen-bearing efficient clusters; (c) metal oxides when combine with AC, and/or CNT, incline to improve the specific superficial area of cross electroactive substance, which will ease the shallow responses, and the medium can partake a decent collaboration with the probe (for its huge surface area),

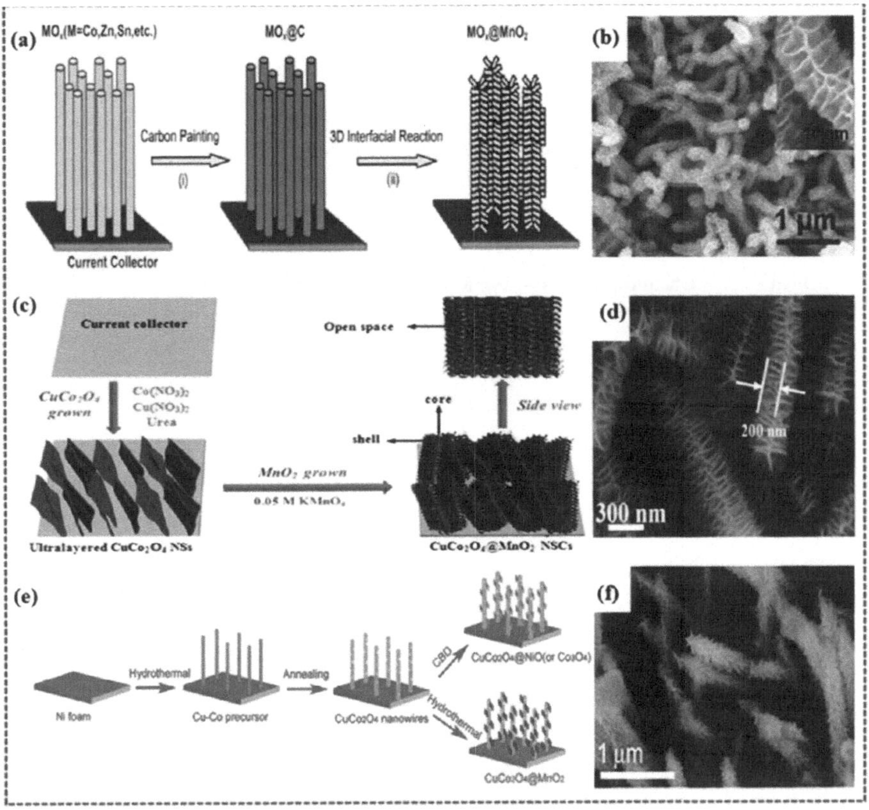

FIGURE 6.8 (a) Diagram illustrating the construction of the Co_3O_4 nanowire@2D MnO_2 ultrathin nanosheet cross core/shell assembly. (b) SEM image of the Co_3O_4 nanowire@2D MnO_2 ultrathin nanosheet mix. Reproduced with permission.[68] (c) Picture of the fabrication process of graded $CuCo_2O_4$@MnO_2 nanostructures. (d) SEM image of 2D $CuCo_2O_4$ nanosheet@MnO_2 nanoflake core/shell arrangements set under diverse investigational circumstances. Reproduced with permission.[69] (e) Pictorial representation for the synthesis of the $CuCo_2O_4$@2D metal oxide nanosheet cross. (f) SEM image of the $CuCo_2O_4$@2D NiO nanosheet amalgam.

Source: Reprinted with permission from Ref. [70]

leading to heightened electrochemical presentations in terms of specific capacity, energy compactness, and power concentration; (d) for advancing the electrical conduction of plain metal oxides, they are typically joint with numerous different CPs to augment the electrical conduction of the metal oxide/polymer crossbred and off course to advance the numerous attributes, for example, flexibility, specific capacity, and comfort of manufacture for real-world applications (Table 6.1).

TABLE 6.1 Summary of Energy Storage Performance of Metal Oxides with Different Hybrid Structures.[53]

Sl No.	Electrode material	Synthesis method	Electrolyte	Specific capacitance	Current density	Potential window (V)	Ref.
1	2D MnO_2/graphene hybrid	Electrostatic co-precipitation method	1 M Na_2SO_4	188 F/g	0.25 A/g	1 V	56
2	2D NiO/graphene hybrid	Chemical precipitation method	2 M KOH	346 F/g	1.5 A/g	0.5 V	57
3	2D MnO_2/graphene hybrid	Solution treatment method	PVA/H_3PO_4 gel electrolyte	267 F/g	0.2 A/g	0.7 V	58
4	2D Co_3O_4/graphene hybrid	Hydrothermal method	6 M KOH	1.75 F/cm^2	1 mA/cm^2	0.5 V	59
5	2D $MnCo_2O_4$/graphene hybrid	Polyol method	2 M KOH	1170 F/g	1 A/g	0.5 V	60
6	2D MnO_2/graphene hybrid	Electrodeposition method	1 M KCl	328 F/g	0.5 mA/cm^2	1 V	71
7	2D $NiCoO_2$/CNT hybrid	Wet chemical method	2 M KOH	1360 F/g	10 A/g	0.5 V	62
8	2D $NiCo_2O_4$/MWCNT hybrid	Microwave followed by calcination method	2 M KOH	1395 F/g	1 A/g	0.4 V	63
9	2D Co_3O_4/PANI/graphene hybrid	Hydrothermal method	6 M KOH	1063 F/g	1 A/g	0.4 V	67
10	2D $CuCo_2O_4$/MnO_2 hybrid	Hydrothermal method	1 M Na_2SO_4	416 F/g	1 A/g	1 V	68

6.4 SUPERCAPACITOR BASED ON METAL DICHALCOGENIDES AND ITS HYBRID

Apart from the meaningly enhanced end proportion and superficial acquaintance from their large-scale stages, 2D TMDs partake graphene-alike reedy constructions (~1 nm profuse, for a single layer) with perfect super brief electron conveyance distance and trivial ion transportation length for rapid electrochemical courses. Researches have recommended that 1H/2H TMDs, for example, shrill MoS_2 layers have analogous supercapacitor presentation to that of CNT arrangements with a theoretic capacity (~1000 F/g) better than graphite.[72,73] Yet, the progress of 2D TMD-founded supercapacitors has not been actually efficacious since the little electrical conduction of maximum steady semiconducting 1H/2H TMDs. In contrast, metallic 1T TMDs are hypothetically metastable and partake the basal level with further chemical bustle over that of the 1H/2H stages.[74–76] They possess very tall electrical conductance (e.g., 10–100 S/cm of single-layered 1T MoS_2 layer at ambiance, 107 times more conducting than 1H MoS_2 single layers) which is at par the excellent RGO sheets (~100 S/cm).[77] Fresh advancements have proposed that constructions of a few metallic 1T TMDs are really steady in convinced electrochemical courses,[78–80] which commits the prodigious upcoming of 2D TMDs in SCs. This capable tender has been confirmed by Chhowalla et al., where single-layered 1T MoS_2 layers (~70% segment proportion, Fig. 6.9(a) were exfoliated as of their wholesale flecks utilizing organolithium.[77] Due to their great electrical conduction, thin supple supercapacitors may be manufactured decently since these 1T MoS_2 panes (piling depth up to 5 μm, Fig. 6.9(b) starved of the count of any obligatory mediators or conducting seasonings (e.g., the flick on polyimide substratum in Fig. 6.9(d).[77] The crammed supercapacitor after this supple probe presented volumetric capacities extending from 400 to 700 F/cm^3 in diverse watery media (H^+, K^+, Na^+, Li^+), laterally with quite high, steady (95% capacitance holding afterward 5000 sequences) energy and power concentrations in organic media (the dappled and tinted area in Fig. 6.9(e). Their CV arcs in the castoff voltage span (watery medium and three probes: −1.0–0 V vs open circuit potential (OCP); biological electrolyte and two rods: 3 V) were all mostly capacitive highlighted,[77] imparting enormous power and energy concentrations (Fig. 6.9e). The energy stowage presentation in organic medium is meaningfully bigger than those of MXenes (2D transition metal carbides, carbonitrides and nitrides), graphene, AC, and few shrill Li–ion batteries (Fig. 6.9e). The effectual storing of organic electrolyte ions (e.g., the tetraethylammonium/ TEA+) and the bigger ions yonder the Li^+ (e.g., Na^+, K^+) was mostly due to the disrupted sheets in loaded leaves with both the interlayer gap about 6.15 and 11.65–12.15 Å which extremely ties the dimensions of these ions (Fig. 6.9c). Succeeding this opinion, other 2D TMDs with great electrical conduction (e.g., 1T WS_2 and 1T VS_2 sheets) and alike big interlayer gapes are likewise predictable to be greatly capable in construction of high-performing supercapacitors.

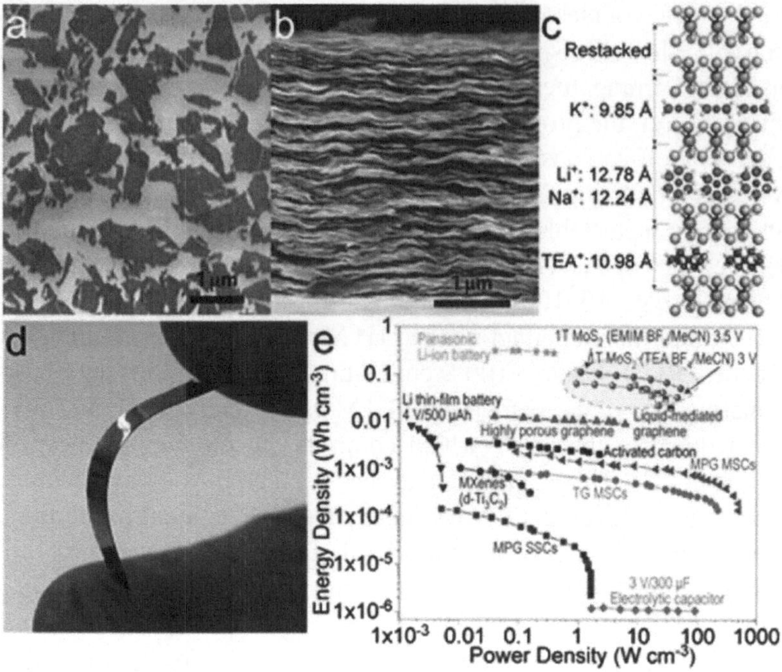

FIGURE 6.9 (a–b) SEM images of the disseminated single-layered nanosheet 1T phase MoS_2: (a) the cross-section of the loaded 1T MoS_2 probe. (c) Diagram displays the interlayer gap of restacked 1T MoS_2 nanosheets (kept as 6.15 Å) earlier and later ions intercalation operation. The K^+ and Na^+/Li^+ ions are harmonized with two and six H_2O molecules, correspondingly. The intercalation of organic electrolyte (TEA+) provided an extra layer parting of 4.83 Å. (d) Picture of the 1T MoS_2 sheets on lithe polyimide substratum. (e) Ragone plot of the bulk power and energy concentrations of diverse ingredients.

Source: Reprinted with permission from Ref. [77]

Above and beyond these 1T TMDs, lately advanced triplex sulfides, the Ni–Co–S,[81–84] might be further stimulating in supercapacitors due to their great constancy, decent redox response places (over that of the respective dual Ni or Co sulfides[81]) and brilliant electrical conduction, for example, the $NiCo_2S_4$ possesses the electrical conduction of about 10^3–10^4 S/cm at ambiance[82,83] which is 1–2 orders more than that of 1T MoS_2 single layers (10–100 S/cm).[77] Supercapacitors are constructed since these triplex sulfides with nano forms (e.g., nanofibers, 3D structures) have been confirmed by diverse research groups.[85–87] Although the manufacture of superthin single incrusted ternary sulfides is yet stimulating, a fresh research on $CoNi_2S_4$ nanosheets (depth: about 20 nm; 2D magnitude: micrometers) founded probe has offered a big capacity of 1117 F/g (at 2 A/g, tripple electrodes method, 6 M KOH) and decent rate presentation (1002 F/g at 10 A/g).[88] An ASC created after these treble sulfides offered great power (e.g., 8 kW/kg at the energy compactness of 55.4 Wh/kg) and outstanding cycling acts (>80% capacity reserved after 10,000 sequences at 10 A/g).[88] Greater capacity (2906 F/g) than the overhead established value of comparable $CoNi_2S_4$ layers (depth: about 45 nm; 2D dimensions: micrometers) founded supercapacitor was similarly attained at a current concentration of 5 mA/cm.[289] These presentations are previously greater than that of their countless complements with conventional arrangements.[88,89]

Like the slim 2D TMD ingredients with deprived electrical conduction (mainly the semiconducting 1H/2H TMDs), cross assemblies after these TMDs and graphene (Fig. 6.10,[90] for example, via physical piling or chemical attachment) might be well-thought-out to hypothetically circumvent numerous faintness of old-style HS, offering better energy and power concentrations, extended cycle lifetime and enhanced protection. Contrasting conventional nanoparticles (where just the topmost exterior takes part in the electrochemical course), the entire 2D TMDs (e.g., single layers), and their extremely active ends in this cross assembly (Fig. 6.10) might underwrite both the Faradaic and non-faradaic capacity, refining ingredients' usage, and dipping cost. Contrasting it with old-style secondary batteries, the close surface redox response is fundamentally dissimilar and related with much greater energy transfer competences. Additionally, 2D TMDs might closely communicate the graphene wall, that would safeguard fast charge transference among the probe and the chemically vigorous site (e.g., lively places on 1H semiconducting TMDs) for big power compactness and better rate presentation. This construction is further capable as soon as extra electrically conductive TMDs nanocrystals (e.g., 1T nanocrystals), chemical alterations, and 3D-G with huge surface parts are presented. Even though cross electrodes since sternly single-layered 2D TMDs and graphene yet requires to be advanced, current information on alike mix constructions have established that there is a prosperous forthcoming for 2D TMDs in supercapacitors.[91–93] For instance, Ajayan et al. have established the mix probe of $CoNi_2S_4$–graphene-layered $MoSe_2$ with graphene and slim $MoSe_2$ leaves are mutually devoted, and extremely conducting $CoNi_2S_4$ nanoparticles on/among the coatings.[93] This probe showed huge capacity (1141 F/g, 1 A/g, tierce electrodes system, in 6 M KOH electrolyte), decent rate (961 F/g at 5 A/g) and cycling presentations (108% capacity holding after 2000 sequences, 20 A/g). The supercapacitor accumulated since this probe also displayed great power concentrations (e.g., ~0.89 and 13.3 kW/kg at energy masses of 38.6 and

23.7 Wh/kg). The part of 2D TMDs in supercapacitors is more reinforced, predominantly when the constituents manufacture on numerous 2D nanostructures has been demonstrated to be progressively fruitful.[94–102]

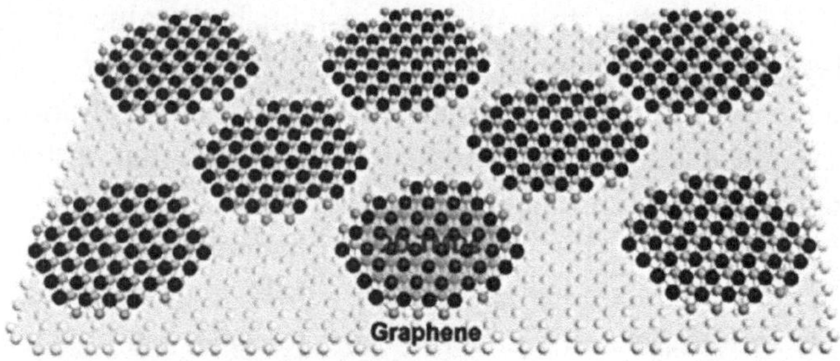

FIGURE 6.10 Diagram demonstrates the cross assembly of 2D TMDs and graphene, where the 2D TMDs closely connect the graphene's surface with their basal planes.

Source: Reprinted with permission from Ref. [90]

6.5 SUPERCAPACITOR BASED ON MXENE AND CROSSES

Since the electrode's substance has a critical part in defining their electrochemical presentation, better materials with exclusive attributes are indispensable for the manufacture of competent probes. Unearthing of hydrous ruthenium oxide[103] presented excellent electrochemical characteristics consenting a capacity of 1000–1800 F/cm^3 for profuse probes (>20 μm)[104] and capacitance of nearly 2800 F/cm^3 for thin flicks (50–100 nm).[105] Nevertheless, its great price and diminished charge stowage competence with the growing probe depth discarded it as an appropriate solid for energy stowing needs.[106] After RuO_2, the encounter of 2D TMDs,[107] metal–organic frameworks (MOFs),[108] graphene,[109] etc., presented energy stowage ability by rescindable superficial redox responses. The incrusted construction of 2D substances offers a bigger surface-to-volume ratio thus guaranteeing larger acquaintance of electrolytes to the face of the electrode. Yet, these substances exhibited restrictions for their poorer conduction, feeble van der Waals attachment and hydrophobicity also. To bargain a substitute, different layered resources were inspected to attain greater capacity. This led to discovery of the 2D constituent MXene with interlayer construction, big face-to-bulk proportion, and plentiful electrochemically vigorous places for energy storing needs.[110,111] The widespread variety of oxidation conditions offered by transition metal atoms deliberates an outstanding faradic charge stowage machinery in MXene founded supercapacitors. In contrast with other pseudocapacitive ingredients, MXenes stock copious charges and transport charges at considerably rapid rate. In its place, the exceptional conduction and hydrophilic theme of MXenes turn it a charming 2D resource for electrochemical energy stowage usages.[112] The practice of primeval MXene as electrode ingredients confines its tender for its relaxed restacking and substandard mechanical characteristics. Thus, the complex of MXene with diverse electrode substances may augment the electrochemical presentation extremely.[113] Owing to excellent elastic moduli, modulable superficial close groups, and bigger length: width relation of MXene, it may be utilized as caulking in numerous complexes. These inimitable attributes of MXene complex warrant scientists to emphasis on manufacturing supple, abiliment, and transportable electrochemical expedients.

6.5.1 MXENE-GRAPHENE FOUNDED SUPERCAPACITORS

The MXene founded fibers created by wet-spinning assemblage technique by MXene panes and RGO liquid crystal displayed an electrical conduction of 2.9×10^4 S/cm with a better areal and bulk capacity of 372.2 mF/cm^2 and 586.4 F/cm^3, correspondingly in PVA/H_3PO_4 gel medium with capacity holding beyond 3000 sequences.[114] Kumar et al. produced $Ti_3C_2T_x$ MXene/graphene/Ni supercapacitor probe with the premium supernatant of HCl/LiF imprinted $Ti_3C_2T_x$ MXene emulsion as the substance of the probe and CVD-developed graphene on Ni-foil as the current accumulator. The manufactured supercapacitor showed a superior gravimetric capacity of ~542 F/g at a probe rate of 5 mV/s, that is, 1.5 times bigger than the capacity attained shorn of graphene.[115]

6.5.2 MXENE-CNT FOUNDED SUPERCAPACITORS

Zhao et al.[116] described that the sandwich-like MXene/CNT complex shaped by discontinuous percolation of MXene and CNT scatterings presented enhanced bulk capacity of 350 F/cm^3 at 5 A/g in contrast with pristine MXene or haphazardly assorted MXene/CNT papers. A greater bulk capacity of 435 F/cm^3 for MXene/rGO complex was detected with outstanding cycling steadiness. Park et al.[117] utilized 100 μm profuse MXene/CNT-positive and MnO_2/CNT-negative probes joint with PVA/LiCl gel medium to arrange an asymmetric yarn supercapacitor that presented a concentration of energy of 100 μWh/cm^2. The $Ti_3C_2T_x$ MXene–CNT compound probe made by photolithography method heralded by blankness purification displayed a superior areal capacity of 61.38 mF/cm^2 at 0.5 mA/cm^2 for CNT intercalation among MXene sheets that deliver a relaxed trail for ion dispersal.[118] The $Ti_3C_2T_x$ MXene–multiwalled CNT compound negative probe and polypyrrole covered MCNT-positive probe accumulated with Na_2SO_4 medium to make an ASC, displays an areal capacity of 0.94 F/cm^2 in a potential opening of 0–1.6 V.[119]

6.5.3 MXENE-CARBON ALLOTROPES-FOUNDED SUPERCAPACITORS

The spray inoculation of $Ti_3C_2T_x$ MXene on nickel–cobalt sulfide ($NiCO_2S_4$) @ carbon cloth (CC) substratum caused $Ti_3C_2T_x/NiCO_2S_4$@ CC composite (TNSC), that showed a gravimetric capacity of 2326 F/g at 1 A/g and superior cyclic firmness of 93.8% at a current compactness of 10 A/g.[120] The electrochemical presentation of the complex probe may be altered by varying the frame load of $Ti_3C_2T_x$ MXene, which underwrites to electrical conduction and bring about the penetrability of ions to nickel–cobalt sulfide. The partially solid-stage supple supercapacitor manufactured by amassing nitrogen-nobbled AC covered carbon cloth (NACC) as the negative and TNSC as the positive electrode in PVA/KOH gel medium displayed a power compactness of 800 W/kg at an energy concentration of 57.5 Wh/kg, indicating a Coulombic efficacy of 99.2 and 90.2% capacity holding beyond 5000 charge–discharge sequences (Fig. 6.11).

6.5.4 MXENE-TMD FOUNDED SUPERCAPACITORS

Rakhi et al.[121] pointed out that the straight chemical creation of ε-MnO_2 whiskers on the surface of MXene leaves heightened the shallow area, thus growing the electrochemical presentation of $Ti_3C_2T_x$/ε-MnO_2 nanocomposite probe in a watery medium, by closely three orders of scale contrasted with pure MXene electrodes. The simplistic hydrothermal creation of $Ti_3C_2T_x$ MXene/MoS_2 nanocomposite combines the upshot of web construction and metallic conduction of $Ti_3C_2T_x$ MXene to impart a greater electrochemical capacity.[122] The net assembly enhanced the approachability of ion electrolytes to the probes and the metallic milieu of $Ti_3C_2T_x$ MXene offered a better electrical conduction that improved the electrochemical action. The SC made by $Ti_3C_2T_x$ MXene/MoS_2 nanocomposite probe stated a precise capacity of 115.2 F/g at 0.5 A/g with a Coulombic efficacy of ~90%.

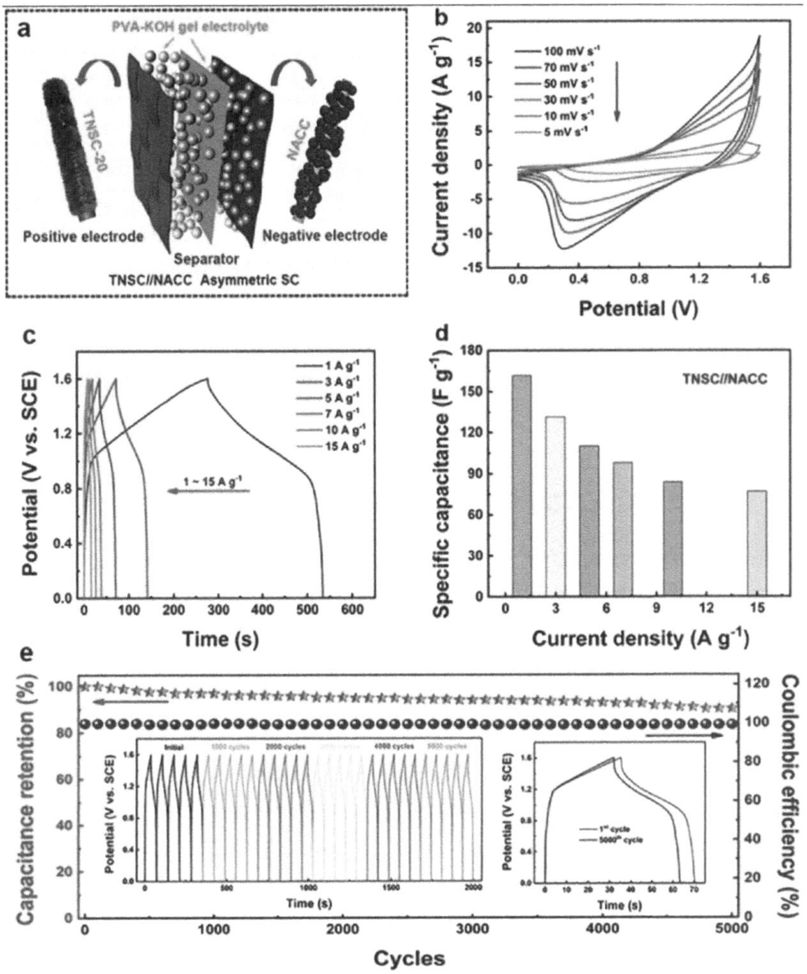

FIGURE 6.11 The electrochemical presentation of asymmetric supercapacitor (ASC) comprising of NACC (positive) and TNSC (negative) electrode in PVA/KOH gel watery electrolyte, (a) construction of ASC, (b) CV arcs at diverse probe rates, (c, d) GCD arcs and specific capacity at diverse current densities, (e) capacity holding and Coulombic efficacy of ASC.

Source: Reprinted with permission from Ref. [120]

6.5.5 MXENE-METAL OXIDE-FOUNDED SUPERCAPACITORS

The compound of Co–Fe oxide with $Ti_3C_2T_x$ MXene displayed an improved bulk capacity of 2467.6 F/cm^3 in 1 M LiCl medium for Co–Fe-oxide insertion which enlarged the layer-to-layer spacing of MXene panes; although utilized as SC probe, it unveiled an areal capacity of 356.4 mF/cm^2 with a capacity preservation of 88.2% post-10,000 sequences.[123] Even after 100 cycles of powered twisting, the complex probe reserved the electrochemical energy stowage steadiness, turning it an appropriate substance for the construction of supple energy stowage device. Co_3O_4 nanoparticles-MXene (Co-MXene) complex created by facetious hydrothermal technique evaded rearrangement of MXene sheets and testified a better mass capacity of 1081 F/g at 0.5 A/g, that is considerably better than the capacity of primeval MXene probe.[124] An ASC was formed using Co–MXene@Nickel foam as the positive and polyaniline (PANI) resultant carbon@carbon fiber paper (PANIC@CFP) as the negative electrode in 6 M KOH solution with a micropermeable polypropylene spacer amid them. The ASC showed a greater voltage space of 1.4 V and mass capacity of 95.71 F/g at 1 A/g. The supreme energy compactness of the ASC was 26.06 Wh/kg at 700 W/kg and 83% capacity holding beyond 8000 sequences at 2 A/g.

6.5.6 MXENE-POLYMER FOUNDED SUPERCAPACITORS

Even though MXenes have exposed superior tenders in numerous electrochemical energy stowage strategies, rebuilding of MXene layers deters ion transference. Also, It was pointed out that the electrochemical presentation of MXene founded probes declined with growing probe depth. These restrictions of MXene probes may be evaded by constructing the complex of MXenes with polymers which would improve the litheness and electrochemical presentation enormously.[125] It was revealed that mono-incrusted MXene leaves had better reachable shallow hydrophilicity and adjustment with polymers for their tall length to width proportion contrasted with multilayered MXene leaves.[126] Thus, multilayered MXenes are delaminated beforehand creating amalgams to augment their electrochemical action. Cheng et al.[127] manufactured zinc hybrid microsupercapacitor (ZHMSC) with capacitor-type MXene/bacterial cellulose (BC)@polypyrrole (PPy) probe and battery-type CNTs@MnO_2 probe in 2 M $Zn(CF_3SO_3)_2$–0.1 M $MnSO_4$/PAM hydrogel. The ZHMSC shared the characteristics of capacitor and battery-type probes and exhibited an areal capacity of 290 mF/cm^2 at 0.75 mA/cm^2, a bigger energy compactness of 145.4 μWh/cm^2, with a superior power concentration of 0.36 mW/cm^2 with 95.8% cycle steadiness over 25,000 sequences (Fig. 6.12). A marvellous upsurge in areal capacity, energy compactness, and power concentration is detected by the integration of conducting PPy polymer contrasted with pure MXene founded ZHMSC. The authors also found the holding of electrochemical presentation over the stretchable deformation up to 400%, earmarking it an apposite substance for the manufacture of stretchy supercapacitors. VahidMohammadi et al.[128]

inspected that the $Ti_3C_2T_x$ MXene compound shaped by onsite polymerizations of PANI on superficial of MXene displayed bulk and mass capacity of 1682 F/cm^3 and 503 F/g, correspondingly. The author also described that the complex probe with an area density of 23.82 mg/cm^2 and a depth of 90 μm exhibited precise capacity of ~336 F/g with 98.3% of capacity preservation after 10,000 sequences. The intercalation of PANI nanoparticles among the interlayers of MXene leads to the pillaring consequence that improves the ion conveyance in MXene probes. This research cemented the track for making thick probes displaying greater capacity.

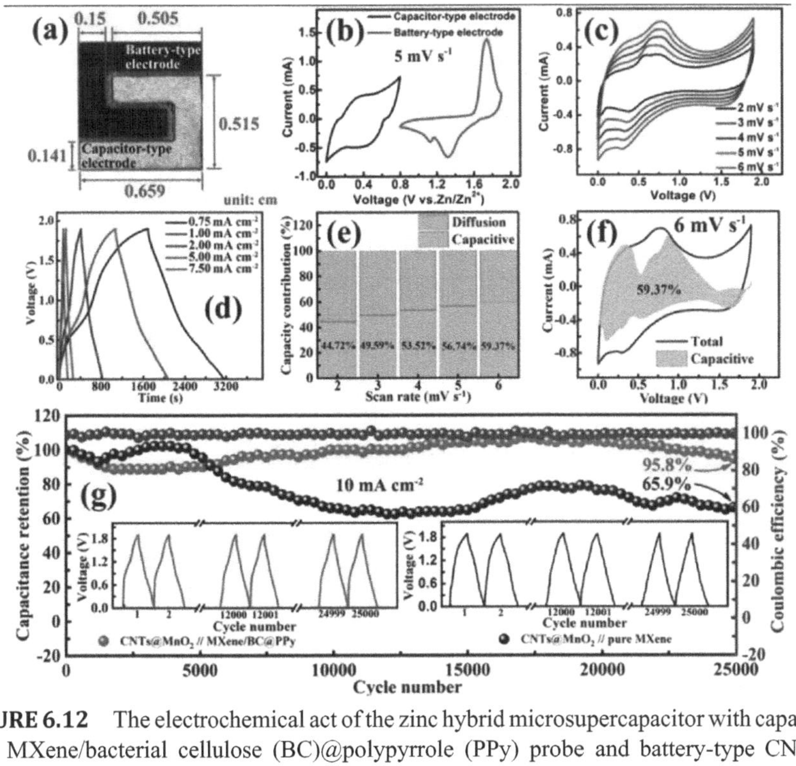

FIGURE 6.12 The electrochemical act of the zinc hybrid microsupercapacitor with capacitor type MXene/bacterial cellulose (BC)@polypyrrole (PPy) probe and battery-type CNTs@ MnO2 electrode. (a) Strategy of ZHMSC. (b) CV arcs of capacitor type and battery type electrodes. (c) CV curve of ZHMSC at changeable sprobe rate. (d) GCD arc at variable current densities, (e) pseudocapacitive influence of capacitor type probe. (f) CV arc demonstrating 59.37% capacitive regulated involvement. (g) Cycle lifespan of ZHMSC.

Source: Reprinted with permission from Ref. [127]

Table 6.2 exemplifies the capacitive presentation of MXenes and a few related complexes-based symmetric and asymmetric supercapacitors.[129]

TABLE 6.2 The Electrochemical Performance of MXene-Based Symmetric and Asymmetric Supercapacitor Electrodes.[129]

MXene supercapacitor electrode (Cathode//Anode)	Electrolyte	Capacitance	Energy density	Power density	Capacitance retention	Ref.
$Ti_3C_2T_x$ MXene//activated carbon	PVA/Na_2SO_4	7.8 mF/cm^2	3.5 mWh/cm^3	100 mW/cm^3	10,000 cycles 91.4%	130
ONCFT//MNCFT	PVA/H_2SO_4	780 mF/cm^2	277.3 $\mu Wh/cm^2$	624 $\mu W/cm^2$	30,000 cycles 90%	131
$Ti_3C_2T_x$/CNF/PC	PVA/KOH	143 mF/cm^2 at 0.1 MA/cm^2	2.4 μ Wh/cm^2	17.5 $\mu W/cm^2$	10,000 cycles 90%	132
$NiCo_2$-LDHs@MXene/rGO// MXene/rGO	2 M KOH	240 F/g at 1 A/g	65.3 Wh/kg	700 W/kg	10,000 cycles 92.8%	133
$Ti_3C_2T_x$-based-Zn ion hybrid (0.851 μm)	PVA/$ZnCl_2$	72.02 mF/cm^2 and 662.53 F/cm^3	0.02 mWh/cm^2	0.50 mW/cm^2	50,000 cycles 80%	134
MXene/BC@PPy//CNTs@MnO_2	2 M $Zn(CF_3SO_3)_2$–0.1 M $MnSO_4$/PAM hydrogel	290 mF/cm^2	145.4 $\mu Wh/cm^2$	0.36 mW/cm^2	25,000 cycles 95.8%	127
MXene/RGo	PVA/H_3PO_4	372.2 mF/cm^2 and 586.4 F/cm^3 at 10 mV/s	13.03 mWh/cm^3	0.59 W/cm^3	3000 cycles 100%	114
RGM-H_2SO_4	1 M H_2SO_4	454.9 F/cm^3 at 10 mV/s	39.4 Wh/L	180 W/L	–	135
V_2O_5//$Ti_3C_2T_x$ MXene (ZIHMSC)	$ZnSO_4$ dissolved PAM hydrogel	129 mF/cm^2 at 0.34 mA/cm^2	48.9 $\mu Wh/cm^2$	673 $\mu W/cm^2$	10,000 cycles 77%	136
Acide treated Ti_3C_2//RGO	PVA/H_2SO_4	256 F/cm^3 at 5 mV/s	58.4 mWh/cm^3 at 20.2 Wh/kg	7466.0 mW/cm^3 at 2583.0 W/kg	10,000 cycles 92.4%	137
MGH//PGH	1 M KCl	137.5 F/g at 10 mV/s	30.3 Wh/kg	1.13 kW/kg	10,000 cycles 81%	138

6.6 THEORETICAL INSIGHT FOR SUPERCAPACITOR

Many nitty-gritties of charge storage mechanism in electrochemical storage devices which are impossible to conceptualize physically, can be theoretically explained using quantum calculations, such as molecular dynamics, Monte Carlo, and DFT. These electrochemical capacitors can accumulate the energy by charge distribution at the interface of electrode materials and electrolyte. We can define two types of ECs, which depend on the charge storage process.[139] Generally, materials hoard charge through the electrostatic adsorption of ions at the interface of electrode and electrolyte without any redox mechanism as illustrated in Figure 6.13(a–e). As h18, h49, and ISV all have lots of charge and due to that they tend to be related to lower capacitance. Here also Figure 6.13d exhibiting correlation between the probability density of average charge with the probability of highly charged atoms. In Figure 6.13e, simulation process of the electrodes (ZTC

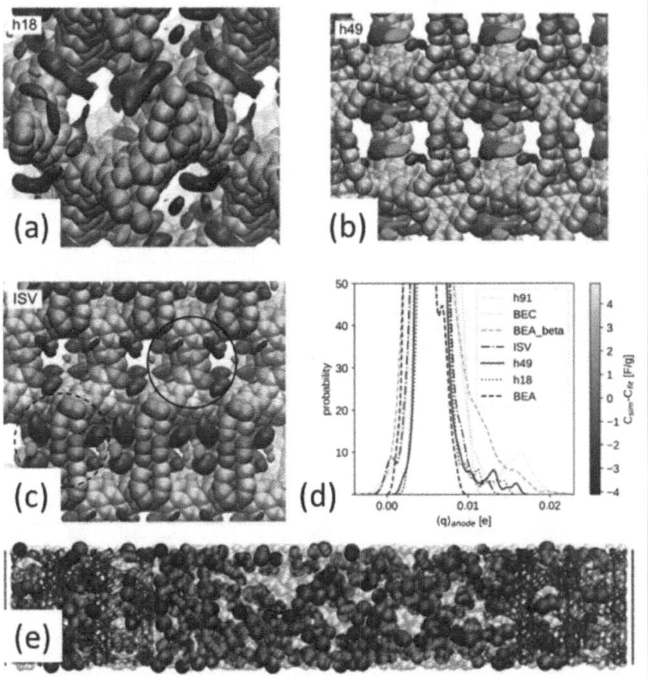

FIGURE 6.13 (a–c) Inspection of atomic charge different h18, h49, and ISV materials simulated for anion. (d) Average charge distribution (probability density) in anode. (e) simulation geometry of Faujasite (FAU) (Scaling of color from 0 (white) to 0.01 (red), total charge indicated by purple color with cut of greater than 0.01e and green show total charge less than 0.1e.

Source: Reprinted with permission from Ref. [139]

electrodes with graphene) within the z-direction where graphene caps the electrodes on both the sides of the capacitor cell has discussed for FAU_1 (Faujasite). Molecular dynamics simulations are good for investigating this type of framework based on redox reactions. However, molecular dynamics simulations usually include electrostatic and van der Waals contacts among molecules.[139] Although, the charge storage mechanism is different for redox type materials and is known as pseudocapacitance phenomenon. They accumulate the charge through redox mechanism at very close surface of the material. The analysis of charge storage mechanism of a supercapacitor basically depends on its energy density. The energy density (*E*) and power density (*P*) can be defined as:

$$E = \frac{CXV^2}{2} \text{ and } P = \frac{V^2}{4R} \tag{6.3}$$

Here *C*, *V*, and *R* are representing capacitance, voltage, and resistance values, respectively. If high energy density is to be obtained, the capacitance and operating potential values are to be maximized.

6.6.1 QUANTUM CAPACITANCE

Quantum capacitance is well-defined as the disparity of electrical charge q commensurating the change of electrochemical potential $\bar{\mu}$, that is, $C_{\bar{\mu}} = \frac{dq}{d\bar{\mu}}$. In the meekest instance, consider a parallel-plate capacitor where one or both of the plates has a low density of states, then the capacitance is not obtained by the usual formulation for parallel-plate capacitors, C_e. As an alternative, the capacitance is inferior, like there was additional capacitor in succession, C_q. This second capacitance, connected to the density of states of the plates, is the quantum capacitance and is characterized by C_q. The correspondent capacitance is known as electrochemical capacitance. $\frac{1}{C_{\bar{\mu}}} = \frac{1}{C_e} + \frac{1}{C_q}$. It is particularly significant for low-density-of-states schemes, for example, a layered electronic scheme in a semiconductor surface or interface or graphene, and may be utilized to concept a pragmatic energy functional of electron density.[140]

When a voltmeter is utilized to amount an electronic expedient, it does not actually quantify the clean electric potential (also termed as Galvani potential). As an alternative, it quantifies the electrochemical potential, known as "fermi level difference," that is the entire free energy alteration per electron, counting not lonely its electric potential energy but of course, all extra forces and effects on the electron (e.g., the dynamic energy in its wave function). For instance, a *p-n* junction in steadiness, there is a Galvani potential (built-in potential) crossways the junction, but the "voltage" across it is zero (in the sagacity that a voltmeter would amount zero voltage). In a capacitor, there is an association among charge and voltage, $Q = CV$. As clarified above, the voltage can be divided into two parts: The Galvani potential, and all different. In an outdated metal–insulator–metal capacitor, the Galvani potential is the lone pertinent input. Consequently, the capacitance may be considered in a forthright way using Gauss's law. But, if one or both of the capacitor plates is a semiconductor, then the Galvani potential is not essentially the solitary significant influence to capacitance. As the capacitor charge upsurges, the negative plate blocks up with electrons, which inhabit higher energy conditions in the band structure, while the positive plate mislays electrons, parting behind electrons with lower energy situations in the band structure. Consequently, as the capacitor charges or discharges, the voltage alters at a diverse rate than the Galvani potential variance. In these circumstances, one cannot compute capacitance just by observing the general geometry and by means of Gauss's law. One must also consider the band-filling/band-emptying consequence associated with the density of states of the plates. The band-filling/band-emptying effect modifies the capacitance, emulating a second capacitance in succession. This capacitance is termed as quantum capacitance, since it is associated with the energy of an electron's quantum wave function. A few experts denote to this identical notion as chemical capacitance, as it is related to the electrons' chemical potential. The philosophies behind quantum capacitance are meticulously linked to Thomas–Fermi screening and band bending.

The quantum capacitance of graphene is pertinent to consider and model gated graphene.[141] It is also applicable for CNTs.[142] In the research of dye-sensitized solar cells, the quantum capacitance of the sintered TiO_2 nanoparticle probe is a vital result.[143] Quantum capacitance can be significant in capacitance–voltage profiling. When supercapacitors are researched in detail, quantum capacitance stages a significant part.[144] Theoretically, it is computed from the density of states using the relation:[145]

$$C_Q = e^2 \int_{-\infty}^{\infty} D(E) F_T (E - e\varphi_G) dE \tag{6.4}$$

where $D(E)$ and ϕ_G are the density of states and electrode potential of the system under consideration, respectively, while $F_T(E)$ is the thermal broadening function due to thermal effect expressed as:

$$F_T(E) = (4K_BT)^{-1} sech^2 (E / 2k_BT) \tag{6.5}$$

Figure 6.14 is to compare the quantum capacitance between pristine VS_4 and VSe_2. and their hybrids when coupled with MXene as a function of electrode potential.[146,147] It is critically evident that hybridization indeed improves specific capacitance in both cases.

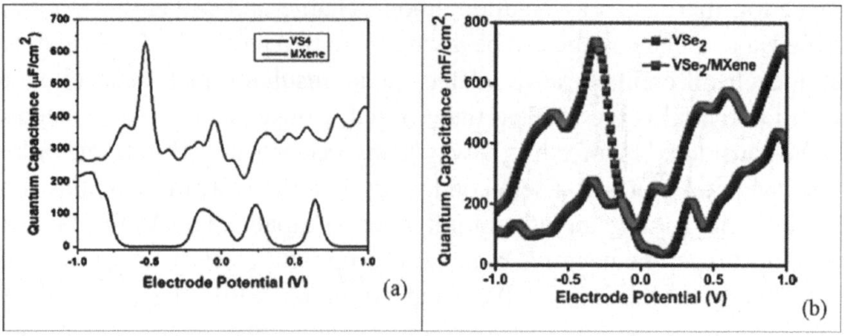

FIGURE 6.14 Variation of QC as a function of electrode potential for pristine. (a) VS4. (b) VSe2 and their respective hybrids with MXene.

Source: Reprinted with permission from Ref. [146,147]

6.6.2 *DIFFUSION ENERGY BARRIER*

The energy barrier is the difference between the energy of the diffusing atom in the transition complex and the energy in the kink site position. High-rate electrode ingredients are more striking for the mandate for rapid charging of EES devices.[148] The most shared technique to advance rate capability is to construct nanosized materials,[149] which feature short diffusion aloofness and big interaction area with the electrolyte. Inopportunely, the low tap density of nanomaterials harshly limits the applied volumetric energy compactness of the strategies. Obviously, micron-sized materials are further favorable for practical energy storage applications. Yet, micron-sized ingredients always transport substandard rate capability for the big diffusion barrier as a consequence of a long diffusion trail and separation of the inner portion of the particles since the electrolyte. As rate capability is mostly limited by the comparatively sluggish solid ion dispersion in the bulk and restricted interfacial ion conveyance amid the active substance and electrolyte, rapid inner dispersion kinetics in the lattice and interfacial ion carriage are powerfully desired to decrease the dissemination barricade and thus advance the rate performance of micron-sized resources. Theoretically, it is thus possible to calculate DEB for the electrolyte ion and a lower barrier indicates improvement in rate capability.

VS_4 parades redox-type capacitive behavior where charge transporters are produced. The movement of the charge carter theaters a significant part in charge stowage presentation. The greater mobility of ions in electrolytes designates the improved energy stowing measurements of supercapacitors. In other words, as per the universal tendency, supercapacitive efficacy of electrolyte ions is improved with an upsurge in the degree of mobility. Besides, the mobility is contrariwise relational to the barrier potential of the system. Hence, lower the diffusion barrier potential, higher is the capacitance of the system. At this time, to deduce the flexibility of K^+ ions in K_2SO_4 solution, the DEB for K^+ ions in VS_4, MXene, VSe_2, and VSe_2/MXene has been displayed in Figure 6.15(a)[146] and (b)[147], respectively. The diffusion energy barrier in VS_4 is much lower (0.17 eV) than that in MXene (0.28 eV). The MXene shows electrical double-layer capacitive behavior for intercalation/de-intercalation of ions. For the VSe_2 case, MXene hybrid appears to offer a lower DEB indicating higher mobility and better charge transfer rate.

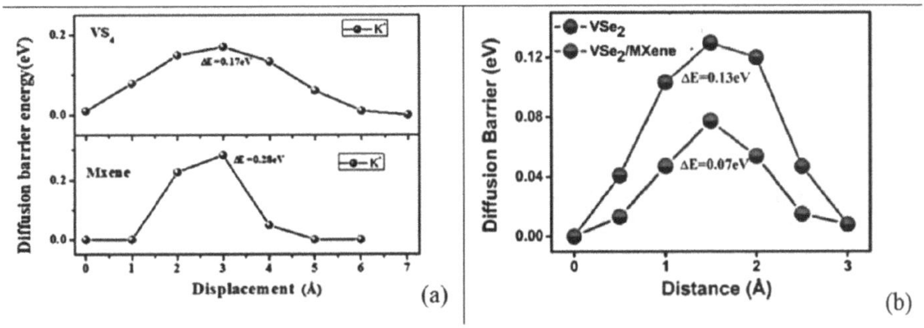

FIGURE 6.15 Diffusion energy barrier for K atom for traveling equivalent positions in (a) pristine VS4 and MXene and (b) pristine VSe2 and VSe2-MXene hybrid.

Source: Reprinted with permission from Ref. [146,147]

6.6.3 *INDUCED VOLTAGE*

The quantum of current over a capacitor is contingent on both the capacitance and how rapidly the voltage is growing or dropping. If the voltage crossways a capacitor speedily rises, a big positive current will be persuaded through the capacitor. A modification in the current passing through the conductor makes (induces) a voltage in that conductor, as well as all neighboring conductors. The induced voltage competes with the alteration in the current that induced the voltage. Inductance is a significance of two rules of physics. First, a continuous current moving through a conductor generates a persistent magnetic field. Secondly, a flexible magnetic field induces a voltage in all close conductors, including the conductor which was utilized to form the magnetic field in the first case. When these two laws are joint, the subsequent effect is inductance. Just like resistors are incorporated to familiarize an anticipated hindrance in a course, and like capacitors are used to present a wanted capacitance, inductors are electrical rudiments used to announce a looked-for amount of inductance into the circuit.

For theoretical calculation of induced voltage, involving supercapacitor, V is defined as[150]

$$V = -\frac{E_{K_{x_1}} - \left(E_{K_{x_2}} + (x_1 - x_2)E_K\right)}{(x_1 - x_2)e} \tag{6.6}$$

where X_1, X_2 are the concentrations of K$^+$ ions and $E_{K_{x_1}}$ is the energy of the system per formula unit for concentration X_1. For example, Pathak et al.[151] computed the average voltage of $NiCo_2O_4$@MWCNT composite with three different K$^+$ ion concentrations, that is, 0.625, 0.25, and 0.5, using the above equation. The middling voltage of the system for diverse K$^+$ ion concentrations is shown in Table 6.3, which is analogous to the foretold values for lithium and sodium ions.[152]

TABLE 6.3 The Average Voltage of the K$^+$ Ion on $NiCo_2O_4$ for Different K$^+$ Ion Concentrations, Where x Represents the Concentrations of the K+ Ion (x: 0.625, 0.25, and 0.5) and $E_{K^+_{x1\,NiCo_2O_4}}$ is the Total Energy of the System per Formula Unit.[151]

$E_{K^+_{x1\,NiCo_2O_4}}$ (eV)	$E_{K^+_{x2\,NiCo_2O_4}}$ (eV)	Voltage (V)
−44.911	−45.398	1.38
−44.853	−45.398	2.08
−44.853	−44.911	3.01

6.7 CONCLUSION AND FUTURE DIRECTIONS

Here in this chapter, a detailed overview about the supercapacitor, role of morphology, nanomaterials that can be used as an electrolyte material, theoretical factors affecting the performance of supercapacitors has been discussed. Supercapacitor, which is a new storage device, is a highly attractive in the field of energy storage or harvesting issues. Research done in the previous years on supercapacitors shows an effective way to achieve a new height in this field but still have some limitations which give a new future direction to the scientist and research communities to overcome these challenges and improving the energy density performance of one of them.

Similarly, finding the ideal parameter to model a supercapacitor for military applications is our future task. Increase the low-rated voltage so we do not need any series combination of supercapacitors for practical applications. The ultimate focus of researchers on supercapacitors in industry and research is to achieve high power density. Such an objective can be achieved not only by the making of supercapacitors but also by the effective way to find electroactive nanomaterials with high electrochemical performance.

KEYWORDS

- **energy storage**
- **electrochemical properties**
- **cycle life**
- **energy density**
- **power density**
- **2D nanostructures**

REFERENCES

1. Tanaka, N. *World Energy Outlook 2010*; International Energy Agency: Paris, 2010.
2. World Gross Electricity Production by Source, 2019—Charts—Data & Statistics [Online]. International Energy Agency (accessed Nov 25, 2021).
3. Powell, C. A.; Bryan, D. M. Materials Challenges in Advanced Coal Conversion Technologies. *MRS Bull.* **2008,** *33* (4), 309–315.
4. Arunachalam, V. S.; Fleischer, E. L. The Global Energy Landscape and Materials Innovation. *MRS Bull.* **2008,** *33* (4), 264–288.
5. Meier, P. J.; Wilson, P. P. H.; Kulcinski, G. L.; Denholm, P. L. US Electric Industry Response to Carbon Constraint: A Life-cycle Assessment of Supply Side Alternatives. *Energy Policy* **2005,** *33* (9), 1099–1108.
6. Ginley, D.; Green, M. A.; Collins, R. Solar Energy Conversion Toward 1 Terawatt. *MRS Bull.* **2008,** *33* (4), 355–364.
7. Holdren, J. P. Energy and Sustainability. *Science* **2007,** *315 (*5813), 737–737.
8. Lee, J. C. Y.; Fields, M. J. An Overview of Wind-Energy-Production Prediction Bias, Losses, and Uncertainties. *Wind Energy Sci.* **2021,** *6* (2), 311–365.
9. Yang, Z.; Zhang, J.; Kintner-Meyer, M. C.; Lu, X.; Choi, D.; Lemmon, J. P.; Liu, J. Electrochemical Energy Storage For Green Grid. *Chem. Rev.* **2011,** *111,* 3577–3613.
10. Lee, B. S.; Gushee, D. E. Electricity Storage: The Achilles' Heel of Renewable Energy. *Chem. Eng. Prog.* **2008,** *104* (3), S29–S32.

11. Burke, A. R&D Considerations for the Performance and Application of Electrochemical Capacitors. *Electrochim. Acta* **2007,** *53* (3), 1083–1091.
12. Pomerantseva, E.; Bonaccorso, F.; Feng, X.; Cui, Y.; Gogotsi, Y. Energy Storage: The Future Enabled by Nanomaterials. *Science* **2019,** *366* (6468), eaan8285.
13. Chen, H.; Cong, T. N.; Yang, W.; Tan, C.; Li, Y.; Ding, Y. Progress in Electrical Energy Storage System: A Critical Review. *Prog. Nat. Sci.* **2009,** *19* (3), 291–312.
14. Aneke, M.; Wang, M. Energy Storage Technologies and Real Life Applications—A State of the Art Review. *Appl. Energy* **2016,** *179*, 350–377.
15. Gür, T. M. Review of Electrical Energy Storage Technologies, Materials and Systems: Challenges and Prospects for Large-scale Grid Storage. *Energy Environ. Sci.* **2018,** *11* (10), 2696–2767.
16. Díaz-González, F.; Sumper, A.; Gomis-Bellmunt, O.; Villafáfila-Robles, R. A Review of Energy Storage Technologies for Wind Power Applications. *Renew. Sustain. Energy Rev.* **2012,** *16* (4), 2154–2171.
17. Liu, C.; Li, F.; Ma, L.-P.; Cheng, H.-M. Advanced Materials for Energy Storage. *Adv. Mater.* **2010,** *22* (8), E28–E62.
18. Long, J. W.; Dunn, B.; Rolison, D. R.; White, H. S. Three-Dimensional Battery Architectures. *Chem. Rev.* **2004,** *104* (10), 4463–4492.
19. Janek, J.; Zeier, W. G. A Solid Future for Battery Development. *Nat. Energy* **2016,** *1* (9), 1–4.
20. Van Noorden, R. A Better Battery. *Nature* **2014,** *507* (7490), 26.
21. Raza, W.; Ali, F.; Raza, N.; Luo, Y.; Kim, K.-H.; Yang, J.; Kumar, S.; Mehmood, A.; Kwon, E. E. Recent Advancements in Supercapacitor Technology. *Nano Energy* **2018,** *52,* 441–473.
22. Zhang, L. L.; Zhao, X. S. Carbon-based Materials as Supercapacitor Electrodes. *Chem. Soc. Rev.* **2009,** *38* (9), 2520–2531.
23. Snook, G. A.; Kao, P.; Best, A. S. Conducting-Polymer-based Supercapacitor Devices and Electrodes. *J. Power Sour.* **2011,** *196* (1), 1–12.
24. Muzaffar, A.; Ahamed, M. B.; Deshmukh, K.; Thirumalai, J. A Review on Recent Advances in Hybrid Supercapacitors: Design, Fabrication and Applications. *Renew. Sustain. Energy Rev.* **2019,** *101,* 123–145.
25. Salanne, M.; Rotenberg, B.; Naoi, K.; Kaneko, K.; Taberna, P.-L.; Grey, C. P.; Dunn, B.; Simon, P. Efficient Storage Mechanisms for Building Better Supercapacitors. *Nat. Energy* **2016,** *1* (6), 1–10.
26. Patra, A.; Namsheer, K.; Jose, J. R.; Sahoo, S.; Chakraborty, B.; Rout, C. S. Understanding the Charge Storage Mechanism of Supercapacitors: In Situ/Operando Spectroscopic Approaches and Theoretical Investigations. *J. Mater. Chem. A* **2021,** *9* (46), 25852–25891.
27. Sahoo, S.; Walke, P.; Kumar Nayak, S.; Sekhar Rout, C.; Late, D J. Recent Developments in Self-powered Smart Chemical Sensors for Wearable Electronics. *Nano Res.* **2021,** *14* (11), 3669–3689.
28. Yu, X.; Yun, S.; Yeon, J. S.; Bhattacharya, P.; Wang, L.; Woo Lee, S.; Hu, X.; Park, H. S. Emergent Pseudocapacitance of 2D Nanomaterials. *Adv. Energy Mater.* **2018,** *8* (13), 1702930.
29. Jiang, Y.; Liu, J. Definitions of Pseudocapacitive Materials: A Brief Review. *Energy Environ. Mater.* **2019,** *2* (1), 30–37.

30. Muzaffar, A.; Basheer Ahamed, M.; Deshmukh, K.; Thirumalai, J. A Review on Recent Advances in Hybrid Supercapacitors: Design, Fabrication and Applications. *Renew. Sustain. Ener. Rev.* **2019,** *101,* 123–145.
31. Isacfranklin, M.; Yuvakkumar, R.; Kumar, P. S.; Thirumal, V.; Ravi, G.; Velauthapillai, D. Hydrogen Free Direct Growth Carbon Nanorod as a Promising Electrode in Symmetric Supercapacitor Applications. *Prog. Org. Coat.* **2021,** *158,* 106379.
32. Mane, V. J.; Kale, S. B.; Ubale, S. B.; Lokhande, V. C.; Lokhande, C. D. Enhanced Specific Energy of Silver-Doped MnO_2/Graphene Oxide Electrodes as Facile Fabrication Symmetric Supercapacitor Device. *Mater. Today Chem.* **2021,** *20,* 100473.
33. Beknalkar, S. A.; Teli, A. M.; Harale, N. S.; Shin, J. C.; Patil, P. S. Construction of IrO_2@ Mn_3O_4 Core-Shell Heterostructured Nanocomposites for High Performance Symmetric Supercapacitor Device. *J. Alloys Compound.* **2021,** *887,* 161328.
34. Mathew, M.; Rout, C. S. Electrochemical Biosensors Based on Ti_3C_2Tx MXene: Future Perspectives for On-Site Analysis. *Curr. Opin. Electrochem.* **2021,** *30,* 100782.
35. Wu, Y.; Chen, H.; Lu, Y.; Yang, J.; Zhu, X.; Zheng, Y.; Lou, G. et al. Rational Design of Cobalt–Nickel Double Hydroxides for Flexible Asymmetric Supercapacitor with Improved Electrochemical Performance. *J. Colloid Interf. Sci.* **2021,** *581,* 455–464.
36. Ghosh, K.; Pumera, M. MXene and MoS3− x Coated 3D-Printed Hybrid Electrode for Solid-State Asymmetric Supercapacitor. *Small Methods* **2021,** *5* (8), 2100451.
37. Shao, Y.; El-Kady, M. F.; Sun, J.; Li, Y.; Zhang, Q.; Zhu, M.; Wang, H.; Dunn, B.; Kaner, R. B. Design and Mechanisms of Asymmetric Supercapacitors. *Chem. Rev.* **2018,** *118* (18), 9233–9280.
38. Brousse, T.; Bélanger, D.; Long, J W. To be or Not to be Pseudocapacitive? *J. Electrochem. Soc.* **2015,** *162* (5), A5185.
39. Tan, H. T.; Sun, W.; Wang, L.; Yan, Q. 2D Transition Metal Oxides/Hydroxides for Energy-Storage Applications. *ChemNanoMat* **2016,** *2* (7), 562–577.
40. Moraes, L. P. R.; Mei, J.; Fonseca, F C.; Sun, Z. Two-dimensional Metal Oxide Nanomaterials for Sustainable Energy Applications. In *2D Nanomaterials for Energy Applications*; Elsevier, 2020; pp 39–72.
41. Jin, H.; Guo, C.; Liu, X.; Liu, J.; Vasileff, A.; Jiao, Y.; Zheng, Y.; Qiao, S.-Z. Emerging Two-dimensional Nanomaterials for Electrocatalysis. *Chem. Rev.* **2018,** *118* (13), 6337–6408.
42. Lee, S.; Jin, X.; Kim, I. Y.; Gu, T.-H.; Choi, J.-W.; Nahm, S.; Hwang, S.-J. Superior Additive of Exfoliated RuO_2 Nanosheet for Optimizing the Electrode Performance of Metal Oxide Over Graphene. *J. Phys. Chem. C* **2016,** *120* (22), 11786–11796.
43. Xia, A.; Yu, W.; Yi, J.; Tan, G.; Ren, H.; Liu, C. Synthesis of Porous δ-MnO_2 Nanosheets and Their Supercapacitor Performance. *J. Electroanal. Chem.* **2019,** *839,* 25–31.
44. Feng, L.; Zhu, Y.; Ding, H.; Ni, C. Recent Progress in Nickel Based Materials for High Performance Pseudocapacitor Electrodes. *J. Power Sour.* **2014,** *267,* 430–444.
45. Low, W. H.; Khiew, P. S.; Lim, S. S.; Siong, C. W.; Ezeigwe, E. R. Recent Development of Mixed Transition Metal Oxide and Graphene/Mixed Transition Metal Oxide Based Hybrid Nanostructures for Advanced Supercapacitors. *J. Alloys Compound.* **2019,** *775,* 1324–1356.
46. Lu, Q.; Chen, J G.; Xiao, J. Q. Nanostructured Electrodes for High-Performance Pseudocapacitors. *Angew. Chem. Int. Ed.* **2013,** *52* (7), 1882–1889.
47. Zheng, J. P.; Jow, T. R. A New Charge Storage Mechanism for Electrochemical Capacitors. *J. Electrochem. Soc.* **1995,** *142* (1), L6.

48. Yu, Z.; Tetard, L.; Zhai, L.; Thomas, J. Supercapacitor Electrode Materials: Nanostructures from 0 to 3 Dimensions. *Energy Environ. Sci.* **2015,** *8* (3), 702–730.
49. Wang, G.; Zhang, L.; Zhang, J. A Review of Electrode Materials for Electrochemical Supercapacitors. *Chem. Soc. Rev.* **2012,** *41* (2), 797–828.
50. Al-Rubaye, S.; Rajagopalan, R.; Dou, S. X.; Cheng, Z. Facile Synthesis of a Reduced Graphene Oxide Wrapped Porous $NiCo_2O_4$ Composite with Superior Performance as an Electrode Material for Supercapacitors. *J. Mater. Chem. A* **2017,** *5* (36), 18989–18997.
51. Patil, S. S.; Dubal, D. P.; Deonikar, V. G.; Tamboli, M. S.; Ambekar, J. D.; Gomez-Romero, P.; Kolekar, S. S.; Kale, B. B.; Patil, D. R. Fern-like rGO/BiVO4 Hybrid Nanostructures for High-Energy Symmetric Supercapacitor. *ACS Appl. Mater. Interfaces* **2016,** *8* (46), 31602–31610.
52. Zhi, M.; Xiang, C.; Li, J.; Li, M.; Wu, N. Nanostructured Carbon–Metal Oxide Composite Electrodes for Supercapacitors: A Review. *Nanoscale* **2013,** *5* (1), 72–88.
53. Kandasamy, M.; Sahoo, S.; Kumar Nayak, S.; Chakraborty, B.; Rout, C. S. Recent Advances in Engineered Metal Oxide Nanostructures for Supercapacitor Applications: Experimental and Theoretical Aspects. *J. Mater. Chem. A* **2021,** *9* (33), 17643–17700.
54. Anasori, B.; Beidaghi, M.; Gogotsi, Y. Graphene–Transition Metal Oxide Hybrid Materials. *Mater. Today* **2014,** *17* (5), 253–254.
55. Wu, Z.-S.; Zhou, G.; Yin, L.-C.; Ren, W.; Li, F.; Cheng, H.-M. Graphene/Metal Oxide Composite Electrode Materials for Energy Storage. *Nano Energy* **2012,** *1* (1), 107–131.
56. Zhang, J.; Jiang, J.; Zhao, X. S. Synthesis and Capacitive Properties of Manganese Oxide Nanosheets Dispersed on Functionalized Graphene Sheets. *J. Phys. Chem. C* **2011,** *115* (14), 6448–6454.
57. Ge, C.; Hou, Z.; He, B.; Zeng, F.; Cao, J.; Liu, Y.; Kuang, Y. Three-dimensional Flower-like Nickel Oxide Supported on Graphene Sheets as Electrode Material for Supercapacitors. *J. Sol-Gel Sci. Technol.* **2012,** *63* (1), 146–152.58. Peng, L.; Peng, X.; Liu, B.; Wu, C.; Xie, Y.; Yu, G. Ultrathin Two-dimensional MnO_2/Graphene Hybrid Nanostructures for High-Performance, Flexible Planar Supercapacitors. *Nano Lett.* **2013,** *13* (5), 2151–2157.
59. Tan, H. Y.; Yu, B. Z.; Cao, L. L.; Cheng, T.; Zheng, X. L.; Li, X. H.; Li, W. L.; Ren, Z. Y. Layer-dependent Growth of Two-dimensional Co_3O_4 Nanostructure Arrays on Graphene for High Performance Supercapacitors. *J. Alloys Compound.* **2017,** *696,* 1180–1188.
60. Shrestha, K. R.; Kandula, S.; Kim, N. H.; Lee, J. H. A Spinel $MnCo_2O_4$/NG 2D/2D Hybrid Nanoarchitectures as Advanced Electrode Material for High Performance Hybrid Supercapacitors. *J. Alloys Compound.* **2019,** *771,* 810–820.
61. Pan, H.; Li, J.; Feng, Y. P. Carbon Nanotubes for Supercapacitor. *Nanosc. Res. Lett.* **2010,** *5* (3), 654–668.
62. Zhou, H.; Zhang, L.; Zhang, D.; Chen, S.; Coxon, P. R.; He, X.; Coto, M.; Kim, H.-K.; Xi, K.; Ding, S. A Universal Synthetic Route to Carbon Nanotube/Transition Metal Oxide Nano-composites for Lithium Ion Batteries and Electrochemical Capacitors. *Sci. Rep.* **2016,** *6* (1), 1–11.
63. Mondal, A. K.; Liu, H.; Li, Z.-F.; Wang, G. Multiwall Carbon Nanotube-Nickel Cobalt Oxide Hybrid Structure as High Performance Electrodes for Supercapacitors and Lithium Ion Batteries. *Electrochim. Acta* **2016,** *190,* 346–353.
64. Barbieri, O.; Hahn, M.; Herzog, A.; Kötz, R. Capacitance Limits of High Surface Area Activated Carbons for Double Layer Capacitors. *Carbon* **2005,** *43* (6), 1303–1310.

65. Zhang, X.; Luo, J.; Tang, P.; Ye, X.; Peng, X.; Tang, H.; Sun, S.-G.; Fransaer, J. A Universal Strategy for Metal Oxide Anchored and Binder-free Carbon Matrix Electrode: A Supercapacitor Case with Superior Rate Performance and high mass loading. *Nano Energy* **2017,** *31,* 311–321.
66. Ramya, R.; Sivasubramanian, R.; Sangaranarayanan, M. V. Conducting Polymers-based Electrochemical Supercapacitors—Progress and Prospects. *Electrochimica Acta* **2013,** *101,* 109–129.
67. Li, S.; Wu, D.; Cheng, C.; Wang, J.; Zhang, F.; Su, Y.; Feng, X. Polyaniline-Coupled Multifunctional 2D Metal Oxide/Hydroxide Graphene Nanohybrids. *Angew. Chem. Int. Ed.* **2013,** *52* (46), 12105–12109.
68. Kuang, M.; Liu, X. Y.; Dong, F.; Zhang, Yu X. Tunable Design of Layered $CuCo_2O_4$ Nanosheets@ MnO_2 Nanoflakes Core–Shell Arrays on Ni Foam for High-performance Supercapacitors. *J. Mater. Chem. A* **2015,** *3* (43), 21528–21536.
69. Liu, J.; Jiang, J.; Cheng, C.; Li, H.; Zhang, J.; Gong, H.; Fan, H. J. Co_3O_4 Nanowire@ MnO_2 Ultrathin Nanosheet Core/Shell Arrays: A New Class of High-performance Pseudocapacitive Materials. *Adv. Mater.* **2011,** *23* (18), 2076–2081.
70. Xu, K.; Ma, S.; Shen, Y.; Ren, Q.; Yang, J.; Chen, X.; Hu, J. $CuCo_2O_4$ Nanowire Arrays Wrapped in Metal Oxide Nanosheets as Hierarchical Multicomponent Electrodes for Supercapacitors. *Chem. Eng. J.* **2019,** *369,* 363–369.
71. Cheng, Q.; Tang, J.; Ma, J.; Zhang, H.; Shinya, N.; Qin, L.-C. Graphene and Nanostructured MnO_2 Composite Electrodes for Supercapacitors. *Carbon* **2011,** *49* (9), 2917–2925.
72. Soon, J. M.; Loh, K. P. Electrochemical Double-layer Capacitance of MoS_2 Nanowall Films. *Electrochem. Solid-State Lett.* **2007,** *10* (11), A250.
73. Cao, L.; Yang, S.; Gao, W.; Liu, Z.; Gong, Y.; Ma, L.; Shi, G. et al. Direct Laser-patterned Micro-supercapacitors from Paintable MoS_2 Films. *Small* **2013,** *9* (17), 2905–2910.
74. Chhowalla, M.; Shin, H. S.; Eda, G.; Li, L.-J.; Loh, K. P.; Zhang, H. The Chemistry of Two-Dimensional Layered Transition Metal Dichalcogenide Nanosheets. *Nat. Chem.* **2013,** *5* (4), 263–275.
75. Voiry, D.; Goswami, A.; Kappera, R.; Kaplan, D.; Fujita, T.; Chen, M.; Asefa, T.; Chhowalla, M. Covalent Functionalization of Monolayered Transition Metal Dichalcogenides by Phase Engineering. *Nat. Chem.* **2015,** *7* (1), 45–49.
76. Tang, Q.; Jiang, D.-E. Stabilization and Band-gap Tuning of the 1T-MoS_2 Monolayer by Covalent Functionalization. *Chem. Mater.* **2015,** *27* (10), 3743–3748.
77. Acerce, M.; Voiry, D.; Chhowalla, M. Metallic 1T Phase MoS_2 Nanosheets as Supercapacitor Electrode Materials. *Nat. Nanotechnol.* **2015,** *10* (4), 313–318.
78. Yin, Y.; Zhang, Y.; Gao, T.; Yao, T.; Zhang, X.; Han, J.; Wang, X.; et al. Synergistic Phase and Disorder Engineering in 1T-$MoSe_2$ Nanosheets for Enhanced Hydrogen-evolution Reaction. *Adv. Mater.* **2017,** *29* (28), 1700311.
79. Piao, M.; Chu, J.; Wang, X.; Chi, Y.; Zhang, H.; Li, C.; Shi, H.; Joo, M.-K. Hydrothermal Synthesis of Stable Metallic 1T Phase WS2 Nanosheets for Thermoelectric Application. *Nanotechnology* **2017,** *29* (2), 025705.
80. Ambrosi, A.; Sofer, Z.; Pumera, M. 2H→ 1T Phase Transition and Hydrogen Evolution Activity of MoS2, MoSe2, WS2 and WSe2 Strongly Depends on the MX2 Composition. *Chem. Commun.* **2015,** *51* (40), 8450–8453.

81. Chen, H.; Jiang, J.; Zhang, Li, Wan, H.; Qi, T.; Xia, D. Highly Conductive $NiCo_2S_4$ Urchin-like Nanostructures for High-rate Pseudocapacitors. *Nanoscale* **2013,** *5* (19), 8879–8883.
82. Xia, C.; Li, P.; Gandi, A. N.; Schwingenschlögl, U.; Alshareef, H. N. Is $NiCo_2S_4$ Really a Semiconductor? *Chem. Mater.* **2015,** *27* (19), 6482–6485.
83. Bouchard, R. J.; Russo, P. A.; Wold, A. Preparation and Electrical Properties of Some Thiospinels. *Inorg. Chem.* **1965,** *4* (5), 685–688.
84. El Radaf, I. M.; Fouad, S. S.; Ismail, A. M.; Sakr, G. B. Influence of Spray Time on the Optical and Electrical Properties of $CoNi_2S_4$ Thin Films. *Mater. Res. Exp.* **2018,** *5* (4), 046406.
85. Shim, J.-J. In Situ Growth of Hierarchical Mesoporous $NiCo_2S_4$@ MnO_2 Arrays on Nickel Foam for High-performance Supercapacitors. *Electrochim. Acta* **2015,** *166,* 302–309.
86. Mei, L.; Yang, T.; Xu, C.; Zhang, M.; Chen, L.; Li, Q.; Wang, T. Hierarchical Mushroom-like $CoNi_2S_4$ Arrays as a Novel Electrode Material for Supercapacitors. *Nano Energy* **2014,** *3,* 36–45.
87. Peng, T.; Qian, Z.; Wang, J.; Song, D.; Liu, J.; Liu, Q.; Wang, P. Construction of Mass-controllable Mesoporous $NiCo_2S_4$ Electrodes for High Performance Supercapacitors. *J. Mater. Chem. A* **2014,** *2* (45), 19376–19382.
88. Li, Z.; Zhao, D.; Xu, C.; Ning, J.; Zhong, Y.; Zhang, Z.; Wang, Y.; Hu, Y. Reduced $CoNi_2S_4$ Nanosheets with Enhanced Conductivity for High-performance Supercapacitors. *Electrochim. Acta* **2018,** *278,* 33–41.
89. Hu, W.; Chen, R.; Xie, W.; Zou, L.; Qin, N.; Bao, D. $CoNi_2S_4$ Nanosheet Arrays Supported on Nickel Foams with Ultrahigh Capacitance for Aqueous Asymmetric Supercapacitor Applications. *ACS Appl. Mater. Interfaces* **2014,** *6* (21), 19318–19326.
90. Lin, L.; Lei, W.; Zhang, S.; Liu, Y.; Wallace, G. G.; Chen, J. Two-dimensional Transition Metal Dichalcogenides in Supercapacitors and Secondary Batteries. *Energy Storage Mater.* **2019,** *19,* 408–423.
91. Muller, G. A.; Cook, J. B.; Kim, H.-S.; Tolbert, S. H.; Dunn, B. High Performance Pseudocapacitor Based on 2D Layered Metal Chalcogenide Nanocrystals. *Nano Lett.* **2015,** *15* (3), 1911–1917.
92. Tu, C.-C.; Lin, L.-Y.; Xiao, B.-C.; Chen, Y.-S. Highly Efficient Supercapacitor Electrode with Two-dimensional Tungsten Disulfide and Reduced Graphene Oxide Hybrid Nanosheets. *J. Power Sour.* **2016,** *320,* 78–85.
93. Shen, J.; Wu, J.; Pei, L.; Rodrigues, M.-T. F.; Zhang, Z. Q.; Zhang, F.; Zhang, X.; Ajayan, P M.; Ye, M. $CoNi_2S_4$-Graphene-2D-$MoSe_2$ as an Advanced Electrode Material for Supercapacitors. *Adv. Energy Mater.* **2016,** *6* (13), 1600341.
94. Lin, L.; Miao, N.; Wen, Y.; Zhang, S.; Ghosez, P.; Sun, Z.; Allwood, D. A. Sulfur-depleted Monolayered Molybdenum Disulfide Nanocrystals for Superelectrochemical Hydrogen Evolution Reaction. *ACS Nano* **2016,** *10* (9), 8929–8937.
95. Lin, L.; Miao, N.; Huang, J.; Zhang, S.; Zhu, Y.; Horsell, D. D.; Ghosez, P.; Sun, Z.; Allwood, D. A. A Photocatalyst of Sulphur Depleted Monolayered Molybdenum Sulfide Nanocrystals for Dye Degradation and Hydrogen Evolution Reaction. *Nano Energy* **2017,** *38,* 544–552.
96. Lin, L.; Xu, Y.; Zhang, S.; Ross, I. M.; Ong, A. C. M.; Allwood, D. A. Fabrication of Luminescent Monolayered Tungsten Dichalcogenides Quantum Dots with Giant Spin-Valley Coupling. *ACS Nano* **2013,** *7* (9), 8214–8223.

97. Tan, C.; Luo, Z.; Chaturvedi, A.; Cai, Y.; Du, Y.; Gong, Y.; Huang, Y.; et al. Preparation of High-percentage 1T-phase Transition Metal Dichalcogenide Nanodots for Electrochemical Hydrogen Evolution. *Adv. Mater.* **2018,** *30* (9), 1705509.
98. Li, H.; Tay, R. Y.; Tsang, S. H.; Zhen, X.; Teo, E. H. T. Controllable Synthesis of Highly Luminescent Boron Nitride Quantum Dots. *Small* **2015,** *11* (48), 6491–6499.
99. Xu, S.; Li, D.; Wu, P. One-pot, Facile, and Versatile Synthesis of Monolayer MoS2/WS2 Quantum Dots as Bioimaging Probes and Efficient Electrocatalysts for Hydrogen Evolution Reaction. *Adv. Func. Mater.* **2015,** *25* (7), 1127–1136.
100. Jeong, S.; Yoo, D.; Ahn, M.; Miró, P.; Heine, T.; Cheon, J. Tandem Intercalation Strategy for Single-layer Nanosheets as an Effective Alternative to Conventional Exfoliation Processes. *Nat. Commun.* **2015,** *6* (1), 1–7.
101. Lin, L.; Zhang, S. Creating High Yield Water Soluble Luminescent Graphene Quantum Dots via Exfoliating and Disintegrating Carbon Nanotubes and Graphite Flakes. *Chem. Commun.* **2012,** *48* (82), 10177–10179.
102. Lin, L.; Xu, Y.; Zhang, S.; Ross, I. M.; Ong, A. C. M.; Allwood, D. A. Fabrication and Luminescence of Monolayered Boron Nitride Quantum Dots. *Small* **2014,** *10* (1), 60–65.
103. Hadz, S.; Angerstein-Kozlowska, H.; Vukovič, M.; Conway, B. E. Reversibility and Growth Behavior of Surface Oxide Films at Ruthenium Electrodes. *J. Electrochem. Soc.* **1978,** *125* (9), 1471.
104. Oh, S. H.; Nazar, L. F. Direct Synthesis of Electroactive Mesoporous Hydrous Crystalline RuO_2 Templated by a Cationic Surfactant. *J. Mater. Chem.* **2010,** *20* (19), 3834–3839.
105. Sassoye, C.; Laberty, C.; Khanh, H. L.; Cassaignon, S.; Boissiere, C.; Antonietti, M.; Sanchez, C. Block-Copolymer-templated Synthesis of Electroactive RuO_2-based Mesoporous Thin Films. *Adv. Func. Mater.* **2009,** *19* (12), 1922–1929.
106. Zheng, J. P.; Cygan, P. J.; Jow, T. R. Hydrous Ruthenium Oxide as an Electrode Material for Electrochemical Capacitors. *J. Electrochem. Soc.* **1995,** 142 (8), 2699.
107. Augustyn, V.; Come, J.; Lowe, M. A.; Kim, J. W.; Taberna, P.-L.; Tolbert, S. H.; Abruña, H. D.; Simon, P.; Dunn, B. High-rate Electrochemical Energy Storage Through Li+ Intercalation Pseudocapacitance. *Nat. Mater.* **2013,** *12* (6), 518–522.
108. Raza, N.; Kumar, T.; Singh, V.; Kim, K.-H. Recent Advances in Bimetallic Metal-Organic Framework as a Potential Candidate for Supercapacitor Electrode Material. *Coord. Chem. Rev.* **2021,** *430,* 213660.
109. Liu, T.; Zhang, L.; Cheng, B.; Hu, X.; Yu, J. Holey Graphene for Electrochemical Energy Storage. *Cell Rep. Phys. Sci.* **2020,** *1* (10), 100215.
110. Naguib, M.; Kurtoglu, M.; Presser, V.; Lu, J.; Niu, J.; Heon, M.; Hultman, L.; Gogotsi, Y.; Barsoum, M. W. Two-dimensional Nanocrystals Produced by Exfoliation of Ti_3AlC_2. *Adv. Mater.* **2011,** *23* (37), 4248–4253.
111. Lukatskaya, M. R.; Mashtalir, O.; Ren, C. E.; Dall'Agnese, Y.; Rozier, P.; Taberna, P. L.; Naguib, M.; Simon, P.; Barsoum, M. W.; Gogotsi, Y. Cation Intercalation and High Volumetric Capacitance of Two-dimensional Titanium Carbide. *Science* **2013,** *341* (6153), 1502–1505.
112. Kshetri, T.; Tran, D. T.; Le, H. T.; Nguyen, D. C.; Van Hoa, H.; Kim, N. H.; Lee, J. H. Recent Advances in MXene-based Nanocomposites for Electrochemical Energy Storage Applications. *Prog. Mater. Sci.* **2021,** *117,* 100733.

113. Huang, W.; Hu, L.; Tang, Y.; Xie, Z.; Zhang, H. Recent Advances in Functional 2D Mxene-based Nanostructures for Next-generation Devices. *Adv. Func. Mater.* **2020,** *30* (49), 2005223.
114. Yang, Q.; Xu, Z.; Fang, B.; Huang, T.; Cai, S.; Chen, H.; Liu, Y.; Gopalsamy, K.; Gao, W.; Gao, C. MXene/Graphene Hybrid Fibers for High Performance Flexible Supercapacitors. *J. Mater. Chem. A* **2017,** *5* (42), 22113–22119.
115. Kumar, S.; Rehman, M. A.; Lee, S.; Kim, M.; Hong, H.; Park, J.-Y.; Seo, Y. Supercapacitors based on Ti_3C_2Tx MXene Extracted from Supernatant and Current Collectors Passivated by CVD-Graphene. *Sci. Rep.* **2021,** *11* (1), 1–9.
116. Zhao, M.-Q.; Ren, C. E.; Ling, Z.; Lukatskaya, M. R.; Zhang, C.; Van Aken, K. L.; Barsoum, M. W.; Gogotsi, Y. Flexible MXene/Carbon Nanotube Composite Paper with High Volumetric Capacitance. *Adv. Mater.* **2015,** *27* (2), 339–345.
117. Park, J. W.; Lee, D. Y.; Kim, H.; Hyeon, J. S.; Andrade, M. J. , Baughman, R. H.; Kim, S. J. Highly Loaded MXene/Carbon Nanotube Yarn Electrodes for Improved Asymmetric Supercapacitor Performance. *MRS Commun.* **2019,** *9* (1), 114–121.
118. Wang, R.; Luo, S.; Xiao, C.; Chen, Z.; Li, H.; Asif, M.; Chan, V.; Liao, K.; Sun, Y. MXene-Carbon Nanotubes Layer-by-Layer Assembly Based on-Chip Micro-supercapacitor with Improved Capacitive Performance. *Electrochim. Acta* **2021,** *386,* 138420.
119. Liang, W.; Zhitomirsky, I. MXene–Carbon Nanotube Composite Electrodes for High Active Mass Asymmetric Supercapacitors. *J. Mater. Chem. A* **2021,** *9* (16), 10335–10344.
120. Li, L.; Fu, J.; Cho, Y.-R.; Moon Yun, J.; Jung, Y. S.; Hun Kwon, S.; Kim, K. H. Hierarchically Layered Nanocomposite Electrodes Formed by Spray-Injected MXene Nanosheets for Ultrahigh-Performance Flexible Supercapacitors. *Appl. Surf. Sci.* **2021,** *549,* 149226.
121. Rakhi, R. B.; Ahmed, B.; Anjum, D.; Alshareef, H. N. Direct Chemical Synthesis of MnO_2 Nanowhiskers on Transition-metal Carbide Surfaces for Supercapacitor Applications. *ACS Appl. Mater. Interfaces* **2016,** *8* (29), 18806–18814.
122. Hou, W.; Sun, Y.; Zhang, Y.; Wang, T.; Wu, L.; Du, Y.; Zhong, W. Mixed-Dimensional Heterostructure of Few-layer MXene Based Vertical Aligned MoS_2 Nanosheets for Enhanced Supercapacitor Performance. *J. Alloys Compound.* **2021,** *859,* 157797.
123. Xie, W.; Wang, Y.; Zhou, J.; Zhang, M.; Yu, J.; Zhu, C.; Xu, J. MOF-derived $CoFe_2O_4$ Nanorods Anchored in MXene Nanosheets for All Pseudocapacitive Flexible Supercapacitors with Superior Energy Storage. *Appl. Surf. Sci.* **2020,** *534,* 147584.
124. Zhang, Y.; Cao, J.; Yuan, Z.; Zhao, L.; Wang, L.; Han, W. Assembling Co_3O_4 Nanoparticles into MXene with Enhanced Electrochemical Performance for Advanced Asymmetric Supercapacitors. *J. Colloid Interf. Sci.* **2021,** *599,* 109–118.
125. Gao, L.; Li, C.; Huang, W.; Mei, S.; Lin, H.; Ou, Q.; Zhang, Y.; et al. MXene/Polymer Membranes: Synthesis, Properties, and Emerging Applications. *Chem. Mater.* **2020,** *32* (5), 1703–1747.
126. Zhan, X.; Si, C.; Zhou, J.; Sun, Z. MXene and MXene-based Composites: Synthesis, Properties and Environment-related Applications. *Nanoscale Horiz.* **2020,** *5* (2), 235–258.
127. Cheng, W.; Fu, J.; Hu, H.; Ho, D. Interlayer Structure Engineering of MXene-based Capacitor-type Electrode for Hybrid Micro-supercapacitor Toward Battery-level Energy Density. *Adv. Sci.* **2021,** *8* (16), 2100775.

128. VahidMohammadi, A.; Moncada, J.; Chen, H.; Kayali, E.; Orangi, J.; Carrero, C. A.; Beidaghi, M. Thick and Freestanding MXene/PANI Pseudocapacitive Electrodes with Ultrahigh Specific Capacitance. *J. Mater. Chem. A* **2018,** *6* (44), 22123–22133.
129. Panda, S.; Deshmukh, K.; Khadheer Pasha, S. K.; Theerthagiri, J.; Manickam, S.; Choi, M. Y. MXene Based Emerging Materials for Supercapacitor Applications: Recent Advances, Challenges, and Future Perspectives. *Coord. Chem. Rev.* **2022,** *462,* 214518.
130. Xie, Y.; Zhang, H.; Huang, H.; Wang, Z.; Xu, Z.; Zhao, H.; Wang, Y.; Chen, N.; Yang, W. High-Voltage Asymmetric MXene-based On-Chip Micro-Supercapacitors. *Nano Energy* **2020,** *74,* 104928.
131. Wang, Y.; Wang, X.; Li, X.; Li, X.; Liu, Y.; Bai, Y.; Xiao, H.; Yuan, G. A High-performance, Tailorable, Wearable, and Foldable Solid-state Supercapacitor Enabled by Arranging Pseudocapacitive Groups and MXene Flakes on Textile Electrode Surface. *Adv. Func. Mater.* **2021,** *31* (7), 2008185.
132. Chen, W.; Zhang, D.; Yang, K.; Luo, M.; Yang, P.; Zhou, X. Mxene (Ti_3C_2Tx)/Cellulose Nanofiber/Porous Carbon Film as Free-Standing Electrode for Ultrathin and Flexible Supercapacitors. *Chem. Eng. J.* **2021,** *413,* 127524.
133. Zheng, J.; Pan, X.; Huang, X.; Xiong, D.; Shang, Y.; Li, X.; Wang, N.; Lau, W.-M.; Yang, H. Y. Integrated $NiCo_2$-LDHs@ MXene/rGO Aerogel: Componential and Structural Engineering Towards Enhanced Performance Stability of Hybrid Supercapacitor. *Chem. Eng. J.* **2020,** *396,* 125197.
134. Li, L.; Liu, W.; Jiang, K.; Chen, D.; Qu, F.; Shen, G. In-situ Annealed Ti_3C_2Tx MXene Based All-solid-state Flexible Zn-ion Hybrid Micro Supercapacitor Array with Enhanced Stability. *Nano-micro Lett.* **2021,** *13* (1), 1–11.
135. Zhang, M.; Cao, J.; Wang, Y.; Song, J.; Jiang, T.; Zhang, Y.; Si, W.; Li, X.; Meng, B.; Wen, G. Electrolyte-Mediated Dense Integration of Graphene-MXene Films for High Volumetric Capacitance Flexible Supercapacitors. *Nano Res.* **2021,** *14* (3), 699–706.
136. Li, X.; Ma, Y.; Yue, Y.; Li, G.; Zhang, C.; Cao, M.; Xiong, Y.; Zou, J.; Zhou, Y.; Gao, Y. A Flexible Zn-ion Hybrid Micro-Supercapacitor based on MXene Anode and V_2O_5 Cathode with High Capacitance. *Chem. Eng. J.* **2022,** *428,* 130965.
137. Zhu, C.; Geng, F. Macroscopic MXene Ribbon with Oriented Sheet Stacking for High-Performance Flexible Supercapacitors. *Carbon Energy* **2021,** *3* (1), 142–152.
138. Sikdar, A.; Dutta, P.; Kumar Deb, S.; Majumdar, A.; Padma, N.; Ghosh, S.; Maiti, U. N. Spontaneous Three-dimensional Self-assembly of MXene and Graphene for Impressive Energy and Rate Performance Pseudocapacitors. *Electrochim. Acta* **2021,** *391,* 138959.
139. Liu, Y. M.; Merlet, C.; Smit, B. Carbons with Regular Pore Geometry Yield Fundamental Insights into Supercapacitor Charge Storage. *ACS Cent. Sci.* **2019,** *5* (11), 1813–1823.
140. Miranda, D. A.; Bueno, P. R. Density Functional Theory and an Experimentally-Designed Energy Functional of Electron Density. *Phys. Chem. Chem. Phys.* **2016,** *18* (37), 25984–25992.
141. Mišković, Z. L.; Upadhyaya, N. Modeling Electrolytically Top-gated Graphene. *Nanoscale Res. Lett.* **2010,** *5* (3), 505–511.
142. Ilani, S.; Donev, L. A. K.; Kindermann, M.; McEuen, P. L. Measurement of the Quantum Capacitance of Interacting Electrons in Carbon Nanotubes. *Nat. Phys.* **2006,** *2* (10), 687–691.
143. Bisquert, J. Chemical Capacitance of Nanostructured Semiconductors: Its Origin and Significance for Nanocomposite Solar Cells. *Phys. Chem. Chem. Phys.* **2003,** *5* (24), 5360–5364.

144. Bueno, P. R. Nanoscale Origins of Super-capacitance Phenomena. *J. Power Sour.* **2019,** *414,* 420–434.
145. Yang, G. M.; Zhang, H. Z.; Fan, X. F.; Zheng, W. T. Density functional Theory Calculations for the Quantum Capacitance Performance of Graphene-based Electrode Material. *J. Phys. Chem. C* **2015,** *119* (12), 6464–6470.
146. Sharma, A.; Patra, A.; Namsheer, K.; Mane, P.; Chakraborty, B.; Rout, C. S. All-Solid-State Asymmetric Supercapacitors Based on VS4 Nano-bundles and MXene Nanosheets. *J. Mater. Sci.* **2021,** *56* (36), 20008–20025.
147. Raj, K. A.; Sree, P. M.; Radhakrishnan, S.; Chakraborty, B.; Rout, C. S. Heterostructured Metallic 1T-VSe_2/Ti_3C_2T x MXene Nanosheets for Energy Storage. *ACS Appl. Nano Mater.* **2022,** *5* (3), 4423–4436.
148. Zhang, J.; Du, C.; Dai, Z.; Chen, W.; Zheng, Y.; Li, B.; Zong, Y.; Wang, X.; Zhu, J.; Yan, Q. NbS2 Nanosheets with M/Se (M=Fe, Co, Ni) Codopants for Li+ and Na+ Storage. *ACS Nano* **2017,** *11* (10), 10599–10607.
149. Zhang, Q.; Wang, W.; Wang, Y.; Feng, P.; Wang, K.; Cheng, S.; Jiang, K. Controllable Construction of 3D-skeleton-carbon Coated Na_3V_2 (PO_4) 3 for High-performance Sodium Ion Battery Cathode. *Nano Energy* **2016,** *20,* 11–19.
150. Jayaseelan, S. S.; Radhakrishnan, S.; Saravanakumar, B.; Seo, M.-K.; Khil, M.-S.; Kim, H.-Y.; Kim, B.-S. Mesoporous 3D $NiCo_2O_4$/MWCNT Nanocomposite Aerogels Prepared by a Supercritical CO_2 Drying Method for High Performance Hybrid Supercapacitor Electrodes. *Colloids Surf. A: Physicochem. Eng. Asp.* **2018,** *538,* 451–459.
151. Pathak, M.; Jose, J. R.; Chakraborty, B.; Rout, C. S. High Performance Supercapacitor Electrodes Based on Spinel $NiCo_2O_4$@ MWCNT Composite with Insights from Density Functional Theory Simulations. *J. Chem. Phys.* **2020,** *152* (6), 064706.
152. Fan, S.; Zou, X.; Du, H.; Gan, L.; Xu, C.; Lv, W.; He, Y.-B.; Yang, Q.-H.; Kang, F.; Li, J. Theoretical Investigation of the Intercalation Chemistry of Lithium/Sodium Ions in Transition Metal Dichalcogenides. *J. Phys. Chem. C* **2017,** *121* (25), 13599–13605.

CHAPTER 7

Two-Dimensional (2D) Materials-Based Nanostructured Hybrid Composite: A Newer Paradigm for Energy Application

NIDHI SHARMA[1], NEETU TALREJA[2], DIVYA CHAUHAN[3], and MOHAMMAD ASHFAQ[4]

[1]*Department of Physics, Dhanauri P.G. College, Dhanauri, Roorkee, Haridwar, Uttarakhand, India*

[2]*Department of Science, Faculty of Science and Technology, Alliance University, Anekal, Bengaluru, Karnataka, India*

[3]*Department of Chemical and Biomedical Engineering, University of South Florida, Tampa, FL, USA*

[4]*Department of Biotechnology & University Centre for Research & Development (UCRD), Chandigarh University, Gharaun, Mohali, Punjab, India*

ABSTRACT

Presently, two-dimensional nanomaterials (2D-NM) uninterruptedly increase the attention of researchers because of their extraordinary properties, mainly high mechanical strength, high surface area, and electrical and optical, which create a promising and appropriate candidates for numerous end applications. The high surface area, atomic thickness, and easily tunable ability might be advantageous to increase their applicability towards end applications. Emphasizing the mechanical, electrical, and thermal properties of 2D-NM, their numerous potential applications, and their characteristic properties affect the applicability of 2D-NM. This mini-review article focuses

Physics and Mechanics of New Materials: Synthesis, Processing, and Emerging Applications. Devrim Balkose, Ana Cristina Faria Ribeiro, Nandakumar Kalarikkal, Ann Rose Abraham, A. K. Haghi (Eds.)

on the synthesis processes of 2D-NM, composition, and crystal structure of the 2D-NM. The role of 2D-NM-based hybrid composite in solar cells and supercapacitor applications on the basis of its characteristics was also discussed. In conclusion, future challenges in the area of the 2D-NM-based hybrid composite were addressed based on their characteristics.

7.1 INTRODUCTION

Nanomaterials (NM) have emerged as next-generation materials that entail a broad spectrum in the form of size (1–100 nm) and dimension (zero [0D], one [1D], two [2D], and three [3D]). The unique characteristics of NM, especially their high surface area, mechanical, electrical, optical, catalytically, and thermal ability, create promising and appropriate candidates for numerous applications such as environmental treatments, energy, electrical, solar cells, agriculture, medicine, drug delivery, and catalyst. Furthermore, the properties of the NM might be simply changed/tuned by changing their surface functional groups, changing synthesis parameters, size, shape, and surface charges.[1–5] NM classified mainly four types based on dimensionality, (1) 0D-NM (particles and clusters), (2) 1D-NM (nanotubes, nanofibers, and nanowires), (3) 2D-NM (plates and sheets), and (4) 3D-NM (bulk powder, and bundle of nanotubes/wires). Recently, 2D-NM have gained significant research interest since the discovery of graphene.[6, 7] Graphene is the first well-known 2D-NM with a zero band gap value attributed to high electron mobility and exceptional conductivity. The extraordinary characteristics such as mechanical, optical, electronic, and thermal transport create appropriate candidates for energy and electronics. Moreover, tunable atomic thickness and high surface area of 2D-NM make them a promising candidate for end applications.[8–13] In one aspect, academicians/scientists continue concentrating on the newer 2D-NM to improve their efficacy toward end applications.

2D-NM such as transition metal dichalcogenides (TMDCs) (WS_2, WSe_2, MoS_2, and $MoSe_2$), graphene, hexagonal boron nitride (h-BN), metal-oxide, transition metal halides (TMHs), graphitic carbon nitride (g-C_3N_4), MXene, layered-double hydroxide (LDHs), and black phosphorus (BP) have been synthesized for countless applications like energy, solar cells, sensor, antibacterial, environmental remediation application, photocatalyst, agriculture, electronics, supercapacitors, biomedical, and wearable device.[14–21]] Interestingly, 2D-NM, mainly thermal, electrical, and mechanical properties, directly affected the potential applicability toward end applications. In this aspect, understanding the 2D-NM properties, especially thermal, electrical, and mechanical properties, might be advantageous to improving the efficacy of the 2D-NM. However, the thickness of the 2D-NM is too thin to absorb adequate light, which limits its effectiveness mainly in photocatalysis, optoelectronics, and optical sensors. Approximately 2.3 and 5.6% light absorption was observed in graphene, and the MoS_2 layer was significantly lower than the bulk materials. Therefore, there is a necessity to be improved optical and optoelectronics properties to fulfill the requirement.[22–25] The development of 2D-NM offers new paradigms due to exceptional potential ability in most areas of research.

Researchers continue focusing on developing the 2D-NM-based hybrid system to improve the electrical, mechanical, and thermal characteristics by incorporating metal, non-metals, and polymers within the 2D-NM. Containing the metal nanoparticles (M-NPs) within the 2D-NM like graphene improved the photon absorption, conductivity, mechanical stability, and thermal properties, thereby improving the applicability for end applications.[7,26,27] Moreover, 2D-NM-based hybrid materials might be tuned according to the applications, thereby easily improving the materials' efficacy via changing the properties of 2D-NM. The materials science, engineering, chemistry, and nanotechnology offer newer paradigms for developing a newer class of materials, namely 2D-NM that contains numerous properties, mainly electrical, thermal, and mechanical properties, which are required for numerous end applications.[28–31] However, there is a knowledge gap between the characteristics and properties of 2D-NM with their chemistry. With the advancement in material science and chemistry, researchers try to develop newer materials 2D-NM for numerous applications, especially environmental remediation, sensor, energy, and drug delivery systems. Moreover, the role of electrical, thermal, and mechanical properties in their application is still desirable to explore for better indulgent and their application in end applications.[32–35] This chapter focuses on synthesizing 2D-NM, mainly electrical, thermal, and mechanical properties. We will also discuss the solar cells and supercapacitor application of the 2D-NM in the context of their characteristics.

7.2 SYNTHESIS OF 2D-NM

The 2D-NM are challenging to synthesize because of their isotropic crystal nature and high surface energy. The 2D-NM and its derivative were synthesized in mainly two types, (1) top-down process, the bulk materials converted into smaller size or nano-sized, and (2) bottom-up process, atoms/molecules built to nano-sized materials. There are mainly two processes used for the synthesis of 2D-NM, (1) exfoliation process and (2) chemical vapor deposition (CVD) process.

7.2.1 EXFOLIATION PROCESS

Graphene and other than 2D-NM were exfoliated using a liquid phase exfoliation process. The liquid-phase exfoliation process produced graphene from graphite using NMP solvent. Usually, bulk graphite is detached into single/multilayer graphene because of generated force. The degree of graphene exfoliation is directly related to dispersion grade and stability. The graphene layers are attached using Van der Waal force and functional group between two-adjacent layers. The exfoliation process is facile and scalable with a high grade of 2D-NM. The high quality and yields of 2D-NM mainly depend on the organic solvents, sonication times, concentration of the materials, integration of surfactant, and polymers.[36–39] Several 2D-NM have been synthesized using the exfoliation process. For instance, MoS_2 nanosheets were synthesized using a liquid phase exfoliation process. The sodium dodecyl benzene sulfonate surfactant in water was used as a solvent for synthesizing MoS_2 nanosheets. The synthesized MoS_2 nanosheets were highly stable, and few-layer MoS_2 nanosheets were controlled with the surfactant concentration.[40] Tan et al. synthesized MoS_2 nanosheets using a liquid phase exfoliation process. The supercritical CO_2-modified *N*-Methyl-2-pyrrolidone (NMP) solvent was used as a solvent. The produced MoS_2 nanosheets are highly stable and single/few-layer nanosheets.[41] Sahoo et al. synthesized MoS_2 nanosheets for photocatalytic applications. The MoS_2 nanosheets were synthesized exfoliation process, and acetone was used as a solvent.[42] Ashfaq et al. synthesized 2D-Cu-WS_2 nanosheets using a simple exfoliation process for biomedical application. WS_2 nanosheets was synthesized using the exfoliation process in NMP solvents.[14] Furthermore, the exfoliation process easily blocks the adjacent graphene layers, thereby easily achieving single/multilayer graphene. Moreover, we can easily control the shape, size, crystal structure, and stability of the 2D-NM using solvent, surfactant, and concentration of materials. Figure 7.1 shows the exfoliation process of the 2D-NM.

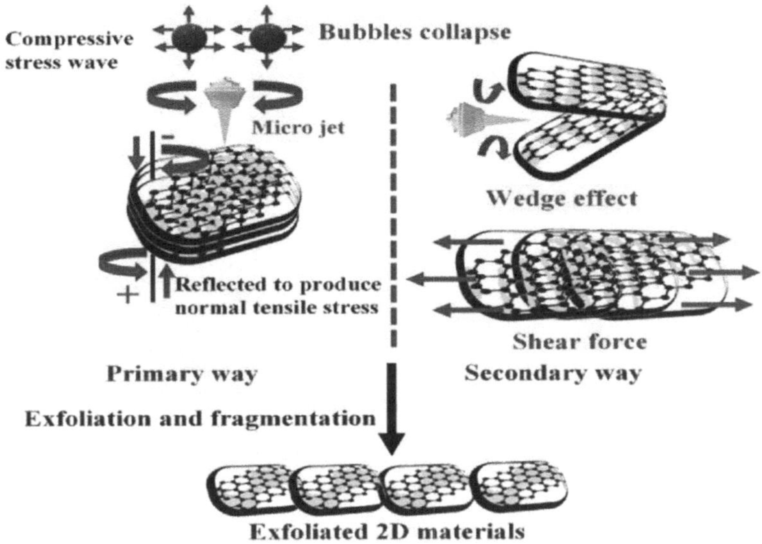

FIGURE 7.1 A schematic illustration of the exfoliation process of 2D-NM.
Source: Reprinted with permission from Ref. [37]

7.2.2 CVD PROCESS

Usually, the CVD process synthesizes numerous nanomaterials like graphene, carbon nanofibers (CNFs), carbon nanotubes (CNTs), and their derivatives. Cu or Ni foil or silicon wafers are used as a substrate for growing nanomaterials, including graphene. The substrate was placed within the CVD reactor, and the precursor was exposed to the substrate surface at high temperatures using inert conditions. The precursor decomposes on the substrate to synthesize 2D-NM.[37,43,44] Moreover, the layers or thickness of the 2D-NM might be controlled by tuning the reaction condition. Figure 7.2 shows the schematic representation of the CVD process.

7.3 PROPERTIES OF 2D-NM

The 2D-NM-like graphene is wide-open, having prodigious potential in numerous end applications due to its extraordinary electrical, mechanical, and thermal properties. These characteristics of 2D-NM might be revolutionized in many areas like solar cells, electronics, sensors, and supercapacitors. Despite this, researchers continuously focus on the amendment of 2D-NM by incorporating metal, non-metals, and polymers to augment the properties of 2D-NM, including mechanical, electrical, and thermal characteristics.[45–47]

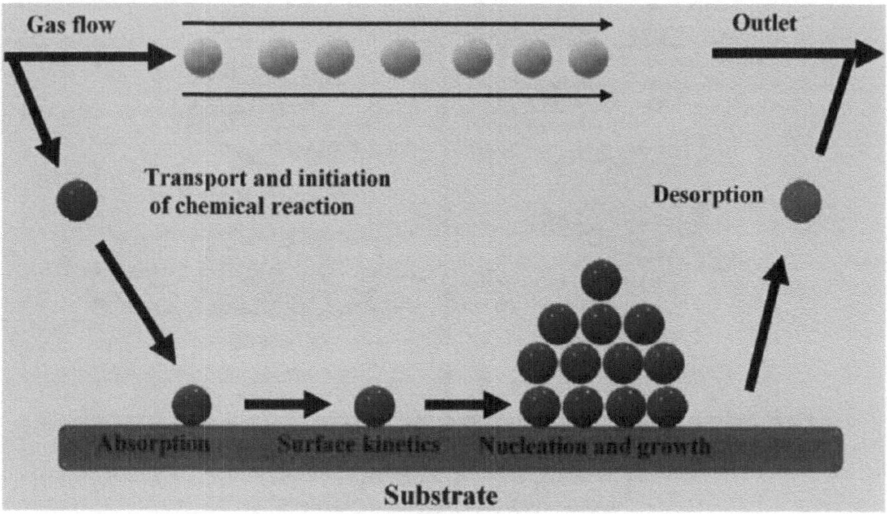

FIGURE 7.2 A schematic representation of the CVD process.

Source: Reprinted with permission from Ref. [37]

7.3.1 THERMAL PROPERTIES

2D-NM, including graphene, is continuously catching researchers' interest because of its thermal properties, mainly thermal conductivity. The thermal properties of 2D-NM mainly depend on several factors such as size, isotopic doping, surface functional group, substrate, and edge configuration. The thermal properties might be easily tuned by modifying these factors. Understanding thermal properties might be beneficial and directly correlate with the application of 2D-NM. Moreover, the NM's electrical conductivity is almost the same as their bulk materials. However, a significant reduction in thermal conductivity of NM compares with that of the bulk materials because of the scattering of photons. With the help of this concept, 2D-NM might be a promising candidate for determining thermal conductivity performance. Research has been done so far on modifying the thermal property of the 2D-NM. For example, Judek et al. focus on theoretical analysis to determine thermal conductivity. The data suggested that the laser beam can distribute power in all directions and provide newer strategies for the Opto-thermal method's theoretical and experimental data acquisition. Moreover, the Opto-thermal method is applicable for 2D-NM that easily characterizes various thermal properties.[48] Feng et al. focus on the metal-doped phosphorene for augmenting thermal stability. The data suggested that the Al doping within the phosphorene increased thermal properties by two times. Moreover, Al doping increases the structural ability of 2D-NM, thereby enhancing thermal stability. The high thermal and structural stability of the 2D-NM have great interest in nanoelectronics applications.[49] Tang et al. synthesized vermiculite sheets (VS) and incorporated them in PEO-LiTFSI to produce a VS-PEO-LiTFSI composite. The data indicate that incorporating PEO-LiTFSI within the VS-PEO-LiTFSI composite considerably increases the thermal stability of the composite.[50] Krishnamoorthy et al. use nonequilibrium molecular dynamics to calculate the thermal stability of $MoSe_2$ and WSe_2 by designing lattices. The data suggested that the doping of metal designs the heterostructures that ultimately increased the thermal stability. Moreover, the thermal stability of the 2D-NM might be controlled by changing the concentration dopant.[51] Tong et al. reported the synthesis of Beryllonitrene and N-doped Beryllonitrene to determine thermal conductivity. The data suggested that doping increases the carrier and electrical density, thereby increasing the electrical and thermal properties of Beryllonitrene.[52] The 2D-NM are mostly semiconductors materials in which transportation of heat occurs by photons. Heat conduction is very efficient because of the carbon atoms that contain the sp2 bond. The graphene's high thermal conductivity is improved by introducing a small gap at the center of the graphene. For the improvement of 2D-NM applicability and its related device performance, the exact mechanism of the thermal properties of the 2D-NM needs to be understood. With this help, we can design the different 2D-NM and their related devices using thermal management.[53–56] The thermal property mainly depends on their atomic structures that might be tuned by incorporating/changing the surface functional group, strains, metal/non-metal doping, structure stability, and lattice design. In general, the literature suggested that the designing of lattice and metal/non-metal doping might improve the structural stability, thereby improving thermal stability of the 2D-NM. With the help of designing lattice, and metal/non-metal doping, we can easily modify the performance of 2D-NM in end application.

7.3.2 ELECTRICAL PROPERTIES

The electrical properties of 2D-NM are mostly ascertained using an electronic band structure that establishes an energy state in which electrons may/may not accompany the crystalline structure. 2D-NM-based devices like solar cells and gas sensors have their electronic characteristics. Numerous studies have been done so far to enhance the electrical properties of 2D-NM. For example, Sreeprasad et al. synthesized Au-incorporated MoS_2 (Au-MoS_2) nanosheets using chemical and micro-wave processes and determined the electrical, thermal, and structural properties. The data indicate that incorporating Au within the MoS_2 significantly improved electrical, thermal, and structural properties. Figure 7.3 shows the conductivity of the MoS_2 and Au-MoS_2. The conductivity of the Au-MoS_2 increased with increasing the temperature. The conductivity of the Au-MoS_2 increased around 10^3-fold compared to that of the MoS_2.[57]

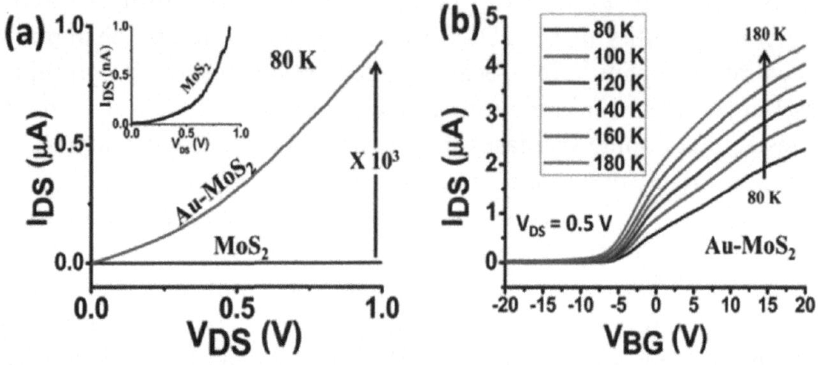

FIGURE 7.3 Conductivity of MoS_2 and Au-MoS_2 (a) at 80 K and (b) at 80–180 K.

Source: Reprinted with permission from Ref. [57]

Yang et al. focus on metal doping within the Sc_2CT_2 to increase the magnetic and electronic properties of Sc_2CT_2-based 2D-NM. The data suggested that the incorporation of transition metals (like Ti, Cr, Mn, and V) might be doped within the Sc_2CT_2-based 2D-NM. The doping of transition metals improved the magnetic and electronic characteristics of the Sc_2CT_2-based 2D-NM.[58] Xiong et al. theoretically examine the role of surface functional groups on Sc_2CT_2 MXene-based 2D-NM. The data suggested that the surface functionalization of Sc_2CT_2 MXene-based 2D-NM shows the high redox ability of photoactive carrier, high absorption, and carrier mobility, thereby high photocatalysis.[59] Vargas-Bernal focuses on the 2D-NM-based gas sensor using the electrical property. The literature suggested that the 2D-NM-based gas sensor is mainly hexagonal crystalline and some materials also have monoclinic, orthorhombic, and triclinic structures. The sensing response is based primarily on the electrical properties of 2D-NM.[60] Figure 7.4 shows the schematic illustration of the gas sensor design based on electrical properties. Postorino et al. understand the effect of biaxial uniform strain on the electrical and structural properties of the MoS_2 and $MoTe_2$-based 2D-NM using the first-principle DFT calculation mainly focused on ionization potential, electron affinity, work function, and band structure. The data suggested that the biaxial uniform strain effectively tuned the structural and electrical properties. Moreover, work function always increases in monolayer 2D-NM, whereas ionization potential and electron affinity show the behavior of materials.[61] Cui et al. focus on tuning the electronic properties of the p- and n-type doped MoSi2N4-based 2D-NM obtained by the first-principal calculation. The data suggested that molecular doping (p- and n-type doping) modifies the electronic properties. The tuning of the electronic properties of p- and n-type doped $MoSi_2N_4$-based 2D-NM might augment the applicability.[62] The above-mentioned data indicate that the 2D-NM have inherent electrical properties. Moreover, electrical properties are easily tuned by structural changes of the 2D-NM. The structural changes by using metals, non-metals, and surface functionalization aided advantageous in numerous applications including in solar cells and sensors.

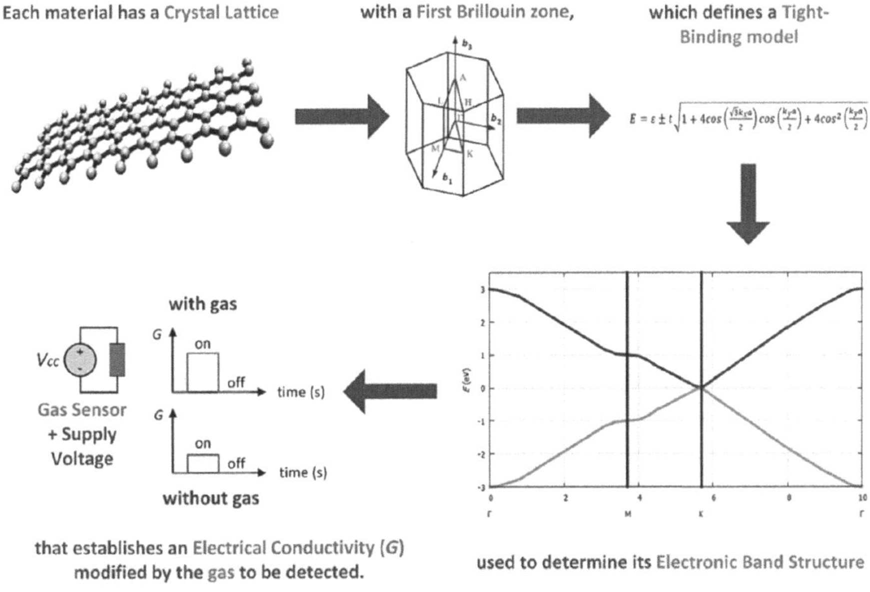

FIGURE 7.4 A schematic illustration of the gas sensor design based on electrical properties.

Source: Reprinted with permission from Ref. [60]. Copyright © 2019. Licensee MDPI, Basel, Switzerland. http://creativecommons.org/licenses/by/4.0/.

7.3.3 MECHANICAL PROPERTIES

The mechanical properties of the 2D-NM provide enthusiastic opportunities for the development/designing of a newer class of 2D-NM that contains high flexibility and in-plane stiffness. The high flexibility and in-plane stiffness of the 2D-NM make them a promising candidate for numerous applications, especially wearable devices. It is important to mention here that the mechanical property of the 2D-NM and their derivative does not directly affect the respective application performance, but it is required for applicability in real-time applications such as wearable electronic devices that have high stiffness and mechanical strength.[56,63,64]

7.4 2D MATERIALS FOR SOLAR CELL APPLICATION

2D-NM-based materials attract interest in solar cell applications due to their superior thermal, mechanical, electronic, and optical properties. Currently, nano-dimensional thin layers are in demand as they can provide flexibility with high performance and mechanical strength. Additionally, these materials are a logical selection for advanced, cost-effective solar cell flexible devices due to their solar energy utilization. Various 2D-NM, like graphene, graphene oxide, TMDs, boron and carbon nitride, metallic nanosheet, etc. are being used in several applications, including solar cells, because of their high conductivity, transparency, and environmental stability. Various functional solar cells like perovskites, organic solar cells, and dye-sensitized solar cells have utilized 2D-NM in the form of electron transport layer, hole transport layer, or absorber layer to achieve high performance. However, commercialization is still a major challenge as the stability of 2D-NM is a concern. Various researchers have reported several methods to get stability and increase the life span of 2D-NM/nanosheets. For example, Krishnamoorthy et al. synthesized graphene-incorporated WS_2 nanosheets using a hydrothermal process. The coated composite demonstrates excellent power conversion efficiency (PCE) of 9.6%, with a fill factor of 0.66.[65] Zhang et al. synthesized Cu_2ZnSnS_4-based nanosheets having high energy (002) facets. These exposed facets demonstrate good photoelectrochemical performance with a PCE of 6.68%. The design of an eco-friendly Cu-based wurtzite structure is a good candidate for solar cell application.[66] Figure 7.5 shows the schematic illustration of synthesized Cu_2ZnSnS_4-based nanosheets, XRD, and Raman spectra.

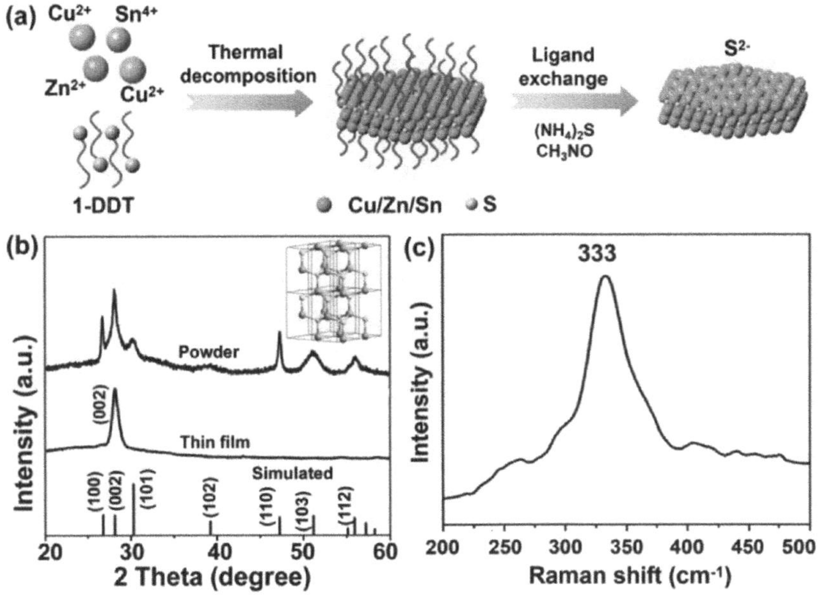

FIGURE 7.5 (a) A schematic illustration of synthesized Cu_2ZnSnS_4-*based* nanosheets. (b) XRD spectra, and (c) Raman spectra.

Source: Reprinted from Ref. [66]. http://creativecommons.org/licenses/by/4.0/.

Fang et al. synthesized a CoS nanosheet using a hydrothermal process. The synthesized nanosheet was dispersed with PEDOT.PSS to form a suspension of CoS/PEDOT PSS. The fabricated device demonstrates a power efficiency of 11.2%.[67] Mahesh et al. synthesized 2D-ZnO nanosheets (2D-ZnO-NS) incorporated with beta-substituted porphyrins photosensitizer. The 2D-ZnO-NS was produced by using the hydrothermal process. The synthesized nanohybrid shows a power efficiency of 0.16%, higher than the 2D-ZnO-NS.[68] Another study by Jiang et al. synthesized S-doped Co_3O_4/graphene using a wet chemical process. The synthesized material shows a power efficiency of 8.24% using visible light irradiation.[69] Huang et al. synthesized ultrathin $FeSe_2$ nanosheets using the phosphine-free hot-injection method. The power efficiency was 7.53%, comparable to the Pt-based device.[70] Sheng et al. synthesized 2D-TiO_2-nanosheet using the self-assembly method for the photovoltaic device. The data suggested that the 2D-TiO_2-nanosheet-based dye-sensitized solar cells have a power efficiency of 8.28%.[71] Dharamalingam et al. synthesized MoS_2 nanosheet and applied it as a dye-sensitized solar cell application. The PCE for the nanosheet was found to be 3.81% which is higher than the bulk MoS_2.[72] Jiang et al. synthesized Co_3S_4 ultrathin nanosheet/reduced-GO (Co_3S_4-rGO) composites as counter electrodes for dye-sensitized solar cell application. The synthesized nanosheet's power efficiency was 8.08%.[73] Another study by Afzali et al. synthesized 2H–TaS_2 nanosheets using the microwave irritation method. The synthesized nanosheet was used as an electron transport layer in perovskites solar cells. The synthesized nanosheet demonstrates an excellent efficiency of 15.23%, which is higher than other reported nanoparticles.[74] Table 7.1 shows the different 2D-NM and their application in solar cells. All these studies prove nanosheet-based materials' applicability for solar cell application due to their low band gap and excellent charge transfer properties.

TABLE 7.1 Different 2D-NM and Their Solar Cell Applications.

S. No	2D nanosheet	Synthesis method	Performance	Reference
1.	2D-ZnO nanosheets-porphyrins	Hydrothermal	PCE 0.13%	68
2.	CoS nanosheets-Si/PEDOT:PSS	Hydrothermal	PCE 11.2%	67
3.	$FeSe_2$ Nanosheets	Phosphine-free hot-injection method	PCE 7.53%	70
4.	2D-TiO_2 nanosheet	Molecular self-assembly using the micelle-assisted solution method	PCE 8.28%	71
5.	Cu_2ZnSnS_4 nanosheets	Thermal decomposition	PCE 6.68%	66
6.	S-Co_3O_4/r-GO	Wet chemical	PCE 8.24%	69
7.	Co_3S_4-r-GO	Noninjection hydrothermal process	PCE 8.08%	73
8.	2H-TaS_2 nanosheets	Microwave irradiation method	PCE 15.23%	74
9.	Graphene-WS_2	Hydrothermal	Open circuit voltage (Jsc) of 0.79 mV, Short circuit current (Voc) of 18.6 mA cm^{-2}, Fill factor-0.66, PCE 9.6%	65
10.	MoS_2 nanosheets	Sodium dodecyl sulfate mediated hydrothermal	PCE 3.81%	72

7.5 APPLICATION OF 2D MATERIALS IN SUPERCAPACITORS

Charge storage system, especially supercapacitors (SC), is a significant global concern. Supercapacitors are divided into two parts: (1) electrochemical double layer capacitors (EDLCs) and (2) pseudo capacitors. Electrode–electrolyte double layer is used to store energy in EDLCs. However, pseudo-capacitors utilize redox reactions on the electrode surface for charge storage. Various materials have been utilized for supercapacitor applications, including carbon-based materials, metal oxides/hydroxides, and polymers metal/carbon composites. The prominent carbon-based materials are activated carbon, mesoporous carbon, CNFs, CNTs, and graphene as active electrode materials in EDLCs. At the same time, several metal oxides and conducting polymer composites are used for pseudo capacitors. Carbon-based materials mainly work based on EDLCs mechanism as pore size distribution is an important factor for EDLC. Therefore, these often exhibit very higher power output and better cycling ability. However, EDLCs have lower energy density values than pseudo capacitors since pseudo capacitors involve active redox materials to store charge both on the surface and in the subsurface layer. Recently, 2D materials like graphene, TMDs, and MXene demonstrated excellent charge storage performance due to their high electrical conductivity, surface-to-charge ratio, and flexible functionality. These materials are attracting interest in energy storage applications. Also, high performance and portability are major requirements for next-generation wearable devices.

Several researchers are reporting the application of 2D materials in advance of supercapacitors application. For instance, Huang et al. synthesized polyaniline-graphene-MoS_2 based composite materials (P-G-MoS_2) using a polymerization process for supercapacitor application. The prepared P-G-MoS_2 is effectively used in supercapacitor applications with a specific capacitance of 575 F/g at A/g. Moreover, high specific capacitance is mainly due to the synergetic effects of polymer, graphene, and $MoS_{2.}$[75] Quan et al. synthesized α-Fe_2O_3 nanoplates-rGO (Fe-rGO) composite for supercapacitor application. The material was synthesized using the one-pot hydrothermal synthesis method, which further applied supercapacitors application and showed an excellent 903 F/g at the current density of 1 A/g.[76] Ansari et al. mechanically exfoliate MoS_2 nanosheets and incorporate them with conductive polyaniline (P-MoS_2) for supercapacitor application. The data indicate that the produced P-MoS_2-based composite materials are effectively used for the supercapacitor application with a specific capacitance of 510 F/g. The extraordinary specific capacitance is mainly due to the synergetic effects of conductive polymers and 2D-MoS_2 nanosheets.[77] Navarro-Suárez et al. synthesized titanium carbide and rGO-based 2D-NM composite nanolayers for supercapacitor application. The synthesized composite shows a specific capacitance of 254 F $g-^1$ at a scan rate of 2 mV $s-^1$. However, on further increasing the scan rate upto 100 mV $s-^1$, specific capacitance decreases to 193 F $g-^{1.}$[78] Rajendran et al. synthesized VS_4 nanosheets using a hydrothermal method which demonstrates the specific capacitance of 215 F g^{-1} at the scan rate of 5 mV s^{-1} with excellent retention of 82% within 2000 cycles.[79] Chang et al. synthesized $NiSe_2$ nanosheets with selenium vacancy, which showed excellent performance (328 $Fg-^1$) toward charge storage.[80] Using the nitridation method, Yuan et al. synthesized 2D carbon-wrapped TiN (2D-C-TiN). The wrapped nanolayer showed excellent performance of 238 F/g at a current density of 0.5 A/g and a high-rate capacitance.[81] Geng et al. synthesized B/N codoped 2D porous carbon nanosheets (B/N-C-NS) by simply mixing and a carbonizing mixture of ammonium humate in $H_3BO_{3,}$ which demonstrates excellent specific capacitance at a current density of 0.2 A/g and on further increasing the current density upto 100 A/g, the material can show the capacitance of 125 F/g with cycling retention of 94.8% after 10,000 cycles.[82] Another study by Zhang et al. uses an Nb_2SnC nano layer for supercapacitor application. Synthesized nanolayers show a capacitance of 140 F/g with an energy density of 14.4 Wh $g-^1$ at a power density of 36 W $kg-^1$ with high cycling stability.[83] Naz et al. synthesized 1T-MoS_2 nanosheets on 3D-r-GO (MoS_2-r-GO) using the one-pot hydrothermal process. The synthesized material demonstrates excellent performance of 442 F/g at a current density of 1 A/g with excellent cycle stability and retention.[84] Venkateshalu et al. synthesized 2D-vanadium-based MXene and applied for supercapacitor. The material was synthesized using high-temperature annealing, etching of the MXene layer, and slurry deposition on the CF surface. The material shows a capacitance of 112.8 F/g at a current density of 1.85 mA/cm^2 using an aqueous electrolyte.[85] Figure 7.6 shows the schematic illustration of the 2D-vanadium-based MXene and its supercapacitor application.

Alam et al. synthesized $Co(OH)_2$-$MoSe_2$ based hybrid (C-$MoSe_2$) nanocomposite using single-step hydrothermal reaction and applied for asymmetrical supercapacitor application. The synthesized material shows the specific capacitance of 541.55 F/g at a current density of 1 A/g, while on further increasing the current density of 10 A/g, the specific capacitance decreases upto 397 F/g with an excellent retention capacity of 86.2%, having an energy density of 18.28 Wh/kg at a maximum power density of

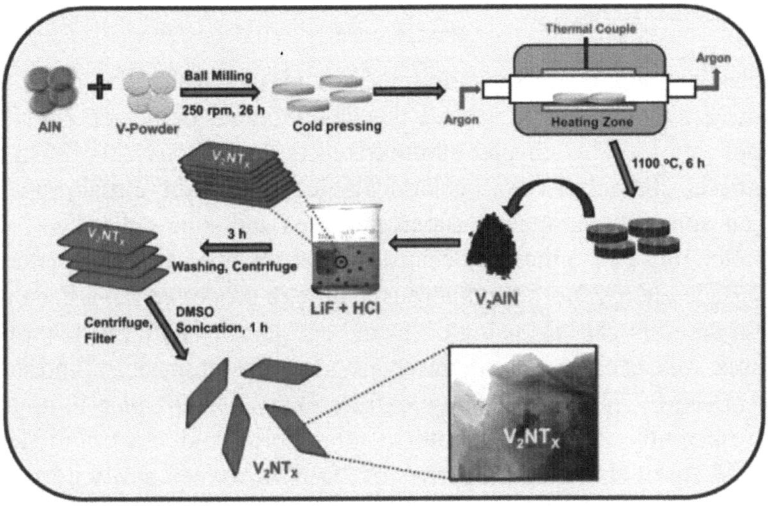

FIGURE 7.6 A schematic illustration of the 2D-vanadium-based MXene and its supercapacitor application.

Source: Reprinted with permission from Ref. [85]. Open Access

12.209 kW kg^{-1}.[86] Table 7.2 summarizes different 2D-NM and their applications in supercapacitors. All these studies prove the excellency of nanosheets toward charge storage; however, the stability of nanosheets is still a concern.

TABLE 7.2 Different 2D-NM and Their Supercapacitor Applications.

S.No.	2D-NM	Synthesis	Capacitance value	References
1.	P-G-MoS_2	Polymerization	575 F/g at A/g	75
2.	Fe-rGO	Hydrothermal	903 F/g at 1 A/g	76
3.	P-MoS_2	Mechanical exfoliation	510 F/g at A/g	77
4.	Carbide-rGO	Etching and liquid phase exfoliation	254 F/g at 2 mV/s 193 F/g at 100 mV/s	78
5.	2D-C-TiN	Nitridation	238 F/g	81
6.	$NiSe_2$	Plasma-assisted dry exfoliation	328 F/g	80
7.	MoS_2-r-GO	Hydrothermal	442 F/g at 1 A/g	84
8.	B/N-C-NS	Carbonization	311 F/g	82
9.	Nb_2SnC	Etching/delamination	140 F/g	83
10.	Vanadium nitride	Etching followed by electro-chemical deposition	112.8 F/g at 1.85 mV/s	85
11.	C- $MoSe_2$	Hydrothermal	541.55 F/g	86
12.	VS_4 nanosheets	Hydrothermal	215 F/g	79

7.6 CONCLUSION

Advancements in characterization tools and increasing the demands of innovative technologies increased the development of newer 2D-NM for end applications. This chapter summarized the numerous 2D-NM in terms of synthesis, characteristics (mechanical, electrical, and thermal properties), and their applications, mainly supercapacitors and solar cells. The 2D-NM are challenging to synthesize because of their isotropic crystal nature and high surface energy. So far, numerous synthetic processes have been developed for creating 2D-NM using CVD and exfoliation. With the help of these processes, we can easily control the shape, size, crystal structure, and stability of the 2D-NM. The 2D-NM show extraordinary performance in numerous areas, especially in supercapacitors, solar cells, and electronic devices. With the help of changing properties of 2D-NM, we can easily improve the performance of 2D-NM in respective applications. The researcher focuses on improving mechanical, electrical, and thermal properties. The data suggested that designing lattice and metal/non-metal doping might improve structural stability, thereby improving thermal, mechanical, and electrical stability. With the help of designing lattice and metal/non-metal doping, we can easily modify the performance of 2D-NM in the end application. The above-mentioned discussion suggested that the characteristic properties of 2D-NM and their hybrid materials might benefit next-generation high-performing devices.

KEYWORDS

- **two-dimensional (2D)**
- **graphene**
- **solar cells**
- **supercapacitors**
- **mechanical**
- **electrical**

REFERENCES

1. Tiwari, J. N.; Tiwari, R. N.; Kim, K. S., Zero-dimensional, One-dimensional, Two-dimensional and Three-dimensional Nanostructured Materials for Advanced Electrochemical Energy Devices. *Prog. Mater. Sci.* **2012,** *57* (4), 724–803. https://doi.org/10.1016/j.pmatsci.2011.08.003.
2. Rizvi, M.; Gerengi, H.; Gupta, P. *Functionalization of Nanomaterials: Synthesis and Characterization, Functionalized Nanomaterials for Corrosion Mitigation: Synthesis, Characterization, and Applications*; American Chemical Society, 2022; pp 1–26. https://doi.org/10.1021/bk-2022-1418.ch001
3. Ashfaq, M.; Talreja, N.; Chauhan, D.; Rodríguez, C. A.; Mera, A. C.; Ramalinga Viswanathan, M. Synthesis of Reduced Graphene Oxide Incorporated Bimetallic (Cu/Bi) Nanorods Based Photocatalyst Materials for the Degradation of Gallic Acid and Bacteria. *J. Indust. Eng. Chem.* **2022,** *110,* 447–455. https://doi.org/10.1016/j.jiec.2022.03.023.
4. Sun, Z.; Talreja, N.; Tao, H.; Texter, J.; Muhler, M.; Strunk, J.; J. C. Catalysis of Carbon Dioxide Photoreduction on Nanosheets: Fundamentals and Challenges. *Angew. Chem. Int. Ed.* **2018,** *57* (26), 7610–7627. https://doi.org/10.1002/anie.201710509.
5. Moraes Silva, S.; Tavallaie, R.; Sandiford, L.; Tilley, R. D.; Gooding, J. J. Gold Coated Magnetic Nanoparticles: From Preparation to Surface Modification for Analytical and Biomedical Applications. *Chem. Commun.* **2016,** *52* (48), 7528–7540. https://doi.org/10.1039/C6CC03225G.
6. Gawande, M. B.; Goswami, A.; Felpin, F.-X.; Asefa, T.; Huang, X.; Silva, R.; Zou, X.; Zboril, R.; Varma, R. S. Cu and Cu-Based Nanoparticles: Synthesis and Applications in Catalysis. *Chem. Rev.* **2016,** *116* (6), 3722–3811. https://doi.org/10.1021/acs.chemrev.5b00482.
7. Jeevanandam, J.; Barhoum, A. Review on Nanoparticles and Nanostructured Materials: History. *Sources Toxic. Regul.* **2018,** *9*, 1050–1074. https://doi.org/10.3762/bjnano.9.98.
8. Sasidharan, V.; Damiri, F.; Talreja, N.; Chauhan, D.; Mangalaraja, R. V.; Berrada, M.; Ashfaq, M. Carbon-Based Nanomaterials: An Efficient Tool for Improving the Nutritional Quality of Crops. In *Metabolic Engineering in Plants*; Aftab, T., Hakeem, K. R., Eds.; Springer Nature Singapore; Singapore, 2022; pp 375–389. https://doi.org/10.1007/978-981-16-7262-0_17.
9. Tao, H.; Gao, Y.; Talreja, N.; Guo, F.; Texter, J.; Yan, C.; Sun, Z. Two-dimensional Nanosheets for Electrocatalysis in Energy Generation and Conversion. *J. Mater. Chem. A* **2017,** *5* (16), 7257–7284. https://doi.org/10.1039/C7TA00075H.
10. Tiwari, S. K.; Sahoo, S.; Wang, N.; Huczko, A. Graphene Research and Their Outputs: Status and Prospect. *J. Sci: Adv. Mater. Devic.* **2020,** *5* (1), 10–29. https://doi.org/10.1016/j.jsamd.2020.01.006.
11. Smith, A. T.; LaChance, A. M.; Zeng, S.; Liu, B.; Sun, L. Synthesis, Properties, and Applications of Graphene Oxide/Reduced Graphene Oxide and Their Nanocomposites. *Nano Mater. Sci.* **2019,** *1* (1), 31–47. https://doi.org/10.1016/j.nanoms.2019.02.004.
12. Mir, S. H.; Yadav, V. K.; Singh, J. K. Recent Advances in the Carrier Mobility of Two-Dimensional Materials: A Theoretical Perspective. *ACS Omega* **2020,** *5* (24), 14203–14211. https://doi.org/10.1021/acsomega.0c01676.
13. Peng, Z.; Chen, X.; Fan, Y.; Srolovitz, D. J.; Lei, D. Strain Engineering of 2D Semiconductors and Graphene: From Strain Fields to Band-structure Tuning and

Photonic Applications. *Light: Sci. Appl.* **2020,** *9* (1), 190. https://doi.org/10.1038/s41377-020-00421-5.

14. Ashfaq, M.; Talreja, N.; Chauhan, D.; Viswanathan, M. R. Synthesis of Cu-doped 2D-WS2 Nanosheet-based Nano-antibiotic Materials for Inhibiting E. Coli and S. aureus Bacterial Strains. *N. J. Chem.* **2022,** *46* (12), 5581–5587. https://doi.org/10.1039/D1NJ05931A.
15. Sultana, A.; Omar, R. A.; Talreja, N.; Chauhan, D.; Mangalaraja, R. V.; Ashfaq, M. Chapter 28—Copper-based Metal-organic Framework for Environmental Applications. In *Copper Nanostructures: Next-Generation of Agrochemicals for Sustainable Agroecosystems*; Abd-Elsalam, K. A., Ed.; Elsevier, 2022, pp 701–717. https://doi.org/10.1016/B978-0-12-823833-2.00007-6.
16. Omar, R. A.; Talreja, N.; Chauhan, D.; Mangalaraja, R. V.; Ashfaq, M. Chapter 14 - Nano Metal-carbon–based Materials: Emerging Platform for the Growth and Protection of Crops. In *Nanotechnology-Based Sustainable Alternatives for the Management of Plant Diseases*; Balestra, G. M., Fortunati, E., Eds.; Elsevier, 2022; pp 341–354. https://doi.org/10.1016/B978-0-12-823394-8.00013-5.
17. Sasidharan, V.; Sachan, D.; Chauhan, D.; Talreja, N.; Ashfaq, M. Three-dimensional (3D) Polymer—Metal–Carbon Framework for Efficient Removal of Chemical and Biological Contaminants. *Sci. Rep.* **2021,** *11* (1), 7708. https://doi.org/10.1038/s41598-021-86661-w.
18. Ahn, S.; Kim, G.; Nayak, P. K.; Yoon, S. I.; Lim, H.; Shin, H.-J.; Shin, H. S. Prevention of Transition Metal Dichalcogenide Photodegradation by Encapsulation with h-BN Layers. *ACS Nano.* **2016,** *10* (9), 8973–8979. https://doi.org/10.1021/acsnano.6b05042.
19. Suzuki, Y.; Watanabe, K. Excitons in Two-dimensional Atomic Layer Materials from Time-Dependent Density Functional Theory: Mono-layer and Bi-layer Hexagonal Boron Nitride and Transition-metal Dichalcogenides. *Phys. Chem. Chem. Phys.* **2020,** *22* (5), 2908–2916. https://doi.org/10.1039/C9CP06034K.
20. Ashfaq, M.; Wongpakham, T.; Talreja, N.; Chauhan, D.; Tharasanit, T.; Srituravanich, W. Synthesis of Polymeric Composite Grafted with Mineral Particles/graphene Oxide-based Biomaterial: A Promising Robust Hemostatic Bandage. *Mater. Today Commun.* **2022,** *33,* 104786. https://doi.org/10.1016/j.mtcomm.2022.104786.
21. Talreja, N.; Ashfaq, M.; Chauhan, D.; Mangalaraja, R. V. Cu-MXene: A Potential Biocide for the Next-generation Biomedical Application. *Mater. Chem. Phys.* **2023,** *294,* 127029. https://doi.org/10.1016/j.matchemphys.2022.127029.
22. Bernardi, M.; Palummo, M.; Grossman, J. C. Extraordinary Sunlight Absorption and One Nanometer Thick Photovoltaics Using Two-Dimensional Monolayer Materials. *Nano Lett.* **2013,** *13* (8), 3664–3670. https://doi.org/10.1021/nl401544y.
23. Samy, O.; Zeng, S.; Birowosuto, M. D.; El Moutaouakil, A. A Review on MoS_2 Properties, Synthesis, Sensing Applications and Challenges. *Crystals.* **2021,** *11* (4). https://doi.org/10.3390/cryst11040355.
24. Singhal, J.; Verma, S.; Kumar, S., The Physio-chemical Properties and Applications of 2D Nanomaterials in Agricultural and Environmental Sustainability. *Sci. Total Environ.* **2022,** *837,* 155669. https://doi.org/10.1016/j.scitotenv.2022.155669.
25. Deng, S.; Sumant, A. V.; Berry, V. Strain Engineering in Two-dimensional Nanomaterials Beyond Graphene. *Nano Today* **2018,** *22,* 14–35. https://doi.org/10.1016/j.nantod.2018.07.001.

26. Derakhshi, M.; Daemi, S. Two-dimensional Nanomaterials Beyond Graphene for Biomedical Applications. **2022,** *13* (1) . https://doi.org/10.3390/jfb13010027.
27. Malik, P.; Gupta, R.; Malik, V.; Ameta, R. K. Emerging Nanomaterials for Improved Biosensing. *Measur. Sens.* **2021,** *16,* 100050. https://doi.org/10.1016/j.measen.2021.100050.
28. Bolotsky, A.; Butler, D.; Dong, C.; Gerace, K.; Glavin, N. R.; Muratore, C.; Robinson, J. A.; Ebrahimi, A. Two-Dimensional Materials in Biosensing and Healthcare: From In Vitro Diagnostics to Optogenetics and Beyond. *ACS Nano*. **2019,** *13* (9), 9781–9810. https://doi.org/10.1021/acsnano.9b03632.
29. Li, X.; Zhu, J.; Wei, B. Hybrid Nanostructures of Metal/Two-dimensional Nanomaterials for Plasmon-enhanced Applications. *Chem. Soc. Rev.* **2016,** *45* (11), 3145–3187. https://doi.org/10.1039/C6CS00195E.
30. Yang, F.; Song, P.; Ruan, M.; Xu, W. Recent Progress in Two-dimensional Nanomaterials: Synthesis, Engineering, and Applications. *FlatChem*. **2019,** *18,* 100133. https://doi.org/10.1016/j.flatc.2019.100133.
31. Zhang, H.; Fan, T.; Chen, W.; Li, Y.; Wang, B. Recent Advances of Two-dimensional Materials in Smart Drug Delivery Nano-Systems. *Bioact. Mater.* **2000,** *5* (4), 1071–1086. https://doi.org/10.1016/j.bioactmat.2020.06.012.
32. Huang, H.; Jiang, R.; Feng, Y.; Ouyang, H.; Zhou, N.; Zhang, X.; Wei, Y. Recent Development and Prospects of Surface Modification and Biomedical Applications of MXenes. *Nanoscale*. **2020,** *12* (3), 1325–1338. https://doi.org/10.1039/c9nr07616f.
33. Jaffar, S. S.; Saallah, S.; Misson, M.; Siddiquee, S.; Roslan, J.; Saalah, S.; Lenggoro, W. Recent Development and Environmental Applications of Nanocellulose-Based Membranes. *Membranes*. **2022,** *12* (3). https://doi.org/10.3390/membranes12030287.
34. Zhang, J.; Chen, H.; Zhao, M.; Liu, G.; Wu, J. 2D Nanomaterials for Tissue Engineering Application. *Nano Res.* **2020,** *13* (8), 2019–2034. https://doi.org/10.1007/s12274-020-2835-4.
35. Hu, T.; Mei, X.; Wang, Y.; Weng, X.; Liang, R.; Wei, M. Two-dimensional Nanomaterials: Fascinating Materials in Biomedical Field. *Sci. Bull.* **2019,** *64* (22), 1707–1727. https://doi.org/10.1016/j.scib.2019.09.021.
36. Li, L.; Zhou, M.; Jin, L.; Liu, L.; Mo, Y.; Li, X.; Mo, Z.; Liu, Z.; You, S.; Zhu, H. Research Progress of the Liquid-Phase Exfoliation and Stable Dispersion Mechanism and Method of Graphene. *Front. Mater.* **2019,** *6*. https://doi.org/10.3389/fmats.2019.00325.
37. Ashfaq, M.; Talreja, N.; Chauhan, D.; Afreen, S.; Sultana, A.; Srituravanich, W. Two-dimensional (2D) Hybrid Nanomaterials for Diagnosis and Treatment of Cancer. *J. Drug Deliv. Sci. Technol.* **2022,** *70,* 103268. https://doi.org/10.1016/j.jddst.2022.103268.
38. Georgakilas, V.; Otyepka, M.; Bourlinos, A. B. ; Chandra, V.; Kim, N.; Kemp, K. C.; Hobza, P.; Zboril, R.; Kim, K. S. Functionalization of Graphene: Covalent and Non-Covalent Approaches, Derivatives and Applications. *Chem. Rev.* **2012,** *112* (11), 6156–6214. https://doi.org/10.1021/cr3000412.
39. Gómez-Santos, G. Thermal Van der Waals Interaction between Graphene Layers. *Phys. Rev. B*. **2009,** *80* (24), 245424. https://doi.org/10.1103/PhysRevB.80.245424.
40. Guan, Z.; Wang, C.; Li, W.; Luo, S.; Yao, Y.; Yu, S.; Sun, R.; Wong, C.-P. A Facile and Clean Process for Exfoliating MoS_2 Nanosheets Assisted by a Surface Active Agent in Aqueous Solution. *Nanotechnology*. **2018,** *29* (42), 425702. https://doi.org/10.1088/1361-6528/aad676.

41. Tan, X.; Kang, W.; Liu, J.; Zhang, C. Synergistic Exfoliation of MoS2 by Ultrasound Sonication in a Supercritical Fluid Based Complex Solvent. *Nanosc. Res. Lett.* **2019,** *14* (1), 317. https://doi.org/10.1186/s11671-019-3126-4.
42. Sahoo, D.; Kumar, B.; Sinha, J.; Ghosh, S.; Roy, S. S.; Kaviraj, B. Cost Effective Liquid Phase Exfoliation of MoS_2 Nanosheets and Photocatalytic Activity for Wastewater Treatment Enforced by Visible Light. *Sci. Rep.* **2020,** *10* (1), 10759. https://doi.org/10.1038/s41598-020-67683-2.
43. Sankararamakrishnan, N.; Chauhan, D.; Dwivedi, J. Synthesis of Functionalized Carbon Nanotubes by Floating Catalytic Chemical Vapor Deposition Method and their Sorption Behavior Toward Arsenic. *Chem. Eng. J.* **2016,** *284,* 599–608. https://doi.org/10.1016/j.cej.2015.08.145.
44. Ashfaq, M.; Verma, N.; Khan, S. Novel Polymeric Composite Grafted with Metal Nanoparticle-dispersed CNFs as a Chemiresistive Non-destructive Fruit Sensor Material. *Mater. Chem. Phys.* **2018,** *217,* 216–227. https://doi.org/10.1016/j.matchemphys.2018.06.056.
45. Derakhshi, M.; Daemi, S.; Shahini, P.; Habibzadeh, A.; Mostafavi, E.; Ashkarran, A. A. Two-Dimensional Nanomaterials Beyond Graphene for Biomedical Applications. *J. Func. Biomater.* **2022,** *13* (1). https://doi.org/10.3390/jfb13010027.
46. Si, A.; Kyzas, G. Z.; Pal, K.; de Souza F. G., Jr. Graphene Functionalized Hybrid Nanomaterials for Industrial-scale Applications: A Systematic Review. *J. Mol. Struc.* **2021,** *1239,* 130518. https://doi.org/10.1016/j.molstruc.2021.130518.
47. Han, X.; Huang, J.; Lin, H.; Wang, Z.; Li, P.; Chen, Y. 2D Ultrathin MXene-Based Drug-Delivery Nanoplatform for Synergistic Photothermal Ablation and Chemotherapy of Cancer. *Adv. Healthc Mater.* **2018,** *7* (9), e1701394. https://doi.org/10.1002/adhm.201701394.
48. Judek, J.; Gertych, A. P.; Świniarski, M.; Łapińska, A. ; Dużyńska, A.; Zdrojek, M. High Accuracy Determination of the Thermal Properties of Supported 2D Materials. *Sci. Rep.* **2015,** *5* (1), 12422. https://doi.org/10.1038/srep12422.
49. Feng, X.; Kulish, V. V.; Wu, P.; Liu, X.; Ang, K.-W. Anomalously Enhanced Thermal Stability of Phosphorene *via* Metal Adatom Doping: An Experimental and First-principles Study. *Nano Res.* **2016,** *9* (9), 2687–2695. https://doi.org/10.1007/s12274-016-1156-0.
50. Tang, W.; Tang, S.; Zhang, C.; Ma, Q.; Xiang, Q.; Yang, Y.-W.; Luo, J. Simultaneously Enhancing the Thermal Stability, Mechanical Modulus, and Electrochemical Performance of Solid Polymer Electrolytes by Incorporating 2D Sheets. *Adv. Energy Mater.* **2018,** *8* (24), 1800866. https://doi.org/10.1002/aenm.201800866.
51. Krishnamoorthy, A.; Baradwaj, N.; Nakano, A.; Kalia, R. K.; Vashishta, P. Lattice Thermal Transport in Two-dimensional Alloys and Fractal Heterostructures. *Sci. Rep.* **2021,** *11* (1), 1656. https://doi.org/10.1038/s41598-021-81055-4.
52. Tong, Z.; Pecchia, A.; Yam, C.; Bao, H.; Dumitrică, T.; Frauenheim, T. Significant Increase of Electron Thermal Conductivity in Dirac Semimetal Beryllonitrene by Doping Beyond Van Hove Singularity. *Adv. Func. Mater.* **2022,** *32* (17), 2111556. https://doi.org/10.1002/adfm.202111556.
53. Moore, A. L.; Shi, L. Emerging Challenges and Materials for Thermal Management of Electronics. *Mater. Today* **2014,** *17* (4), 163–174. https://doi.org/10.1016/j.mattod.2014.04.003.

54. Ali, H. M.; Arshad, A.; Jabbal, M.; Verdin, P. G. Thermal Management of Electronics Devices with PCMs Filled Pin-fin Heat sinks: A Comparison. *Int. J. Heat Mass Transf.* **2018,** *117,* 1199–1204. https://doi.org/10.1016/j.ijheatmasstransfer.2017.10.065.
55. Zhang, Y.-F.; Han, D.; Zhao, Y.-H.; Bai, S.-L. High-performance Thermal Interface Materials Consisting of Vertically Aligned Graphene Film and Polymer. *Carbon* **2016,** *109*, 552–557. https://doi.org/10.1016/j.carbon.2016.08.051.
56. Khan, K.; Tareen, A. K.; Aslam, M.; Mahmood, A.; khan, Q.; Zhang, Y.; Ouyang, Z.; Guo, Z.; Zhang, H. Going Green with Batteries and Supercapacitor: Two Dimensional Materials and Their Nanocomposites Based Energy Storage Applications. *Prog. Solid State Chem.* **2020,** *58*, 100254. https://doi.org/10.1016/j.progsolidstchem.2019.100254.
57. Sreeprasad, T. S.; Nguyen, P.; Kim, N.; Berry, V. Controlled, Defect-Guided, Metal-Nanoparticle Incorporation Onto MoS_2 via Chemical and Microwave Routes: Electrical, Thermal, and Structural Properties. *Nano Lett.* **2013,** *13* (9), 4434–4441. https://doi.org/10.1021/nl402278y.
58. Yang, J.; Luo, X.; Zhang, S.; Chen, L. Investigation of Magnetic and Electronic Properties of Transition Metal Doped Sc2CT2 (T = O, OH or F) Using a First Principles Study. *Phys. Chem. Chem. Phys.* **2016,** *18* (18), 12914–12919. https://doi.org/10.1039/C6CP00138F.
59. Xiong, K.; Wang, P.; Yang, G.; Liu, Z.; Zhang, H.; Jin, S.; Xu, X. Functional Group Effects on the Photoelectronic Properties of MXene (Sc_2CT_2, T = O, F, OH) and Their Possible Photocatalytic Activities. *Sci. Rep.* **2017,** 15095.
60. Vargas-Bernal, R. Electrical Properties of Two-Dimensional Materials Used in Gas Sensors. *Sensors* **2019,** *19* (6). https://doi.org/10.3390/s19061295.
61. Postorino, S.; Grassano, D.; D'Alessandro, M.; Pianetti, A.; Pulci, O.; Palummo, M. Strain-induced Effects on the Electronic Properties of 2D Materials. *Nanomater. Nanotechnol.* **2020,** *10* 1847980420902569. https://doi.org/10.1177/1847980420902569.
62. Dong, Z.; Cui, H.; Zhang, H.; Wang, F.; Zhan, X.; Mayer, F.; Nestler, B.; Wegener, M.; Levkin, P. A. 3D Printing of Inherently Nanoporous Polymers *via* Polymerization-induced Phase Separation. *Nat. Commun.* **2021,** *12* (1), 247. https://doi.org/10.1038/s41467-020-20498-1.
63. Castro-Gutiérrez, J.; Celzard, A.; Fierro, V. Energy Storage in Supercapacitors: Focus on Tannin-Derived Carbon Electrodes. *Front. Mater.* **2020,** *7*.
64. Pomerantseva, E.; Gogotsi, Y. Two-dimensional Heterostructures for Energy Storage. *Nat. Energy* **2017,** *2* (7), 17089. https://doi.org/10.1038/nenergy.2017.89.
65. Krishnamoorthy, D.; Prakasam, A. Graphene Hybridized with Tungsten Disulfide (WS2) Based Heterojunctions Photoanode Materials for High Performance Dye Sensitized Solar Cell Device (DSSCs) Applications. *J. Clust. Sci.* **2021,** *32* (3), 621–630. https://doi.org/10.1007/s10876-020-01828-1.
66. Zhang, X.; Xu, Y.; Zhang, J.; Dong, S.; Shen, L.; Gupta, A.; Bao, N. Synthesis of Wurtzite Cu2ZnSnS4 Nanosheets with Exposed High-Energy (002) Facets for Fabrication of Efficient Pt-Free Solar Cell Counter Electrodes. *Sci. Rep.* **2018,** *8* (1), 248. https://doi.org/10.1038/s41598-017-18631-0.
67. Fang, X.; Song, T.; Liu, R.; Sun, B. Two-Dimensional CoS Nanosheets Used for High-Performance Organic–Inorganic Hybrid Solar Cells. *J. Phys. Chem. C* **2014,** *118* (35), 20238–20245. https://doi.org/10.1021/jp506345a.

68. Mahesh, A. Photovoltaic Performance of ZnO Nanosheets Solar Cell Sensitized with Beta-Substituted Porphyrin. *J. Nanomater.* **2011,** *2011*, 301873. https://doi.org/10.1155/2011/301873.
69. Jiang, T.; Yin, N.; Bai, Z.; Dai, P.; Yu, X.; Wu, M.; Li, G. Wet Chemical Synthesis of S Doped Co_3O_4 Nanosheets/Reduced Graphene Oxide and Their Application in Dye Sensitized Solar Cells. *Appl. Surf. Sci.* **2018,** *450,* 219–227. https://doi.org/10.1016/j.apsusc.2018.04.148.
70. Huang, S.; He, Q.; Chen, W.; Qiao, Q.; Zai, J.; Qian, X. Ultrathin $FeSe_2$ Nanosheets: Controlled Synthesis and Application as a Heterogeneous Catalyst in Dye-Sensitized Solar Cells. *Chemistry* **2015,** *21* (10), 4085–4091. https://doi.org/10.1002/chem.201406124.
71. Sheng, L.; Liao, T.; Kou, L.; Sun, Z. Single-crystalline Ultrathin 2D TiO_2 Nanosheets: A Bridge Towards Superior Photovoltaic Devices. *Mater. Today Energy* **2017,** *3,* 32–39. https://doi.org/10.1016/j.mtener.2016.12.004.
72. Dharamalingam, K.; Arjun Kumar, B.; Ramalingam, G.; Sasi Florence, S.; Raju, K.; Senthil Kumar, P.; Govindaraju, S.; Thangavel, E. The Role of Sodium Dodecyl Sulfate Mediated Hydrothermal Synthesis of MoS2 Nanosheets for Photocatalytic Dye Degradation and Dye-sensitized Solar Cell Application. *Chemosphere* **2022,** *294,* 133725. https://doi.org/10.1016/j.chemosphere.2022.133725.
73. Jiang, T.; Yang, S.; Dai, P.; Yu, X.; Bai, Z.; Wu, M.; Li, G.; Tu, C. Economic Synthesis of Co_3S_4 Ultrathin Nanosheet/Reduced Graphene Oxide Composites and Their Application as an Efficient Counter Electrode for Dye-sensitized Solar Cells. *Electrochim. Acta* **2018,** *261,* 143–150. https://doi.org/10.1016/j.electacta.2017.12.121.
74. Afzali, M.; Mostafavi, A.; Shamspur, T. Improved Perovskite Solar Cell with 2H–TaS_2 Nanosheets as an Electron Transport Layer Using Microwave Irradiation. *J. Alloys Compd.* **2020,** *817*, 152742. https://doi.org/10.1016/j.jallcom.2019.152742.
75. Huang, K.-J.; Wang, L.; Liu, Y.-J.; Wang, H.-B.; Liu, Y.-M.; Wang, L.-L. Synthesis of Polyaniline/2-Dimensional Graphene Analog MoS_2 Composites for High-Performance Supercapacitor. *Electrochim. Acta* **2013,** *109,* 587–594. https://doi.org/10.1016/j.electacta.2013.07.168.
76. Quan, H.; Cheng, B.; Xiao, Y.; Lei, S. One-pot Synthesis of α-Fe_2O_3 Nanoplates-reduced Graphene Oxide Composites for Supercapacitor Application. *Chem. Eng. J.* **2016,** *286,* 165–173. https://doi.org/10.1016/j.cej.2015.10.068.
77. Ansari, S. A.; Fouad, H.; Ansari, S. G.; Sk, M. P.; Cho, M. H. Mechanically Exfoliated MoS_2 Sheet Coupled with Conductive Polyaniline as a Superior Supercapacitor Electrode Material. *J. Colloid Interface Sci.* **2017,** *504,* 276–282. https://doi.org/10.1016/j.jcis.2017.05.064.
78. Navarro-Suárez, A. M.; Maleski, K.; Makaryan, T.; Yan, J.; Anasori, B.; Gogotsi, Y. 2D Titanium Carbide/Reduced Graphene Oxide Heterostructures for Supercapacitor Applications. *Batter. Supercaps* **2018,** *1* (1), 33–38. https://doi.org/10.1002/batt.201800014.
79. Rajendran, K.; Nagamony, P.; Annal Therese, H.; Template-Free Synthesis of 2D-Pristine VS_4 Nanosheets for Supercapacitor Applications. *Physica Status Solidi.* **2022,** *219,* 2200203. https://doi.org/10.1002/pssa.202200203.
80. Chang, A.; Zhang, C.; Yu, Y.; Yu, Y.; Zhang, B. Plasma-Assisted Synthesis of $NiSe_2$ Ultrathin Porous Nanosheets with Selenium Vacancies for Supercapacitor. *ACS Appl. Mater. Interfaces.* **2018,** *10* (49), 41861–41865. https://doi.org/10.1021/acsami.8b16072.

81. Yuan, W.; Cheng, L.; Wu, H.; Zhang, Y.; Lv, S.; Guo, X. One-step Synthesis of 2D-layered Carbon Wrapped Transition Metal Nitrides from Transition Metal Carbides (MXenes) for Supercapacitors with Ultrahigh Cycling Stability. *Chem. Commun.* **2018,** *54* (22), 2755–2758. https://doi.org/10.1039/C7CC09017J.
82. Geng, Q.; Huang, G.; Liu, Y.; Li, Y.; Liu, L.; Yang, X.; Wang, Q.; Zhang, C. Facile Synthesis of B/N Co-doped 2D Porous Carbon Nanosheets Derived from Ammonium Humate for Supercapacitor Electrodes. *Electrochim. Acta* **2019,** *298* 1–13. https://doi.org/10.1016/j.electacta.2018.12.038.
83. Zhang, Z.; Huang, X.; Lu, W.; Qiu, X.; Ma, T.; Xia, W. Synthesis of 2D Layered Nb_2SnC at Low Sintering Temperature and its Application for High-performance Supercapacitors. *J. Solid State Chem.* **2020,** *288,* 121425. https://doi.org/10.1016/j.jssc.2020.121425.
84. Naz, R.; Imtiaz, M.; Liu, Q.; Yao, L.; Abbas, W.; Li, T.; Zada, I.; Yuan, Y.; Chen, W.; Gu, J. Highly Defective 1T-MoS_2 Nanosheets on 3D Reduced Graphene Oxide Networks for Supercapacitors. *Carbon* **2019,** 152, 697–703. https://doi.org/10.1016/j.carbon.2019.06.009.
85. Venkateshalu, S.; Cherusseri, J.; Karnan, M.; Kumar, K. S.; Kollu, P.; Sathish, M.; Thomas, J.; Jeong, S. K.; Grace, A. N. New Method for the Synthesis of 2D Vanadium Nitride (MXene) and Its Application as a Supercapacitor Electrode. *ACS Omega* **2020,** *5* (29), 17983–17992. https://doi.org/10.1021/acsomega.0c01215.
86. Alam, A.; Saeed, G.; Lim, S. One-step Synthesis of 2D–2D $Co(OH)_2$–$MoSe_2$ Hybrid Nanosheets as an Efficient Electrode Material for High-performance Asymmetric Supercapacitor. *J. Electroanal. Chem.* **2020,** *879,* 114775. https://doi.org/10.1016/j.jelechem.2020.114775.

CHAPTER 8

Bismuth Ferrite/Barium Titanate ($BiFeO_3/BaTiO_3$) Thick Films for High-Energy Applications: A Review

DHANYA RAJ, T. V. VENKIDESH, and M. V. MANJUSHA

Department of Physics, the Cochin College, Kochi, Ernakulam, Kerala, India

ABSTRACT

Previous studies have focused primarily on lead-based fuels. Research over the past decades has shown that lead-containing ferroelectric components possess high energy storage densities and high energy efficiencies. However, they consist of solids of complex composition. These are very toxic substances and using them harms nature. The main focus of subsequent research has been to find memories that are easy to manufacture, naturally fit, and cheap to manufacture. A new material is needed that overcomes all these shortcomings. Single-component lead-free perovskites have been investigated by researchers for their application in thick-film dielectric capacitors, where we can see the importance of nanotechnology. Among them, the most well-known thick films are barium titanate ($BaTiO_3$) and bismuth ferrite ($BiFeO_3$). Both these materials have very good dielectric properties. A material with a high dielectric breakdown electric field, excellent impact resistance, and relative permittivity has low permanent polarization and high saturation polarization. These are the ideal conditions for highly efficient energy storage devices. Bismuth ferrite exhibits a largely saturated polarization, allowing it to be tailored in terms of electrical properties via strain or chemical doping. Barium titanate has excellent dielectric properties. Its relative dielectric constant is very large and the value of the

Physics and Mechanics of New Materials: Synthesis, Processing, and Emerging Applications. Devrim Balkose, Ana Cristina Faria Ribeiro, Nandakumar Kalarikkal, Ann Rose Abraham, A. K. Haghi (Eds.)

breakdown field is large; hence, it is extensively used in multi-layer ceramic capacitors (MLCC). The development of lead-free dielectric capacitors with high energy storage density and efficiency can be highly promoted using $BiFeO_3/BaTiO_3$ bilayer thick films, and this review discusses the progress in this field. Bilayer thick films are deposited onto $Pt/Ti/SiO_2/(100)$ Si substrate using $LaNiO_3$ buffer. As a result of this thick film's low remnant polarization and largely saturated polarization, it exhibits a slim polarization electric field hysteresis loop that enhances its energy storage capability. Many studies on bilayer films have shown that they have excellent energy storage properties. In this researched field, many papers have come out on how to effectively use bismuth ferrite—barium titanate bilayer thick films; this paper aims to summarize the reading of these papers.

8.1 INTRODUCTION

Energy is a very important and necessary factor in our daily life. In recent decades, the world's population has grown, and so have people's needs. Population growth, massive climate change, depletion of fossil fuels, and pollution of natural resources require the development of environment friendly and clean renewable energy sources. Reducing CO_2 emissions and fossil fuel use requires more reliance on renewable and sustainable energy sources such as solar, wind, tidal, and hydropower, which are only intermittently available. Energy storage and energy conversion technology are the biggest challenges of this century. It also opens up promising possibilities for the construction and development of energy storage systems. Electrochemical energy storage (EES) plays an important role in the energy storage technologies available to us. Electrochemical capacitors and rechargeable batteries are the most promising energy storage devices. Their energy production and power density are very high and their cycle stability is excellent.[1-5]

Electrochemical energy storage technology includes lead-acid batteries. Lead-acid battery technology has long been used for energy storage. They are very cheap to manufacture and easy to maintain, but they pollute the environment, and have very low energy density and life expectancy. A major shift in energy storage was brought about by the advent of lithium-ion batteries. Their main advantages are high energy density, self-discharge properties, low maintenance, and load characteristics. However, the production cost is very high and ageing is also a drawback. Studies over the past decades have shown that lead-containing ferroelectric films have very high energy efficiency and energy storage density. However, increasing environmental concerns from their use have led to a search for storage devices that are suitable for manufacturing by reducing their environmental impact.

Recently, many researchers around the world are working to create energy storage devices that are harmless to nature, easy to manufacture, and lead-free. This is where nanotechnology applications come into play. Improved ionic transport and electronic conductivity are prominent features of nanomaterials compared to existing batteries and supercapacitors using nanomaterials. Electroceramics are a class of materials with unique electrical, optical, and magnetic properties used in communications, power conversion, energy storage, and many other electronic applications. Dielectrics, ferroelectrics, piezoelectric, pyroelectrics, magnetic ceramics, and ceramic ionic conductors are all variants of electroceramics. Dielectric capacitors have a very large electrostatic storage capacity and can release stored energy very quickly. The most popular materials for dielectric applications are barium titanate BT-($BaTiO_3$), lead zirconate titanate PZT-($Pb(Zr, Ti)O_3$), and bismuth ferrite BF-($BiFeO_3$). Of these, BT is used for manufacturing multilayer ceramic capacitors, PZT is used for pyro, piezo, and ferroelectric applications, and BF is used for ferroelectric relaxor applications.

Bismuth ferrite/barium titanate ceramic is an electroceramic developed to replace lead zirconate titanate. Many such publications have shown that the $BiFeO_3/BaTiO_3$ system is a highly reliable and efficient ceramic energy storage system. This review focuses on the study of energy storage properties applied to $BiFeO_3/BaTiO_3$ bilayer thick films, their applications, and new developments. In this study, we found that the $BiFeO_3/BaTiO_3$ bilayer thick films have improved energy density and charge–discharge efficiency superior to single-layer thin films of this material. Therefore, the $BiFeO_3/BaTiO_3$ bilayer thick film is a great innovation in the fabrication of lead-free energy storage devices.[6-10]

8.2 HIGH-ENERGY APPLICATIONS OF BFO/BTO

Nanoscience and nanotechnology are growing areas of science. The very rapid development of nanoscience and nanotechnology allows us to expand into all industries. The possibilities it brings us are not limited to any particular one. Today, nanotechnology is successfully used in all fields, and energy storage is one of the key areas. Rational material structure design using nanotechnology has been suggested by Lele Peng et al. to improve the overall performance of electronic energy storage devices. Nanomaterials with particle sizes in the hundreds of nanometer range offer excellent ion storage specific surface area (SSA), excellent ion exchange, surface redox reactions, and shorter ion transport distances than bulk materials. These criteria improve attributes such as capacity and rate capability. As a 2D nanomaterial, this material has excellent charge storage capability due to its extremely high SSA and high aspect ratio.[11-15]

Dielectrics are materials with very high electrical resistance. Their specific electrical resistance is close to 108 Ωm. The formation or reorientation of an electric field creates lattice polarization that can be used to store electrical energy. When a dielectric is placed under an electric field, the electric field is developed by the polarization of the atoms and molecules in the material, which oppose the applied electric field. For these reasons, dielectric capacitors can deliver charge very quickly. This distinguishes dielectric materials from conventional batteries. In conventional batteries, the release of charge is slow due to the use of chemical reactions, the dielectric has a longer life compared to conventional batteries, and the short life of conventional batteries is due to irreversible chemical reactions. The energy density stored or released in a dielectric can be determined from its characteristic dielectric displacement versus the electric field curve.

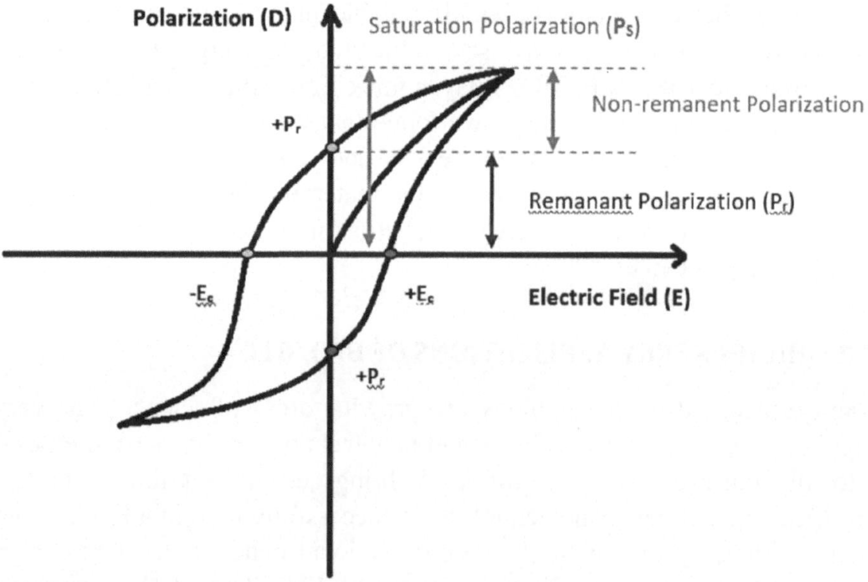

FIGURE 8.1 Schematic illustration of ferroelectric hysteresis loop showing total polarization in a ferroelectric material.

The stored or charged energy density is W_c and is given by

$$W_c = \int_0^{D_s} E\,dD \tag{8.1}$$

The released or discharged energy density will be W_d, and it is given by

$$W_d = \int_{D_r}^{D_s} E\,dD \tag{8.2}$$

where D_s is the saturated or maximum electrical displacement and D_r is the residual electrical displacement.

This gives you energy efficiency,

$$\eta = \frac{W_d}{W_c} \tag{8.3}$$

For an ideal linear dielectric with relative permittivity ε_r,

$$D = \varepsilon_0\ \varepsilon_r\ E \text{ and } D_r = 0 \tag{8.4}$$

Energy density is

$$W_c = \frac{1}{2}\varepsilon_0\varepsilon_r E_{max}{}^2 \tag{8.5}$$

$$W_d = W_c = \frac{1}{2}\varepsilon_0\varepsilon_r E_{max}{}^2 \tag{8.6}$$

Efficiency $\eta = 1$

where ε_0 is the vacuum permittivity and E_{max} is the maximum applicable electric field, which increases with the ideal breakdown field E_b. For an ideal dielectric $E_{max} \approx E_b$. For nonlinear dielectrics, such as ferroelectrics, we usually do not have $D \propto E$ or $D_r = 0$, so $W_c > W_d$. Therefore, $\eta < 1$. However, eq 8.5 can be used to estimate the energy density W_c of nonlinear dielectrics.

The polarization is given by

$$P = \varepsilon_0\ \varepsilon r\ E \tag{8.7}$$

So, for dielectric ceramic materials to achieve good electrical density, the material must have a high εr, a low remanent polarization P_r, a high saturation polarization P_s, and higher breakdown strength (BDS). A key determinant of the electrical density properties of a good electrical energy storage system is the material's BDS.

Bismuth ferrite ($BiFeO_3$) is one of the most widely studied multiferroic magnetoelectric material, rich in dielectric and ferroelectric properties, and research into ceramic capacitors made of it is increasing day by day. R.K. Kotnala et al. proposed that it is mainly used for energy storage applications due to its high polarization and ferroelectric properties. Its Curie temperature is very high, around $T_c = 1043$ K, and its Neel temperature is around $T_n = 647$ K. Due to the high Curie temperature, the ferroelectricity of this material is not lost even at high temperatures. Energy storage applications can be used for spintronics applications at room temperature. Bismuth ferrite is different from other materials for several reasons. At room temperature, bismuth ferrite has a rhombohedral structure with point group R3C ($\alpha = \beta = \gamma = 59.4°$, a = b = c = 5). This material can achieve ferroelectricity due to the 6s lone pair of Bi and magnetic ordering due to the partially filled d orbitals of Fe.

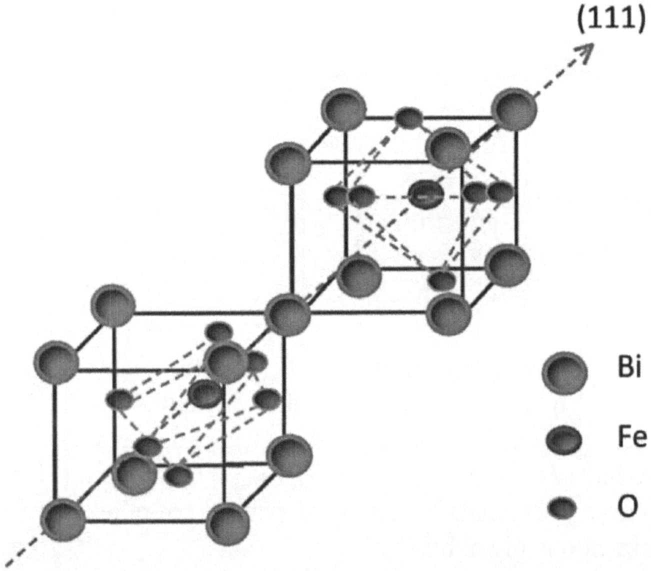

FIGURE 8.2 Schematic view of the R3C structure built up from two cubic perovskite $BiFeO_3$ unit cells.

Barium titanate ($BaTiO_3$) is a material that is primarily used effectively in the electronic device and ceramics industries; its ferroelectric, thermoelectric, and piezoelectric properties make it a useful material in these areas. Barium titanate is a very strong dielectric material with a perovskite structure and is used in the electronics industry for the production of transducers, actuators, capacitors, and high-k dielectrics. Due to its high dielectric constant, ceramic dielectric capacitors, capacitors, and ferroelectric memories can be made. These uses make barium titanate a more attractive material. Barium titanate has a large bandgap (~3.3 eV), which is larger than the energy corresponding to the wavelength of 405 nm. If the optical bandgap can be reduced, this material can be effectively used for ferroelectric photovoltaic effect (FEPV) applications in the visible region while maintaining ferroelectric properties. The multilayer structure exhibits superior ferroelectric properties compared to the single barium titanate starting material.

The launch of the bismuth ferrite-barium titanate (BF-BT) thick-film system is not limited to piezoelectric applications but is aimed at deriving a lead-free system that is equally applicable to all applications. It is an alternative to PZT. Dawei Wang et al. suggested that $BaTiO_3$ and $BiFeO_3$ each have a T&R structure at room temperature and the latter has a high critical temperature of about T_c = 825°C. In the BF-BT system, there are multiple ion species occupying the A and B sites, as opposed to only the B sites in PZT.

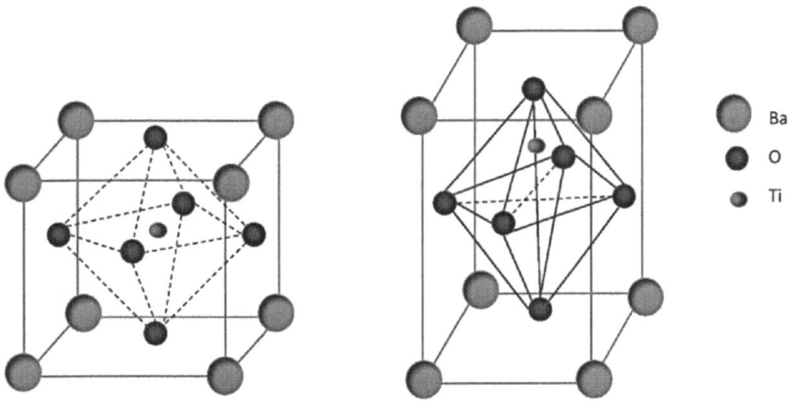

FIGURE 8.3 Schematic illustration of the perovskite structure of $BaTiO_3$. (a) Cubic lattice (above Curie temperature, >120°C). (b) Tetragonal lattice (above Curie temperature, <120°C).

In the BF-BT system, Ba is a large blocking ion that facilitates the displacement of the Bi lone pair species. This is ideal for high energy density storage, electrocaloric and pyroelectric applications. This makes the BF-BT system a promising candidate for effective lead-free ceramics that can be used for centuries. Studies have shown that ferroelectric random-access memories (FeRAM) and microelectromechanical systems (MEMS) can be designed using BF-BT films, and the temperature dependence of ε_r between the films and the bulk of the same composition has been investigated. It is observed that the dielectric anomaly of the film is wider than that of the bulk. This is due to the higher volume fraction of grain boundaries compared to the bulk and the difference in thermal expansion between the foil and the substrate. Increasing the number of layers in the BF-BT film system (monolayer, bilayer, multilayer) significantly changes the dielectric and energy storage properties.

Liu Menglin et al. fabricated $BiFeO_3/BaTiO_3$ thick bilayer films with a thickness of about 1 μm. They fabricated this bilayer thick film by the magnetron sputtering method, $BiFeO_3/BaTiO_3$ bilayer thick film was fabricated on a Pt/Ti/SiO_2/(100) Si substrate buffered with $LaNiO_3$ layer, and the ceramic target was $BaTiO_3$, $Bi_{1.05}FeO_3$, and $LaNiO_3$. It was used in the deposition process and carried in a multi-target magnetic sputtering system. $LaTiO_3$, $BaTiO_3$, and $BiFeO_3$ layers were then sequentially sputtered with a gas mixture of Ar and O_2, and the deposited multilayer thick films were cooled with pure oxygen. A circular top electrode was used for electrical characterization. Temperature is measured using an evaporated shadow mask.

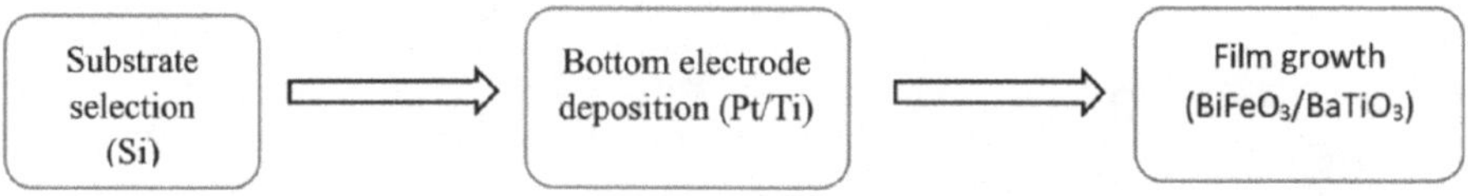

FIGURE 8.4 Schematics of experimental procedure.

X-ray diffraction (XRD) analysis was performed to determine the crystal orientation and phase structure characteristics of the $BiFeO_3/BaTiO_3$ double-layer thick film. Field-emission scanning thermal microscopy (SEM) and scanning probe microscopy were used to determine the cross-sectional morphology of the film. Scanning probe microscopy was used to examine the surface morphology, and its dielectric properties were determined by Menglin Liu et al. It was done with a high-precision digital bridge. The ferroelectric hysteresis loop (P-V) and leakage current (I-V) of the bilayer film at room temperature were determined using a Radiant Precision Premium ferroelectric tester. From the XRD analysis performed to study the microstructure and crystallographic orientation, we were able to analyze the XRD peaks of the bilayer thick films. It is dominated by the preferred (001) peak of the bulk perovskite structure and shows no crystals or secondary impurity phase. Menglin Liu et al. confirmed that the presence of the (100)-$LaNiO_3$ buffer layer facilitated the growth of the (001) oriented $BiFeO_3/BaTiO_3$ bilayer film. XRD investigations revealed that all the grains in the upper $BiFeO_3$ layer were densely packed with an average size of 200–300 nm.

Liu Menglin et al. were able to determine the energy storage density W_c and energy efficiency of Au/$BiFeO_3$/$BaTiO_3$/$LaNiO_3$/Pt/T capacitors from the hysteresis loop of the polarization electric field (P-E). They found a large maximum polarization value of about $P_m \sim 65\ \mu C/cm^2$ and a small remanent polarization value of $P_r \sim 7.8\ \mu C/cm^2$ for the bilayer membrane, compared to a reduced field coercive field $E_c \sim 15\ 2kV/cm^2$. From these characteristics, the film is known to have excellent energy storage properties.

Compared with pure $BaTiO_3$ deposited under the same conditions, we find that the energy storage density W_c of the bilayer film reaches 71 J/cm^3. There is an improvement from 100% to 101%, but the energy efficiency remains at the same level (60%–70%). This W_C improvement is attributed to a significantly improved maximum polarization and a high dielectric strength comparable to that of pure $BaTiO_3$ films. They found that the "slimming" of these P–E loops in bilayer films was due to competition between space-charge effects and charge coupling between the layers. Therefore, the improved energy storage ability of $BiFeO_3/BaTiO_3$ bilayer films may be attributed to the combination of the space-charge effect and charge-coupling effect between the layers, which is dominant at low electric fields. The interfacial energy barrier between the $BiFeO_3$ and $BaTiO_3$ layers reduces the leakage current and exposes the films to large electric fields, thus making it possible to achieve high energy densities in the $BiFeO_3/BaTiO_3$ bilayer films.

8.3 CONCLUSION

As there are many studies currently being conducted on the need for $BiFeO_3/BaTiO_3$ bilayer lead-free thick films in energy storage applications, we wanted to review such studies. Compared with the $BiFeO_3/BaTiO_3$ monolayer, we find that the bilayer has much better energy charge density and energy efficiency. The reason for this intervention is that it is the combined effect of interlayer space charge and interlayer charge coupling. It is also a characteristic that these properties increase as the number of layers increases. The dielectric constant Er is 33% higher than the pure $BaTiO_3$ film and 70% higher than the $BiFeO_3$ film. Your leakage current is very low. Therefore,

$BiFeO_3/BaTiO_3$ bilayer films exhibit excellent and efficient dielectric performance and energy storage properties, making the films promising materials for lead-free thick-film capacitor applications and also in the range of electrocaloric and pyroelectric applications.

KEYWORDS

- **energy storage**
- **$BiFeO_3$**
- **$BaTiO_3$**
- **ferroelectric**
- **bilayer**
- **thick film**

REFERENCES

1. Deka, B.; Cho, K.-H. $BiFeO_3$-Based Relaxor Ferroelectrics for Energy Storage: Progress and Prospects. *Materials* **2021,** *14* (23), 7188. https://doi.org/10.3390/ma14237188
2. Kholodkova, A. A.; Danchevskaya, M. N.; Ivakin, Y. D.; Muravieva, G. P.; Smirnov, A. D.; Tarasovskii, V. P.; Ponomarev, S. G.; Fionov, A. S.; Kolesov, V. V. Properties of Barium Titanate Ceramics based on Powder Synthesized in Supercritical Water. *Ceram. Int.* **2018,** *44* (11), 13129–13138. https://doi.org/10.1016/j.ceramint.2018.04.135
3. Lin, Y.; Li, D.; Zhang, M.; Zhan, S.; Yang, Y.; Yang, H.; Yuan, Q. *Excellent Energy Storage Properties Achieved in $BaTiO_3$-based Lead-Free Relaxor Ferroelectric Ceramics via Domain Engineering on the Nanoscale.* n.d.; 7.
4. Liu, M.; Zhu, H.; Zhang, Y.; Xue, C.; Ouyang, J. Energy Storage Characteristics of $BiFeO_3/BaTiO_3$ Bi-Layers Integrated on Si. *Materials* **2016,** *9* (11), 935. https://doi.org/10.3390/ma9110935
5. Liu, T.; Wang, W.; Qian, J.; Li, Q.; Fan, M.; Yang, C.; Huang, S.; Lu, L. Excellent Energy Storage Performance in $Bi(Fe_{0.93}Mn_{0.05}Ti_{0.02})O_3$ Modified $CaBi_4Ti_4O_{15}$ Thin Film by Adjusting Annealing Temperature. *Nanomaterials* **2022,** *12* (5), 730. https://doi.org/10.3390/nano12050730
6. Liu, Z.; Zhang, Z.; Zhang, F.; Li, Y. Enhanced Energy Storage Properties of Barium Titanate Ceramics Made from Surface Modified Particles. *J. Adv. Dielectr.* **2013,** *3* (3), 1350023. https://doi.org/10.1142/S2010135X13500239
7. Sharma, S.; Tomar, M.; Puri, N. K.; Gupta, V. $BiFeO_3/BaTiO_3$ Multilayer Structures for Solar Energy Harvesting Application. *Energy Harvest. Syst.* **2016,** *3* (3), 237–243. https://doi.org/10.1515/ehs-2016-0001

8. St. Peterburg State University, 7/9, Universitetskaya nab., 199034 St. Petersburg, Russia; Sarnatsky, V. M.; Vinokurov, N. A.; St. Peterburg State University, 7/9, Universitetskaya nab., 199034 St. Petersburg, Russia, Murlieva, Zh. Kh., Dagestan State University, 43, Gadzhiyev st., 367000 Makhachkala, Russia, Alikhanov, N. M.-R.; Dagestan State University, 43, Gadzhiyev st., 367000 Makhachkala, Russia. Magnetic and Electrical Characteristics of Bismuth Ferrite, Depending on the Impurities, Method of Preparation and Size of the Nanoparticles. *J. Nanoelectron. Phys.* **2016,** *8* (3), 03008-1-03008-4. https://doi.org/10.21272/jnep.8(3).03008
9. Veerapandiyan, V.; Benes, F.; Gindel, T.; Deluca, M. Strategies to Improve the Energy Storage Properties of Perovskite Lead-Free Relaxor Ferroelectrics: A Review. *Materials* **2020,** *13* (24), 5742. https://doi.org/10.3390/ma13245742
10. Verma, N.; Singh, S.; Yadav, B. C. Experimental Investigations on Barium Titanate Nanocomposite Thin Films as an Opto-electronic Humidity Sensor. *J. Exp. Nanosci.* **2014,** *9* (7), 669–677. https://doi.org/10.1080/17458080.2012.689433
11. Wang, D.; Wang, G.; Murakami, S.; Fan, Z.; Feteira, A.; Zhou, D.; Sun, S.; Zhao, Q.; Reaney, I. M. $BiFeO_3$-$BaTiO_3$: A New Generation of Lead-free Electroceramics. *J. Adv. Dielectr.* **2018,** *8* (6), 1830004. https://doi.org/10.1142/S2010135X18300049
12. Wang, Y.; Yao, M.; Ma, R.; Yuan, Q.; Yang, D.; Cui, B.; Ma, C.; Liu, M.; Hu, D. Design Strategy of Barium Titanate/Polyvinylidene Fluoride-based Nanocomposite Films for High Energy Storage. *J. Mater. Chem. A* **2020,** *8* (3), 884–917. https://doi.org/10.1039/C9TA11527G
13. Zou, K.; Dan, Y.; Xu, H.; Zhang, Q.; Lu, Y.; Huang, H.; He, Y. Recent Advances in Lead-free Dielectric Materials for Energy Storage. *Mater. Res. Bull.* **2019,** *113,* 190–201. https://doi.org/10.1016/j.materresbull.2019.02.002
14. Korkmaz, S.; Kariper, A. $BaTiO_3$-based Nanogenerators: Fundamentals and Current Status. *J. Electroceramics* **2022,** https://doi.org/10.1007/s10832-021-00266-3
15. Shima, H.; Naganuma, H.; Okamura, S. Optical Properties of Multiferroic $BiFeO_3$ Films. In *Materials Science—Advanced Topics*; Mastai, Y., Ed.; IntechOpen, **2013.** https://doi.org/10.5772/54908

SECTION 3

Case Studies in Thermophysics and Applied Mechanics for Engineering Technology

CHAPTER 9

Investigating the Adhesion Problem of a Double-Sided Tape and Car Bumper Packing Materials

DEVRIM BALKÖSE, GIZEM CIHANOĞLU, SEMA KIRKÖSE, MERVE KARABIYIK, and AYBIKE NIL OLCAY

İzmir Institute of Technology, Urla İzmir, Turkey

ABSTRACT

Pressure-sensitive adhesives (PSAs) are viscoelastic materials with permanent stickiness, and can adhere strongly to solid surfaces upon application of slight contact pressure under a relatively short contact time. The adhesion problem of double-sided adhesive to packing films of car bumpers was investigated by analyzing the flexible adhesive layer and two alternative packing films. The films were blue in color and one of them was opaque and the other one was transparent. Scanning electron microscopy (SEM), Fourier transfer infrared (FTIR) spectroscopy were used to characterize the samples. Water contact angle and roughness of surfaces were also determined. It was shown that the film with high surface roughness and high water contact angle adhered better to the flexible adhesive layer.

9.1 INTRODUCTION

Pressure-sensitive adhesives (PSAs) are a class of materials that have the ability to adhere rapidly to any surface under a minimum pressure load. The term pressure-sensitive defines adhesives which are permanently sticky in the dry condition at temperature of room and gently adhere to different

Physics and Mechanics of New Materials: Synthesis, Processing, and Emerging Applications. Devrim Balkose, Ana Cristina Faria Ribeiro, Nandakumar Kalarikkal, Ann Rose Abraham, A. K. Haghi (Eds.)

ranges of various surfaces without the need of hand pressure.[1] Silicone pressure-sensitive adhesives were proposed in 60s. They revealed acceptable ranges of properties and are extensively used.[2] The nano filler montmorillonite was adjoined to aforesaid adhesives. Different types of loadings of this component were studied and their effects on properties such as tack, viscosity, adhesion, cohesion at different temperatures, and the maximum ranges of working temperature were examined. The influence of CNCs (cellulose nanocrystals) as a property changer for PSAs (pressure-sensitive adhesives) was studied in detail by Dasterji et al.[3]

To produce the high-performance semistructural double-sided adhesive tape, the 2-ethylhexyl acrylate (2-EHA) and AAC (acrylic acid), along with the syrup, that was prepared by UV irradiation, as suggested by Kim were used.[4] The thickness impacts along with different inorganic filler silica contents, and filler size on the semistructural properties of acrylic double-sided adhesive tape were studied in detail in reference.[4] Silicone, acrylic, rubber, epoxy, hybrid polymers are used as base polymers in pressure-sensitive adhesives.[5]

Surface development, like wettability along with cohesion and adhesion, play an eminently dominant role in producing adhesive joints, as they influence the capability of generating such a joint and the quality.[6] Adhesion is the attraction force among substrate and adhesive, which relies on the substrate along with the application conditions like: time period of pressure, along with the temperature during application pressure. Cohesion explains the inner strength of an adhesive. It is appropriate for applications where the force is considered as parallel to the bonding area. Tack is the dominant property of a PSA and is defined as wetting capability or the adhesive capability to form a bond of measurable strength to another surface under conditions of low contact pressure and short contact time. Considerations of surface energies are usually regarded as being fundamental in understanding adhesion.[6] Wetting is in fact a surface phenomenon which rests in replacing the surface of the solid and the liquid with a boundary surface, distinguished by specific tension (σ), which results from the difference in the surface tension among the solid and liquid in the gaseous medium.[7] It is a procedure that explains the diffusion of a liquid (adhesive) over a solid surface (substrate), creating close contact between them. The proper wettability of a surface is essential for a proper adhesive bonding. Wettability is a key problem in the case of forming adhesive joints (due to direct effects of adhesion, by increasing or decreasing adhesion forces). One of the normal methods to examine the wettability of a surface or material is the use of contact angle. As a matter of fact, the contact point of the liquid and the solid surface is considered as the angle formed between the wetted solid surface and the tangent to the wetting liquid surface. Any time, wetting the surface of a solid, the value of contact angle could be in the range of lower than 90°. Any time, however, the contact angle θ is equal to 0°, then we have to expect that the liquid spreads over the surface evenly.

Sowa et al.[8] studied the adhesion of solvent-borne acrylic pressure-sensitive adhesives (PSA) cross-linked using crosslinking agent aluminum acetylacetonate (AlACA). Again, they used contact angle measurements to obtain surface free energy. Contact angle values for polymer increase when polarity of liquid increases. They concluded that adhesion is essentially related to surface-free energy.[8] Correlation between surface tension, work of adhesion in liquid metals/ceramic systems was investigated by Hadef et al.[9] They derived the formulas for the prediction of surface tension and work of adhesion of such metal/ceramic interfaces depending on the elastic and acoustic characteristics. Work of adhesion and surface tension of soft polymeric materials were investigated by Tian et al.[10] The effects of substrate surface properties on the tack performance of acrylic pressure-sensitive adhesives (PSAs) were investigated by Kowalski and Czech.[11]

In the automotive industry, double-sided tape was used in the car bumper packaging process, and in this process the surface of the packaging material is adhered to each other via a double-sided tape. In this paper, the characterization of the two alternative packing films and the pressure-sensitive adhesive was aimed at using Fourier transform infrared spectroscopy (FTIR) and scanning electron microscopy (SEM). The contact angle of water on the surfaces and the surface roughness values were also measured to explain the problem of the debonding of the adhesive and the films.

9.2 EXPERIMENTAL METHODS AND MATERIALS

9.2.1 MATERIALS

The materials used in the present study are taken from the packing industry which packs car bumpers in plastic envelopes. In Figure 9.1, the packing process of a car bumper is shown. As seen in Figure 9.1a, the envelope consists of plastic film coated inside with plastic bubble-foamed material. One surface of the adhesive stuck on the polymer material of the envelope by the workers, while the other side of the adhesive stuck on the carrier film of the adhesive. The carrier film was removed after placing the bumper in the envelope as seen in Figure 9.1b and the lid of the envelope is folded and the adhesive was pressed manually to close the envelope as seen in Figure 9.1c. The blue plastic film of the envelope was made from two alternative materials from the market. The first one was an opaque blue film supplied by Elif Plastik and the second one is a transparent blue film supplied by Naksan. The companies claimed that they were identical in chemical composition and thickness. The car bumpers packed in envelopes are placed in containers and exported to different countries. While the envelopes with opaque films were stable, the envelopes using transparent films were self-opened during the transportation leading to a false impression of reused car bumpers. Thus, the reason of the sticking problem of the envelopes by the double-sided adhesive is investigated in the present study.

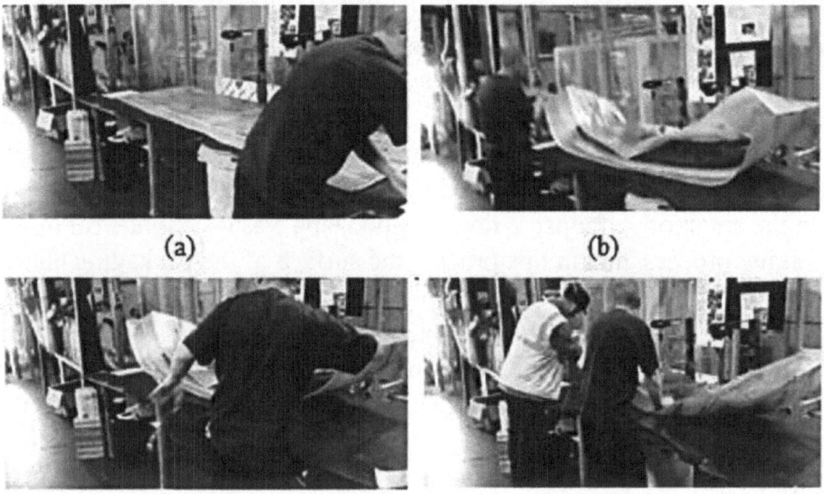

(a) (b) (c) (d)

FIGURE 9.1 The packing process of car bumper: (a) the plastic envelope, (b) car bumper placed in the envelope, (c) placing double-sided band to close the envelope, (d) closing the envelope by manual pressing.

The double-sided adhesive and the blue polymer films which are adhered with the double-sided adhesive are shown in Figure 9.2.

9.2.2 METHODS

In this study, Shimadzu-8400S FTIR spectrophotometer was used as an instrument for the transmission technique of infrared spectroscopy measurement.

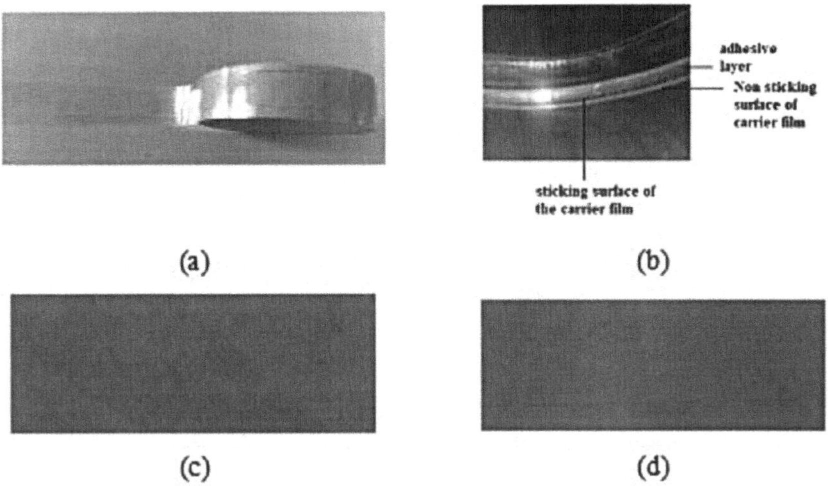

FIGURE 9.2 (a) A roll and (b) crossection of double-sided adhesive, (c) opaque, (d) transparent blue film.

Digilab Excalibur Series FTIR spectrophotometer was used as an instrument for the specular reflection technique using PIKE-80 Spec accessory.

The microstructures of samples were investigated with scanning electron microscopy (SEM), FEI Quanta250, with a field emission gun. Before viewing the samples, sputter coater was used to coat the surface of samples with this layer of gold at 15 mA under vacuum condition at $8 \times 10{-}^2$ mbar during 75 s to increase surface conductivity. In order to examine the interface between blue film and the adhesive of double-sided film, the morphological analysis was carried out. Before the investigation of morphology of samples, all samples were cut using scissors after immersing in liquid nitrogen. The interfaces between double-sided adhesive and opaque blue film and double-sided adhesive and transparent film were also examined by scissors cutting of liquid nitrogen immersed samples.

Mitutoyo surftest SJ-301 profile meter was used to determine the surface roughness of opaque and transparent films. Arithmetical mean roughness (R_a), maximum height (R_y), ten-point mean roughness (R_z), and root mean square roughness (R_q) were reported as the results of roughness measurement.

Attension Optical Tensiometer was used as water contact angle measurement of opaque and transparent blue films and the adhesive layer of the pressure-sensitive adhesive.

9.3 RESULTS AND DISCUSSION

9.3.1 FTIR SPECTROSCOPY

9.3.1.1 TRANSMISSION SPECTRA OF TRANSPARENT AND OPAQUE FILMS

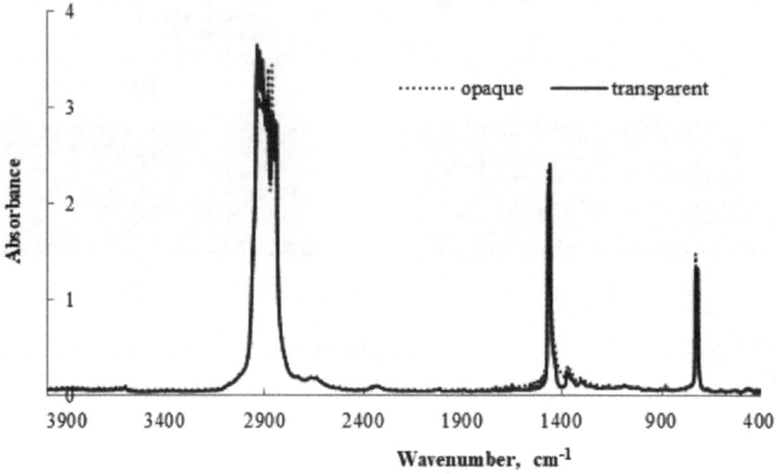

FIGURE 9.3 Transmission FTIR spectrum of tranparent and opaque alterrnative packing films.

In Figure 9.3, for both transparent and opaque films polyethylene characteristic peaks at 1463, 1377 and also between 731 and 720 cm^{-1} can be seen clearly. It can be said that both films were low density polyethylene (LDPE) because the individual peak at 1377cm^{-1} can be observed clearly. If it was not observed in the spectrum, then the material would be high density polyethylene (HDPE).[12]

9.3.1.2 SPECULAR REFLECTANCE SPECTRA OF TRANSPARENT AND OPAQUE FILMS

In the spectra in Figure 9.4 obtained by using the specular reflectance technique of the transparent and opaque blue films, peaks were obtained at the same frequencies but the peak intensities in the transparent film were higher than the opaque film. In addition to this, the peaks in the spectrum of the bubble film were observed at nearly the same wave numbers with both transparent and opaque films, but intensity was higher than the blue films. Furthermore, in the specular reflectance spectrum of transparent opaque and bubble films, the characteristic peaks of the LDPE were observed at

1463 and 720 cm^{-1} like in their transmission spectra. And also the prominent peaks between 2950 and 2800 cm^{-1} were observed clearly for opaque film like in the ATR spectrum of Gulmine et al's[12] and in the reference LDPE specular reflectance spectrum.[13]

Additionally, in the opaque blue film fringes were observed clearly when compared with the transparent blue film. The fringe pattern takes place due to constructive and destructive inference caused by reflection from both sides of the thin film. This effect is clearly seen in the reflection–absorption spectrum of polyethylene film in Figure 9.4.

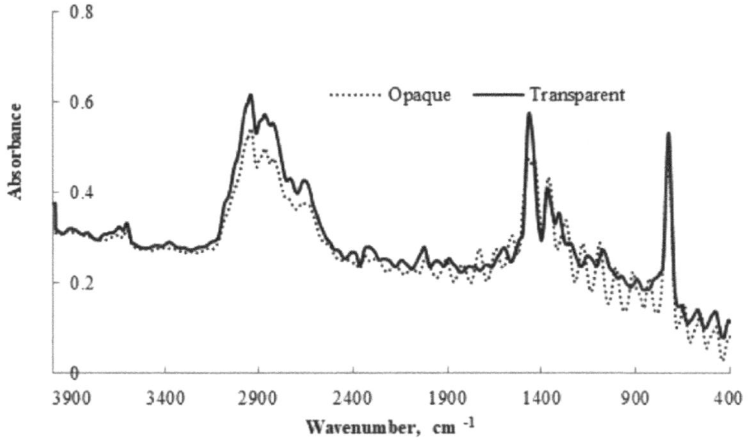

FIGURE 9.4 Specular reflectance spectra of opaque and transparent films.

Therefore, the thickness calculation was done for only opaque blue film by using eq 9.1 described in reference[13] for the polyethylene.

$$b = \frac{1}{2n} x \frac{N}{(\upsilon_1 - \upsilon_2)} \quad (9.1)$$

where b is film thikness, n refractive index of sample, N number of fringes within a given spectral region, υ_1 and υ_2: start and end points in the spectrum cm–^{1}.

In the specular reflectance spectrum of the opaque film, in Figure 9.4 for the region between 2013 and 1556 cm–^{1} there are 6 fringes. Taking the refractive index of the polyethylene as 1.5[14] the film thickness is calculated as 43.9 μm using eq 9.1. For the region between 1271 and 817 cm–^{1} the film thickness is found as 44.0 μm using the same equation. These results show that the thickness of the opaque blue film was found as approximately 44 μm.

9.3.1.3 TRANSMISSION SPECTRUM OF ADHESIVE LAYER OF DOUBLE-SIDED FILM

According to Figure 9.5, the peaks that belong to both polystyrene (PS) and polybutadiene (PB) were observed in the transmission spectrum of double-sided film.[15] The peak at 3647cm$-^1$ was attributed to the presence of isolated hydroxyl groups. The peak at 1942 cm$-^1$ occurs owing to the PS unit. The unique peaks at 1601 and 1580 cm$-^1$ were caused by the aromatic C=C stretching vibration in the PS unit. Additionally, the peak at about 1666 cm$-^1$ assigned to stretching vibrations ν (C=C), at between 1467 and 1435 cm$-^1$ attributed to bending vibrations δ (CH_2) and also at 970 cm^1 implied to =CH- groups in PB unit. The results indicate that the adhesive part of the double-sided tape seems to be made from the styrene–butadiene copolymer because it contains the unique peaks of both of them. However, the isolated OH group peak at 3647 cm$-^1$ and stretching vibrations ν (C=O) at 1743 cm$-^1$ indicated the presence of other additives in the adhesive layer of the double-sided film.

The specular reflectance FTIR spectrum of the adhesive layer of the double-sided film is seen in Figure 9.6. When this figure was examined, it was seen that the specular reflectance spectrum of this sample was nearly the same with the transmission spectrum. The only difference observed between them is that in the specular reflectance method, the intensity of the peaks was obtained as higher than in the transmission technique. On the other hand, the characteristic peaks of styrene-butadiene copolymer were clearly observed in the spectrum above when this method was applied to the sample like the transmission method.

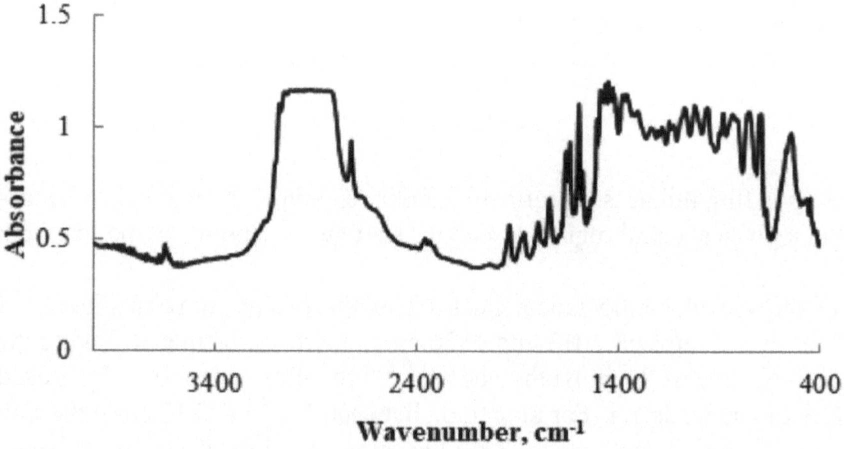

FIGURE 9.5 Transmission spectrum of adhesive layer of the double-sided tape.

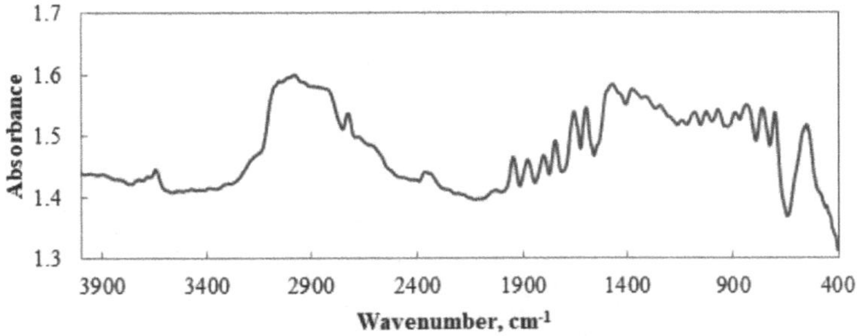

FIGURE 9.6 Specular reflectance FTIR spectrum of the adhesive layer of the double-sided film.

9.3.2 *MORPHOLOGY OF THE FILMS BY SEM*

The cross sections and surfaces of the films were examined by SEM. The films were scissor-cut after immersing in liquid nitrogen to prevent any deformation such as fibrillation or buckling. The film thickness of the opaque blue film is measured as 55.95 μm by SEM as seen in Figure 9.7b.

The surface of the films was examined just after coating them with gold. As seen in Figures 9.7c and 9.7d, the surface of the opaque film is not smooth. In addition to irregularities created by the polymer phase, some inorganic crystals with sharp edges were present making the surface rougher.

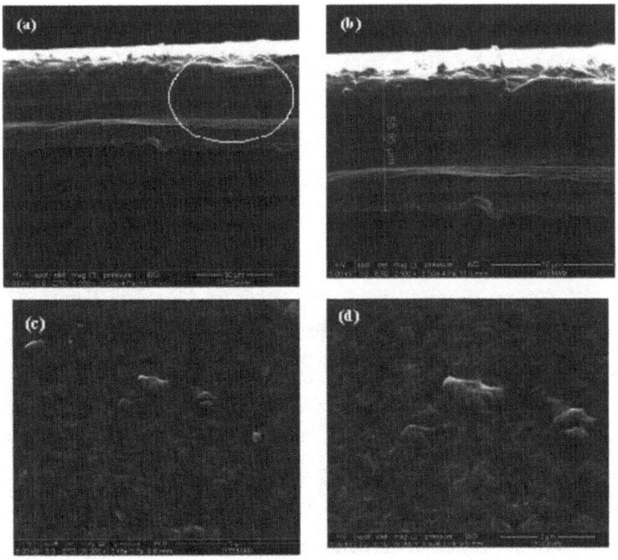

FIGURE 9.7 The SEM images of a. the cross section of opaque blue film prepared breaking in liquid nitrogen at a. 1500× b. 2500× magnification and the top view of the opaque film at 25,000× and 50,000× magnification.

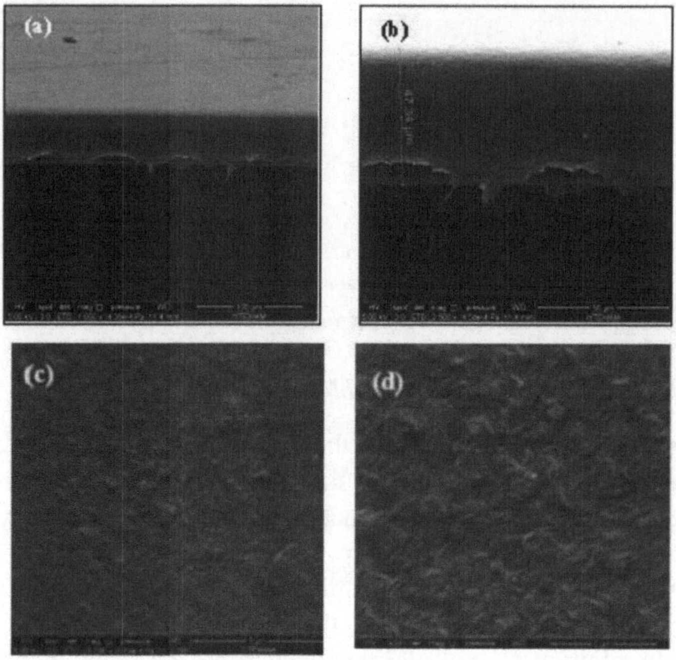

FIGURE 9.8 The SEM images of the cross section of transparent blue film prepared breaking in liquid nitrogen at (a) 1000×, (b) 2500× magnification and the top surface of transparent blue film (c) at 25,000× (d) 50,000× magnification.

The film thickness of the transparent film was measured as 47.34 μm as seen in Figure 9.8b. The surface of the transparent film was not very smooth as seen in Figure 9.8c and 9.8d. However, the surface roughness was only due to the polymer phase. No inorganic particles with sharp edges were present on the surface.

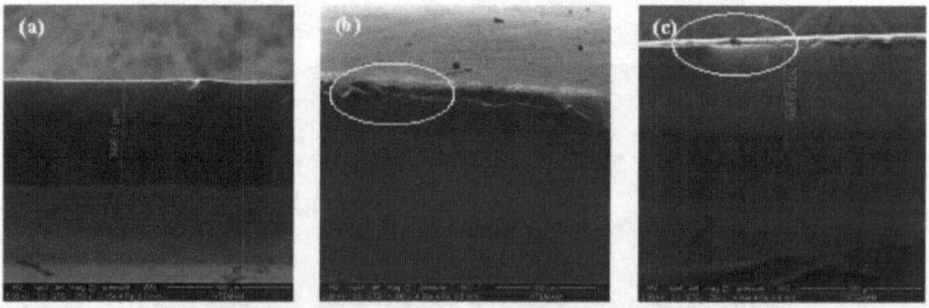

FIGURE 9.9 The SEM observations of the cross section of (a) adhesive of double-sided film at 250 ×, (b) the interface between opaque film and adhesive double-sided film at 100× , (c) the inteface between transparent film and double-sided film at 250 × magnification prepared by breaking in liquid nitrogen.

The cross section of the adhesive double film in Figure 9.9a indicated that it has 386 μm thickness. The interfaces of the opaque and transparent films to double-sided film were also examined by scanning electron microscopy. The films were sticked to adhesive double film manually and then immersed in liquid nitrogen and cut with scissors to observe the sticking surfaces by SEM. As seen in Figure 9.9b the sticking interface of the opaque film was not smooth, it has many irregularities making the sticking area larger. The sticking surface of the transparent film had small empty spaces between the film and sticking band.

9.3.3 SURFACE ROUGHNESS OF THE FILMS

The surface profiles of the opaque and transparent films are seen in Figure 9.10. The surface roughness values determined by two trials are tabulated in Table 9.1. When the average of the surface roughness parameters of two trials was considered for opaque blue film, the arithmetic mean roughness, R_a, was calculated as 0.21 μm. The maximum height, R_y, was calculated as 1.18 μm. Ten-point mean roughness, R_z, was found by using formulas as 0.80 and the root mean square roughness, Rq, was obtained as 0.27. All values for first and second trials were close to each other.

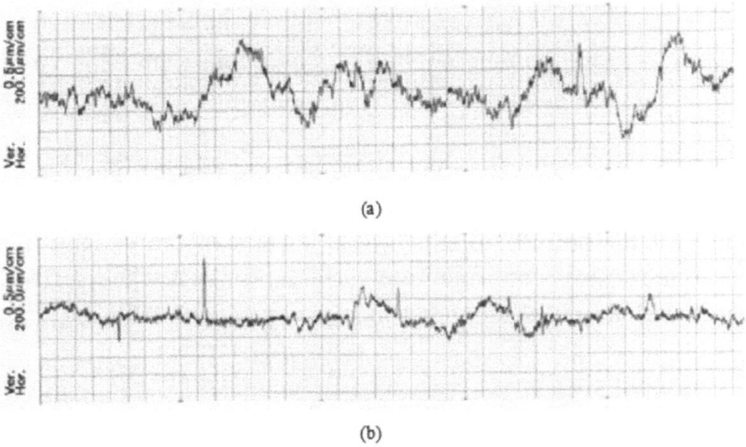

FIGURE 9.10 Surface roughness profile of (a) opaque film, (b) transparent film.

When the surface roughness profiles of transparent blue film were examined, the vertical and horizontal scales of trials were the same. Therefore, the surface profile of each trial of the transparent blue films was smoother than the opaque blue film. However, at the first trial the maximum height, *Ry*, of the peak was found as 0.77 and at the second trial the *Ry* was obtained as 0.68. At the same time, the surface roughness parameters were found as almost the same. When the average of the surface roughness parameters of two trials was investigated for a transparent blue film, the arithmetic means roughness, R_a, was calculated as 0.105 μm. The maximum height, *Ry*, was calculated as 0.725 μm. Ten-point mean roughness, R_z, was found by using formulas as 0.43 and the root means square roughness, R_q, was obtained as 0.135. When compared between the transparent and opaque blue films, all the values of the surface roughness parameters of the transparent blue film were about half of the surface roughness parameters of the opaque blue film. Therefore, the results confirmed the surface profile that the surface of the transparent blue film was smoother than the opaque blue film.

Roughness can promote adhesion. When the overall surface roughness parameters were investigated, the double-sided tape could stick to opaque blue film rather than transparent film properly.

TABLE 9.1 Surface Roughness Values of Opaque and Transparent Films.

Roughness,	Opaque film			Transparent film		
μm	1st trial	2nd trial	Average	1st trial	2nd trial	Average
R_a	0.21	0.21	0.21	0.12	0.09	0.11
R_y	1.31	1.05	1.18	0.77	0.68	0.73
R_z	0.84	0.75	0.80	0.44	0.42	0.43
R_q	0.28	0.26	0.27	0.16	0.11	0.14

When compared between the transparent and opaque blue films, all the values of the surface roughness parameters of the transparent blue film were about half of the surface roughness parameters of the opaque blue film. Therefore, the results confirmed the surface profile that the surface of the transparent blue film was smoother than the opaque blue film.

9.3.4 WATER CONTACT ANGLE OF THE OPAQUE AND TRANSPARENT FILMS AND ADHESIVE LAYER

Contact angle of water droplets on the polymer surfaces was measured from right and left of the droplets as seen in Figure 9.11. Each experiment was repeated twice and their average is shown in Table 9.2. The contact angle of water droplets was 83.6°, 74.1°, and 93.7° for opaque film, transparent film, and adhesive layer surfaces, respectively.

The water contact angle of the polyethylene surfaces would give information about the corona treatment state of polyethylene surfaces.[16] While the water contact angle of untreated polyethylene surface was 101.7°, by corona treatment periods of 0.5, 1, and 5 s it is lowered to 83.5°, 68.0°, and 53°, respectively.[16] Thus, the surface of polyethylene films was corona treated and the treatment period of the transparent film should be longer than that of the opaque film.

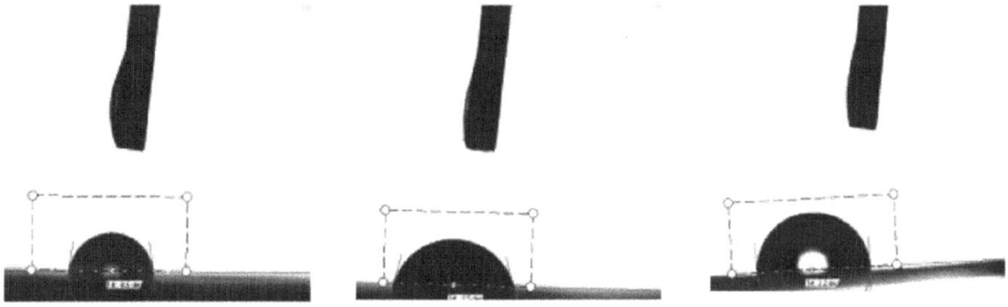

(a) (b) (c)

FIGURE 9.11 Water droplets on the surface of the (a) opaque film, (b) transparent film, and (c) adhesive layer of the double-sided film.

Adhesion between opaque film and adhesive elastomer is known as better than adhesion between transparent film and adhesive elastomer. The opaque film and adhesive have closer contact angle values (83.6 and 93.7°) than that of transparent film and adhesive elastomer (74.1 and 93.7°). Thus, opaque film and the adhesive stick each other more strongly than that of transparent film and adhesive since their surfaces are more similar to each other.

TABLE 9.2 Determined Contact Angles for Layers of Packaging, Adhesive Elastomer, and Adhesive Carrier.

Measurement direction	Contact angle (°)				
	Left 1	Right 1	Left 2	Right 2	Average
Opaque film	82.29	82.14	85.84	84.24	83.6
Transparent film	75.12	73.71	74.69	72.88	74.1
Adhesive layer	91.53	90.58	96.5	96	93.7

The experimental results showed that opaque and transparent films were LDPE as indicated by FTIR spectroscopy and their thicknesses were

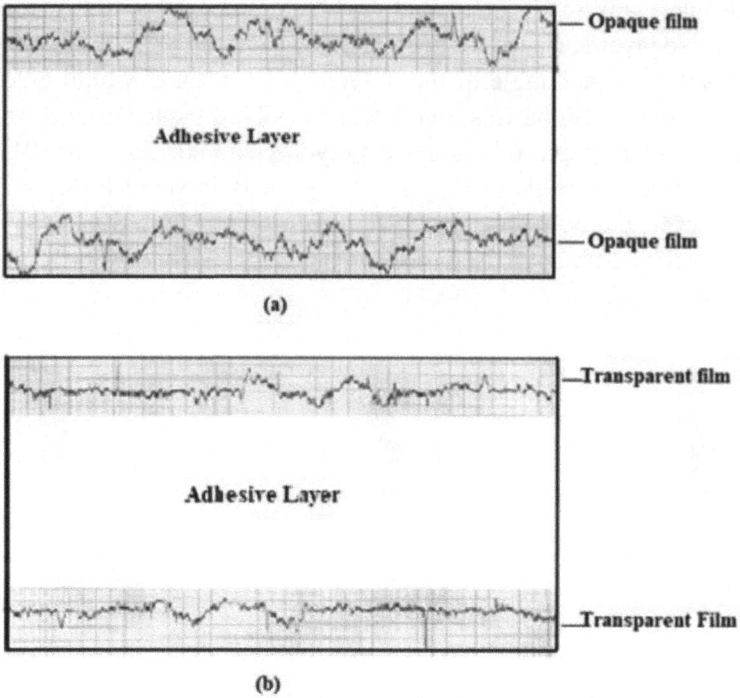

FIGURE 9.12 Representative figures for the adhesion of two (a) Opaque polymer surfaces, (b) Transparent polymer surfaces by the double-sided adhesive.

55.95 μm 47.43 μm respectively as measured by SEM. The film thickness of the opaque film was also measured as 44.0 μm by using the fringe peaks observed in the specular reflection spectrum. A lower thickness value was measured by specular reflection. For transparent film the refraction fringes were not observed clearly; thus, its thickness was not measured by specular reflection. The opaque film had sharp-edged mineral filler creating roughness at its surface as observed in its SEM photographs. The adhesive layer of the double-sided film was polstyrene-polybutadien copolymer with other additives; however, no mineral filler was present in its structure. The thickness of the adhesive layer was measured as 386 μm by SEM. The thin film and adhesive interfaces observed by SEM indicated the presence of void spaces between transparent film and adhesive. The adhesion between opaque film and adhesive was better since no void spaces were present at the interphase. The higher surface roughness of the opaque film made better adhesion to adhesive layer as shown by the representative figures in Figure 9.12. The contact area between the adhesive and the films was higher for the rougher surface of the opaque film. The closer contact angle of water on porous films surface and adhesive indicated that they had similar surfaces and better wetting of the polymer film by the adhesive layer occurred. Further work should be done about the rheological and viscoelastic properties of the adhesive, since they have a very important role in peel strength and shear strength of the bonded interface.[17]

9.4 CONCLUSION

The problem of opening the sealed packages of car bumpers during transportation was investigated in the present study. FTIR spectroscopy showed the two alternative packaging films, one is opaque and other transparent, were polyethylene. The adhesive layer of the pressure-sensitive adhesive was styrene-butadiene copolymer with additives. The surface of opaque film had sharp-edged mineral particles as seen in their SEM micrographs creating a higher average surface roughness value (0.21 μm) than that of the transparent film (0.11 μm). The contact angles of water on the opaque film and the transparent film were 83.6 and 74.1°. The surfaces of both films should have been corona treated since the untreated surface of polyethylene has a contact angle of 101.7°. The surface of the adhesive is more hydrophobic than polyethylene films since its water contact angle is 83.5°. Thus, adhesion of opaque film to the adhesive was stronger than the transparent film since their surfaces are more similar.

Thus, the self-opening of the packages having a transparent polyethylene envelope was due to their smooth surface and lower water contact angle values. If some mineral filler was added during the production process of transparent films, the problem of the opening of the packages during transportation would be solved. However, further quantitative studies could be made measuring flexibility and stickiness of the adhesive layer.

KEYWORDS

- **car bumper**
- **polymer film envelope**
- **double-sided adhesive**
- **surface roughness**
- **surface tension**

REFERENCES

1. Czech, Z.; Milker, R. Development Trends in Pressure-sensitive Adhesive Systems. *Mater. Sci.-Pol.* **2015,** *23* (4), 1015–1022.
2. Antosik, A. K.; Mozelewska, K.; Zbigniew Czech, Z.; Piątek-Hnat, M. Influence of Montmorillonite on the Properties of Silicone Pressure Sensitive Adhesives: Preparation of a Double-sided Tape Based on the Best Composition. *Silico*n **2020,** *12* (8), 1887–1893.
3. Dastjerdi, Z.; Cranston, E. D.; Dubé M. A. Pressure Sensitive Adhesive Property Modification Using Cellulose Nanocrystals. *Int. J. Adhes. Adhes.* **2018,** *81,* 36–42.
4. Kim, D. B. Effects of Tape Thickness and Inorganic Fillers on the Adhesion Properties of Double-sided Acrylic Adhesive Tape by Ultraviolet Curing. *Polymer-Korea* **2014,** *38* (3), 397–405.
5. Mapari, S.; Mestry, S.; Mhaske, S. T. Developments in Pressure-sensitive Adhesives: A Review. *Polym. Bull.* **2021,** *78* (7), 4075–4108.
6. Rudawska, A. *Adhesive Properties*; INTECH Open Access Publisher, 2012.
7. Hay, K.; Dragila, M.; Liburdy, J. Theoretical Model for the Wetting of a Rough Surface. *J. Colloid Interface Sci.* **2008,** *325* (2), 472–477.
8. Sowa, D.; Czech, Z.; Byczyński, Ł. Peel Adhesion of Acrylic Pressure-sensitive Adhesives on Selected Substrates Versus Their Surface Energies. *Int. J. Adhes. Adhes.* **2014,** *49,* 38–43.
9. Hadef, Z.; Doghmane, A.; Kamli, K.; Hadjoub, Z. Correlation Between Surface Tension, Work Of Adhesion In Liquid Metals/Ceramic Systems, And Acoustic Parameters. *Prog. Phys. Metals* **2018,** *19* (2), 168–184.
10. Tian, Y.; Ina, M.; Cao, Z.; Sheiko, S. S.; Dobrynin, A. V. How to Measure Work of Adhesion and Surface Tension of Soft Polymeric Materials. *Macromolecules* **2018,** *51* (11), 4059–4067.
11. Kowalski, A.; Czech, Z. The Effects of Substrate Surface Properties on Tack Performance of Acrylic Pressure-Sensitive Adhesives (PSAs). *Int. J. Adhes. Adhes.* **2015,** *60,* 9–15.
12. Gulmine, J. V.; Janissek, P. R.; Heise H. M.; Akcelrud, L. Polyethylene Characterization by FTIR. *Polym. Test.* **2002,** *21* (5), 557–563.
13. https://www.piketech.com/files/user-manuals/10Spec_Manual.pdf. Date of entry (accessed Sept 28, 2022).
14. http://www.filmetrics.com/refractive-index-database/Polyethylene/PE-Polyethene. Date of Entry (accessed Sept 29, 2022).
15. Munteanua, S. B.; Vasile, C. Spectral and Thermal Characterization of Styrene Butadiene Copolymers with Different Architectures. *J. Optoelectron. Adv. Mater.* **2005,** *7* (6), 135–3148.
16. De Geyter, N.; Morent, R.; Leys C. Surface Characterization of Plasma-modified Polyethylene by Contact Angle Experiments and ATR-FTIR Spectroscopy. *Surf. Interf. Anal.* **2008,** *40,* 608–611.
17. Deng, X. H. Progress on Rubber-based Pressure-sensitive Adhesives. *J. Adhes.* **2018,** *94* (2), 77–96.

CHAPTER 10

Coherent Analysis of Fickian Diffusion in Mixtures of Cyclohexane, Toluene, and Methanol

C. I. A. V. SANTOS[1], A. MIALDUN[2], M. C. F. BARROS[1], V. SHEVTSOVA[3], and A. C. F. RIBEIRO[1]

[1]*CQC-IMS, Department of Chemistry, University of Coimbra, Coimbra, Portugal*

[2]*MRC, CP165/62, Université libre de Bruxelles, Brussels, Belgium*

[3]*Mechanical and Manufacturing Department, Mondragon University, Mondragon, Spain; Ikerbasque, Basque Foundation for Science, Bilbao, Spain*

ABSTRACT

Diffusion coefficients provide important information of different aspects of mass transport that are extremely important to understand the behavior of complex solutions. The nonideal mixture of cyclohexane, toluene, and methanol has remarkable interest since it exhibits a miscibility gap in ambient conditions.

This work presents a parametric description of the Fickian diffusion data for the binary subsystems of the mixture composed of toluene, cyclohexane, and methanol in different steps. Determination of binary Fickian diffusion coefficients through various methods, namely optical beam deflection, optical digital interferometry, Taylor dispersion technique among others, was carried out in stationary and transient regimes. Results are presented together with a summary of the available literature data, compiling a database. From

Physics and Mechanics of New Materials: Synthesis, Processing, and Emerging Applications. Devrim Balkose, Ana Cristina Faria Ribeiro, Nandakumar Kalarikkal, Ann Rose Abraham, A. K. Haghi (Eds.)

this database, the values of mutual diffusion coefficients at infinite dilution limits were determined for all three binary mixtures. After fixing the limiting values, correlations for mutual diffusion in concentrated binary solutions, consistent with the diluted limits, were established, providing a model that allows knowledge of the diffusion coefficient in concentration regions not covered by the measurements. The ternary Fickian diffusion coefficients were experimentally reported using the Taylor dispersion technique, and qualitative analysis is done, using the obtained binary dependencies as boundary conditions.

10.1 INTRODUCTION

Accurate knowledge of the thermophysical properties of working fluids in energy and process engineering is necessary for the design of efficient processes and the corresponding apparatus. Diffusion in liquid systems is of substantial importance in studies involving mass transport in liquids. While such diffusion is, in most cases, the summation of various diffusion mechanisms including, for example, molecular diffusion or thermal diffusion, it is, in itself, important for calculations in many mass transfer operations. In addition, the availability of reliable experimental data on molecular diffusion can contribute to fundamental studies on liquid structure, theories of the liquid state, and aspects of nonideal behavior.

To understand and model the various mass transfer processes, e.g., like absorption, evaporation, or liquid–liquid extraction,[1–3] it is essential not only to evaluate the diffusion coefficients but also to comprehend their composition dependence (of all the components present in the solution). While the diffusion coefficients for binary liquid mixtures are more widely studied by the scientific community,[4–8] most mixtures present in nature or in technical applications are multicomponent. The measurement of diffusion coefficients for ternary or higher order mixtures is much more complex when compared to binary mixtures, which leads to a small number of publications by the scientific community.[4,9–12] One reason for this difficulty can be the coupled diffusion effects where a gradient of one species induces a diffusive flux of another species.

The diffusion coefficients in liquids can be determined by various techniques such as: diaphragm cell,[13] conductimetric and optical cells,[14,15] dynamic light scattering[16–19] and Taylor dispersion technique.[9,18,20,21] This last one becomes one of the most used methods for the measurement of mutual diffusion coefficients (D_{ik}) due to its simplicity, quickness, and consistent results.

In this work, we give special attention to a nonideal ternary liquid mixture: cyclohexane + toluene + methanol. It has exceptional interest because it exhibits a very complex behavior as well as a miscibility gap in ambient conditions.[22] This ternary mixture, as well as its binary limits, has been already well characterized both experimentally[10,23,24] and theoretically[25–27] motivated by the experiment DCMIX (diffusion coefficients in mixtures) performed on-board the International Space Station (ISS).[11] Besides, the knowledge on continuous variation of the diffusive properties in this system is very much in demand for the interpretation of the results of space experiments DCMIX2[11] which target different aspects of mass transport in this ternary system.

It is customary in multicomponent diffusion studies to provide the results of measurements as a data tabulated for a set of isolates state points. At the same time, for the practical use of the data it may be required the knowledge of the corresponding property in concentration regions not covered by the measurements. Consequently, it demands some interpolation efforts which are not as easy even for binary mixtures. The condition provides the development of a methodology allowing an accurate and consistent parameterization of the diffusion data in multicomponent mixtures.

The aim is to contribute with data on the mutual diffusion coefficients for the systems of interest of the second series of the DCMIX experiment (DCMIX2) and at the same time provide, to the scientific community, information on the behavior of mass transport in this ternary system, specifically in understanding the coupled flows of solutes occurring in solution, far and close to the demixing zone. The present work provides ternary diffusion data for twenty compositions (from which five are part of the original mixtures studied in DCMIX2[10]) and relates them with others already published in literature.[23,24] Also, the eigenvalues of the ternary diffusion matrix were determined to check the accuracy of the tested results.

The authors also propose the following way of establishing the parametric description of the diffusion data (with the structure of the work reproducing these steps): (1) creation of a database by summarizing literature data and our own experimental measurements; (2) determination of the mutual diffusion coefficients at infinite dilution limits for all binaries; (3) establishing correlations for mutual diffusion in concentrated binary solutions, consistent with the diluted limits. The model established here, using the obtained binary dependencies as boundary conditions, can be used for the qualitative analysis of ternary diffusion coefficients.

10.2 DIFFUSION BASIC CONCEPTS

The slope of concentration inside a mixture (not to consider either the convection or migration) provides a flux of matter in the opposite direction. This, in fact, appears from incidental variation in the condition of molecules in space. This in fact is designated by diffusion. This is called an irreversible process. In the real solution, the slope of chemical potential is called the true virtual force producing diffusion. In ideal solutions, that force can be evaluated by the slope of the concentration at constant temperature. For such a reason, it is more appropriate that we apply the law of thermodynamics for irreversible processes along with Fick's laws. According to[28,29] this can be considered valid approaches to describe the isothermal diffusion.

Mutual diffusion coefficient, *D,* could be considered in a binary system, in terms of the concentration gradient by a phenomenological relationship

$$J = -D\, grad\, c \tag{10.1}$$

where

J = the flux of matter across a proper chosen reference plane (per area unit and per time unit),

c = the concentration of solute (in moles per volume unit at the point considered);

And eq 10.1 could be considered to measure D.

The diffusion coefficient could be examined using Fick's second law

$$\frac{\partial c}{\partial t} = div\left(D\, grad\, c\right) \tag{10.2}$$

Base on eqs 10.1 and 10.2, the acceptable approach for the measurements of diffusion coefficients is divided into two main groups: steady and unsteady states. Diffusion is a three-dimensional phenomenon but experimentally the diffusion phenomenon is restrained to a one-dimensional. In eq 10.2 the resolution for a unidimensional process is simpler if we consider D as a constant, but this approximation is applicable only when there are small differences of concentration, which is the case in the experimental method here used (Taylor dispersion technique).[30] In these conditions, it is legitimate to consider that our measurements of differential diffusion coefficients obtained by the above method are parameters with a well-defined thermodynamic meaning.[28,30]

For a ternary mixture, the molar diffusion fluxes can be written as

$$J_1 = -\left(D_{11}\nabla C_1 + D_{12}\nabla C_2\right) \tag{10.3}$$

$$J_2 = -\left(D_{21}\nabla C_1 + D_{22}\nabla C_2\right) \tag{10.4}$$

where J_1, J_2, ∇c_1, and ∇c_2 are the molar fluxes and the gradients in the concentrations of solutes 1 and 2, respectively. D_{ij} symbolize the elements of the diffusion coefficient matrix. In a ternary mixture, these equations describe two coupled mass fluxes, assuming that the third component is the reference (corresponding to what would be the solvent). The diffusion coefficients D_{11} and D_{22} provide the flux of each solute obtained by its own concentration gradient. Information about coupled diffusion is provided by cross-diffusion coefficients D_{12} and D_{21}. A positive D_{ik} cross-coefficient ($i \neq k$) shows that the co-current coupled transport of solute i from zone of higher to lower concentrations of solute k. It should be noted that the negative D_{ik} coefficient describes counter-current coupled transport of solute i from zone of lower to higher concentration of solute k.

10.3 EXPERIMENTAL

10.3.1 MATERIALS

The reagent grade solvents toluene (mass fraction purity 0.9985, density 0.8669 g cm−$^{-3}$, CAS Number: 110-82-7) and cyclohexane (mass fraction purity 0.9999, density 0.779 g cm^{-3}, CAS Number: 110-82-7) were supplied by Acros Organics. Methanol (mass fraction purity 0.9999, density 0.7918 g cm^{-3}, CAS Number: 67-56-1) was supplied by Fisher Scientific. All liquids used as received, without considering any purification.

Mixtures were prepared in mass fraction, by weighing each component using a Mettler H80 analytical balance with resolution 0.1 mg/160 g. Each mixture was freshly prepared and de-aerated for about 30 min before each set of runs. Component 1 was cyclohexane (Chex), component 2 was toluene (Tol), and component 3 (reference or solvent) was methanol (MeOH).

10.3.2 DIFFUSION MEASUREMENTS

10.3.2.1 TAYLOR DISPERSION METHOD

Taylor dispersion method for diffusion measurements in multicomponent systems[5,18,20,21] is considered on the dispersion of low amounts of solution injected into laminar carrier streams of solution of a slightly different composition flowing through a long capillary tube. The assembly schematics for the Taylor dispersion technique is present in Figure 10.1. For each each run, a pulse of mixture was injected using a 6-port injection valve (Rheodyne, model 5020) to a laminar carrier stream with no similar composition. A flow rate of 0.30 cm^3 min–[1] was considered using metering pump (Gilson model Minipuls 3) to give storage times of about 3.8×10^3 s. The dispersion tube (length 30.480 (±0.001) m, r = 0.322 mm) and the injection valve were hold at 298.15 K (±0.01 K) in an air thermostat. Dispersion of the injected samples was studied by differential refractometer (Waters model 2410) at the outlet of the dispersion tube. Detector voltages, $V(t)$, were studied at 5 s intervals with a digital voltmeter (Agilent 34401 A) obtained with an IEEE interface.

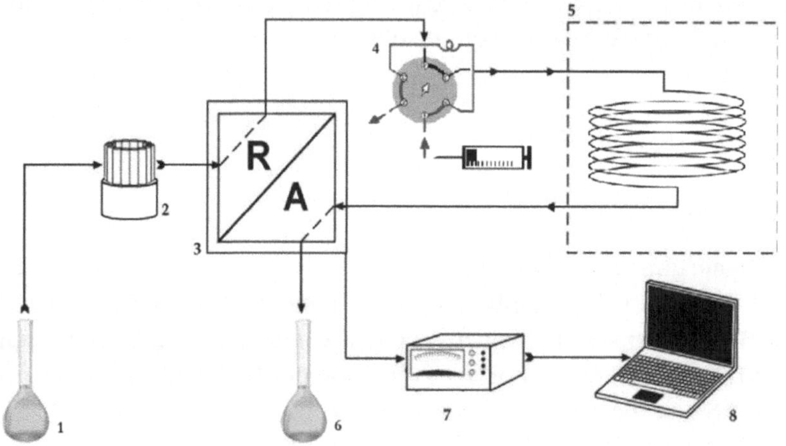

Legend: 1—Solution reservoir, 2—Pump, 3—Refractometer, 4—Injection valve, 5—Thermostatized box with coiled dispersion tube, 6—Waste, 7—Voltímeter, 8—Computer

FIGURE 10.1 Schematic of the experimental setup for the Taylor dispersion technique. The flow direction is shown by arrows.

Binary diffusion coefficients are calculated from the dispersion equation

$$V(t) = V_0 + V_1 t + V_{max}(t_R / t)^{1/2} \exp -12D(t - t_R)^2 / r^2 t \tag{10.5}$$

being the additional fitting parameters: t_R, the mean sample retention time, V_{max}, the peak height, V_0, the baseline voltage, and V_1, the baseline slope.

The determination of the ternary diffusion coefficients was done by adopting the procedure presented by Leaist and Hao.[31] To raise the consistency of the data evaluation, four runs with different Δc_i were performed, with both positive and negative amplitudes of the signals,[10] and the corresponding signal profiles analyzed. In total, eight peaks of four different injection samples were fitted, two by two, and the average of the attained values was considered to obtain the four coefficient values.

The D_{ik} coefficients defined by eqs 10.3 and 10.4 were evaluated by fitting the ternary dispersion equation

$$V(t) = V_0 + V_1 t + V_{max}\left(t_R/t\right)^{1/2}\left[W_1 \exp\left(-\frac{12D_1(t-t_R)^2}{r^2 t}\right) + (1-W_1)\exp\left(-\frac{12D_2(t-t_R)^2}{r^2 t}\right)\right] \quad (10.6)$$

The two pairs of refractive-index profiles, D_1 and D_2, represent the eigenvalues of the matrix of the ternary D_{ik} coefficients. W_1 and $1 - W_1$ are the normalized pre-exponential factors.

The eigenvalues, D_1 and D_2, for the 2 × 2 matrix of the ternary diffusion coefficients, D, are given by

$$D_1 = \left\{D_{11} + D_{22} + \left(D_{11} - D_{22}\right)\left[1 + \left(4D_{12}D_{21}/\left(D_{11} - D_{22}\right)^2\right)\right]^{1/2}\right\}/2 \quad (10.7)$$

$$D_2 = \left\{D_{11} + D_{22} - \left(D_{11} - D_{22}\right)\left[1 + \left(4D_{12}D_{21}/\left(D_{11} - D_{22}\right)^2\right)\right]^{1/2}\right\}/2 \quad (10.8)$$

where the determinant of the matrix is

$$D = D_{11}D_{22} - D_{12}D_{21} \quad (10.9)$$

and the normalized weighting factors

$$W_1 = \frac{\left(a + b\alpha_1\right)\sqrt{D^{(1)}}}{\left(a + b\alpha_1\right)\sqrt{D^{(1)}} + \left(1 - a - b\alpha_1\right)\sqrt{D^{(2)}}} \quad (10.10)$$

$$W_2 = 1 - W_1$$

defined in terms of the a and b parameters

$$a = \frac{D_{11} - D^{(1)} - \left(\frac{R_1}{R_2}\right)D_{12}}{D^{(2)} - D^{(1)}} \quad (10.11)$$

$$b = \frac{D_{22} - D_{11} + \left(\frac{R_2}{R_1}\right)D_{21} - \left(\frac{R_1}{R_2}\right)D_{12}}{D^{(2)} - D^{(1)}} \quad (10.12)$$

The ratio of the detector sensitivity to the components (10.1) and (10.2) of the solution is given by

$$\frac{R_2}{R_1} = \frac{\partial V / \partial C_2}{\partial V / \partial C_1} \tag{10.13}$$

and α_1 is the fraction of the initial refractive index change contributed by the component (10.1), given by

$$\alpha_1 = \frac{R_1 \Delta c_1}{R_1 \Delta c_1 + R_2 \Delta c_2} \tag{10.14}$$

In a typical experiment, small volumes, ΔV, of a solution with composition $(c_1 \pm \Delta c_1)$ and/or $(c_2 \pm \Delta c_2)$ are injected into carrier solutions of composition (c_1, c_2) at time $t = 0$.[9,18]

After the injection, the flow is dispersed by the combined action of molecular diffusion and the parabolic velocity profile of the carrier. At the end of the tube, the radially averaged concentrations of the dispersed solutes, relative to the background concentrations in the carrier stream, are given by the expression

$$\Delta C_i(t) = C_i(t) - \bar{C}_i = \sum_{k=1}^{2} \frac{2A_{ik}}{\pi r^3 U} \left(\frac{3D_i}{\pi t_R} \right)^{1/2} exp\left[-\frac{12 D_i U (t - t_R)^2}{r^2 L} \right] \tag{10.15}$$

A nonlinear least-squares method was used to fit eq 10.5 to dispersion profiles measured for two or more different values of α_1 for each carrier stream. The adjustable least-squares incorporated a, b, $D^{(1)}$, $D^{(2)}$, and, for each profile, the retention time t_R, the peak height ΔV_{max}, the baseline V_0, and the baseline slope V_1. The ternary mutual coefficients were calculated from the fitted parameters by using the expressions[32]

$$D_{11} = D_1 + \frac{a(1-a-b)}{b} \left(D^{(1)} - D^{(2)} \right) \tag{10.16}$$

$$D_{12} = \frac{R_2}{R_1} \frac{a(1-a)}{b} \left(D^{(1)} - D^{(2)} \right) \tag{10.17}$$

$$D_{21} = \frac{R_1}{R_2} \frac{(a+b)(1-a-b)}{b} \left(D^{(2)} - D^{(1)} \right) \tag{10.18}$$

$$D_{22} = D_2 + \frac{a(1-a-b)}{b} \left(D^{(2)} - D^{(1)} \right) \tag{10.19}$$

10.3.2.2 CONVECTIONLESS OPTICAL METHODS

Optical beam deflection technique (OBD) and optical digital interferometry (ODI)[6,33,34] are nonisothermal methods, normally employed for the measurement of the Soret effect (thermodiffusion) but also provide information for the diffusion coefficient.[70–73] These methods employ a classical Soret cell with relatively small height (typically 1 mm in OBD). The ODI is an instrument designed to study heat and mass transfer in liquids and consists of two principal parts: the optical cell[35] and an interferometric system in combination with equipment for digital recording and processing the phase information. Figure 10.2 shows a cross section of the diffusion cell (Fig. 10.2a) and a sketch of the ODI setup (Fig. 10.2b). This technique allows performing and reproducing experiments without disturbing the media. It also allows measuring not only Δc between the hot and cold plates, but also the concentration and temperature distributions along the diffusion path, giving the possibility to adopt this technique for some other purposes like the study of hydrodynamic fluctuations. The reliability of the experiments can be improved by using statistical means.[36]

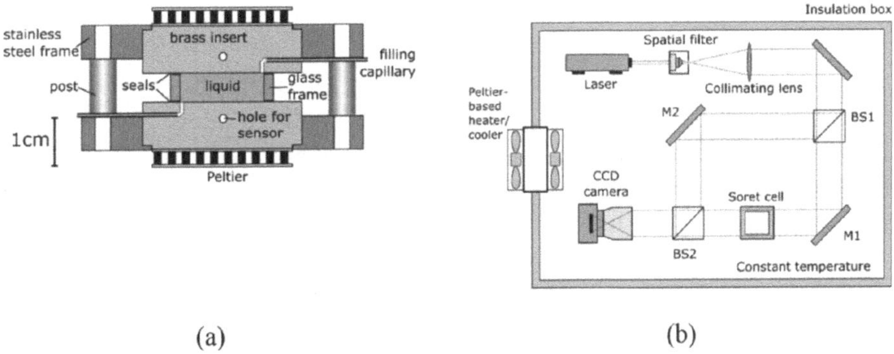

FIGURE 10.2 Optical digital interferometry setup; cross section of the Soret cell (a) and top view of the entire setup (b).

Source: Reprinted from Ref.[67]. https://creativecommons.org/licenses/by/4.0/

The experiments are performed in a transparent cubic cell filled initially by a homogenous mixture. The thermal slope is charged in fact by using heating and cooling the top and bottom walls of the cell, respectively. The geometrical temperature differences activate mass transfer through the Soret effect. An expanded collimated laser beam traverses the complete cell perpendicular to the temperature slope. Both temperature and composition differences grant to the geometrical circulation of the refractive index that modulates the wavefront of the emerging optical beam. The use of an interferometric technique allows the final wave-front shape to be accurately determined by comparison with undisturbed shape. The interferometer (with a Mach–Zehnder interferometer scheme) coupled with a digital recording and processing is used for a very accurate determination of a phase shift.[6,34] Each interferogram is reconstructed by performing a 2D fast Fourier transform (FFT) of the fringe image, filtering a selected band of the spectrum, performing an inverse 2D FFT of the filtered result and phase unwrapping. Knowledge of phase shifts gives information about the local gradients of the refractive index. Then the gradients of composition inside the fluid are calculated from refractive index gradients and local temperature both for steady state and dynamic regimes. Thus, optical digital interferometry enabled measurements of Δc between two arbitrary points.

The light source is a He–Ne laser with the wavelength of $\lambda = 632.8$ nm. An expanded and collimated laser beam is split into a reference and an objective arm, with the cell assembly placed within the latter. Then, both beams are deflected by mirrors and merged at the second beam splitter. One of the beams is inclined with respect to the other to create a narrow fringe pattern. The resulting interferogram is recorded by a CCD camera with a sensor of 1280×1024 pixels. The resolution of the imaging system is ≈50 pixels/mm. To reconstruct a wave front (spatial distribution of optical phase), the Fourier transform method is applied to obtain the phase distribution from a single interferogram. To be precise, two interferograms are used in processing, as the interference pattern of interest is always processed against a reference interferogram taken before the refractive index change.

To ensure the thermal stability of the interferometer during an experiment (up to 2–3 days), the whole setup is placed inside a box of thermal insulation material and equipped with air-to-air cooling/heating assembly based on a Peltier element and driven by a dedicated PID controller. A set of shields is inserted in the box to prevent air motion over optical paths. The temperature inside the box is kept equal to the mean temperature of the liquid with residual fluctuations of ±0.02 K.

The unique feature of this method is that it traces the transient path of the system over the entire two-dimensional cross section of the cell. The quantity obtained after processing an interferogram is a spatial phase distribution $\phi(x, z, t)$, which is transferred into a concentration distribution by the equation

$$\frac{\lambda}{2H}\Delta\varphi(x,z,t) = \left(\frac{\partial n}{\partial c}\right)_{p,T} \Delta c(x,z,t) + \left(\frac{\partial n}{\partial T}\right)_{p,c} \Delta T(x,z,t) \qquad (10.20)$$

where the second term is defined during the initial step of the experiment.[37] Consequently, Soret and diffusion coefficients are simultaneously extracted by fitting of $c(x, z, t)$ from eq 10.20, averaged in the x-direction, with the one-dimensional analytical solution[35]

$$c(z,t)=c_0+c_0\left(1-c_0\right)S_T\Delta T\left[\frac{1}{2}-\frac{z}{H}-\frac{4}{\pi^2}\sum_{n,0dd}^{\infty}\frac{1}{n^2}\cos\times\left(\frac{n\pi z}{H}\right)exp\left(\frac{n^2\pi^2}{H^2}Dt\right)\right] \quad (10.21)$$

The counterflow diffusion cell (Fig. 10.3) for the measurement of diffusion by interferometric probing[26,38] is a small cubic cell (see Fig. 10.3(a)) made of a brass disc of 25 mm diameter and 5 mm thickness where a rectangular opening of 10 mm × 5 mm was cut, providing a volume to be filled with liquid. The disk is inserted between two glass windows with 25.4 mm diameter and 2 mm thickness. The whole assembly is mounted inside an optical holder for 1-inch lenses. Four channels aimed at injecting and withdrawing the liquid from the cell are made of stainless-steel capillaries with 0.58 mm inner diameter, and those are equipped with valves to simplify injection and block the liquid inside the cell before the experiment starts. Thus, the total diffusion path is $H = 10.0$ mm, the height of the interface is $b = 5.0$ mm, and the length of the optical path in the liquid is $L = 5.30$ mm.

(a) (b)

FIGURE 10.3 The schematics of the counterflow diffusion cell (a) and a photo of the cell (b).

A similar cell had been developed and successfully applied for accurate diffusion measurements by Komiya et al. in 2006.[39]

The observation of the concentration variation inside liquid mixtures can be done with different interferometric schemes, being the most precise the phase-shifting technique.[40] This cell was tested with two types of interferometer schemes, easy to implement and that provide complete phase information from a single snapshot: the classical Mach–Zehnder scheme (as described above) and the common-path scheme. Both are based on an identical set of components and differ only by the positioning of the cell with respect to the interferometer. In the Mach–Zehnder interferometer, the cell is located within an object arm inside the interferometer. In the common-path version, the cell is placed in front of the interferometer. The arms of the interferometer in this case are aligned in a special way, to create a small spatial vertical shift between both. The last scheme is particularly useful for diffusion experiments, as it allows resolution of steep refractivity gradients.[41] The overall implementation of the interferometer is the same as that described before for the ODI setup. The cell and the interferometer were placed inside a thermally insulated box, with controlled temperature of 298 ± 0.1 K.

When performing an experiment, the cell was filled through the bottom, and allowed to attain thermal and optical stabilization of the interferometer for 8–10 hours. Then another liquid was injected from the top. When the interface approached the outlets, injection from the bottom inlet was also activated. Both liquids flow for some time into the inlets and outlets to homogenize concentrations in the bulk of the two cell sections. Later in time, the injection speed was abruptly increased to make the interface sharp, then the injection was stopped, all the valves were closed, and image recording was initiated. The image acquisition step was variable, often at the beginning of the experiment (each 10 s) and seldom by the end of the experiment (once per 300 s). Then the set of images was analyzed to extract refractive index profiles. And finally, diffusion coefficients were evaluated by fitting an analytical solution to the obtained experimental data. Details on image processing and data extraction are explained in detail elsewhere.[38,42]

10.4 EXPERIMENTAL RESULTS

10.4.1 BINARY DIFFUSION COEFFICIENTS AT INFINITE DILUTION

The infinite dilution diffusion coefficients presented in this work were collected from the literature data for each binary subsystem under study, toluene + cyclohexane, toluene + methanol, and methanol + cyclohexane. For each literature value, the table presents the experimental method used, the assessed diffusion coefficient, and the corresponding literature source.

10.4.1.1 TOLUENE–CYCLOHEXANE DIFFUSION COEFFICIENTS AT INFINITE DILUTION

The data for the diffusion coefficients at infinite dilution for the binary system toluene + cyclohexane, measured by different experimental methods, are collected in Table 10.1.

TABLE 10.1 Infinite Dilution Diffusion Coefficients, $D^{\infty}/10^{-9}$ m^2 s^{-1}, in the Binary System Toluene + Cyclohexane.

Method/source	*x* Tol	D^{∞}_{Tol}	*x* Tol	D^{∞}_{Chex}
Diaphragm cell, mutual diffusion[43]	0.0	1.569	1.0	2.420
Diaphragm cell, intradiffusion[44]	0.0	1.74**	1.0	2.17**
Diaphragm cell, intradiffusion[45]			1.0	2.10
Taylor dispersion, mutual diffusion[46]	0.0*	1.715	1.0*	2.237
Nuclear magnetic resonance, intradiffusion[47]	0.05	1.28	0.99	2.24
Molecular dynamic simulation, mutual diffusion[25]	0.0	1.79	1.0	2.24
Diffusion cell/interferometry, mutual diffusion (present work)	0.0021	1.74	0.9976	2.16
Average		1.71		2.22

Notes: *Data obtained by extrapolation from close concentrations, **Data recalculated from the product of diffusion and viscosity, D has units 10^{-9} m^2 s^{-1}

As far as the data is available, we indicate an exact mean concentration at which the measurement has been done. In some cases (marked in the table by asterisk) the limiting coefficients were obtained by extrapolating data from diluted solutions of close concentrations.

The final values have been obtained by averaging all available data, with one exception of D^{∞}_{12} coefficient measured by Merzliak *et al.*,[47] that is an obvious outlier.

10.4.1.2 TOLUENE–METHANOL DIFFUSION COEFFICIENTS AT INFINITE DILUTION

The data for the diffusion coefficients at infinite dilution for the binary system toluene + methanol, measured by different experimental methods, are collected in Table 10.2.

TABLE 10.2 Infinite Dilution Diffusion Coefficients, $D^{\infty}/10^{-9}$ m^2 s^{-1}, in the Binary System Toluene + Methanol.

Method/source	***x* Tol**	D^{∞}_{Tol}	***x* Tol**	D^{∞}_{MeOH}
Diaphragm cell, mutual diffusion[48]	0.0	2.56	1.0	3.62
Diaphragm cell, intradiffusion[45]			1.0	3.99
Taylor dispersion, mutual diffusion[49]	0.0	2.42		
Molecular dynamic simulation, mutual diffusion[50]	0.0	2.30	1.0	3.76
Diffusion cell/interferometry, mutual diffusion (present work)	0.0007	2.43	0.9873	3.88
Average		2.43		3.81

Notes: *D* has units 10^{-9} m^2 s^{-1}

10.4.1.3 METHANOL-CYCLOHEXANE DIFFUSION COEFFICIENTS AT INFINITE DILUTION

The data for the diffusion coefficients at infinite dilution for the binary system methanol + cyclohexane, measured by different experimental methods, are collected in Table 10.3.

TABLE 10.3 Infinite Dilution Diffusion Coefficients, $D^{\infty}/10^{-9}$ m^2 s^{-1}, in the Binary System Methanol-Cyclohexane.

Method/source	***x* MeOH**	D^{∞}_{MeOH}	***x* MeOH**	D^{∞}_{Chex}
Open-ended capillary, intradiffusion[51]	0.0*	4.02		
Flow cell, mutual diffusion[52]			1.0*	2.23
Diffusion cell/interferometry, mutual diffusion[50]	0.0*	2.31	1.0*	2.37
Diffusion cell/interferometry, mutual diffusion[26]	0.0065	2.495		
Diffusion cell/interferometry, mutual diffusion (present work)	0.0013	3.45		
Average		3.07		2.30

Notes:*Data obtained by extrapolation from close concentrations. *D* has units 10^{-9} m^2 s^{-1}

10.4.2 MUTUAL DIFFUSION COEFFICIENTS FOR BINARY SYSTEMS

10.4.2.1 TOLUENE–CYCLOHEXANE MUTUAL DIFFUSION COEFFICIENTS

Searching the literature for a data on mutual diffusion in the binary mixture of toluene-cyclohexane resulted in several data sources with full concentration coverage and some data with several state points, or even with individual measurements. The eldest and most complete data[43] were obtained by diaphragm cell, and self-cited in a later publication.[53] Other full concentration range measurements in this binary system were made in 1995 by the Taylor dispersion technique.[46] The original work is difficult to access, but the data is available in Winkelmann's compilation.[54] Another dataset covering complete concentration span was obtained by Guevara-Carrion et al. by molecular dynamic simulations.[25] This set is also accounted in our database since the authors demonstrated an excellent agreement of their calculations with experiments on an extended set of binary mixtures.

Using optical methods, where this binary had been measured as a subset of the ternary mixture with n-heptane by the use of a Gouy interferometer and a diffusion cell, a less complete dataset is also available in Winkelmann's book.[54] A few state points of the system were assessed by Wittko et al. with the thermodiffusion-forced Rayleigh scattering technique when measuring an isotopic effect in the Soret coefficient, data is available graphically[55] and tabulated numerically in this author PhD thesis.[56] Several measurements of concentrated solutions have been published in a collective work by Lapeira et al.,[50] where two different techniques were applied, the sliding symmetric tubes technique (a version of the diffusion cell), and the optical beam deflection technique, in which the diffusion coefficient is determined alone with the Soret coefficient.

Some individual data points can be found in a study using Raman spectroscopy, where the equimolar mixture has been measured at different temperatures.[57] Another point of 0.4 mass fraction of toluene was assessed by two different methods where, again, a purely diffusive technique of a counter-flow cell has been compared with a result of thermodiffusion measurement implemented in space.[58]

In Figure 10.4 all the toluene–cyclohexane mutual diffusion coefficients data available on literature are presented. As expected, due to the large quantity of different techniques presented in the literature data, a large dispersion in the diffusion coefficients can be observed, although they present a similar behavior.

Plotting the full concentration dependency allows additionally confirming the correctness of the limiting values of this system. Comparing them with the infinite dilution diffusion coefficients average obtained in section 10.4.1.1, we realize that these values are quite consistent.

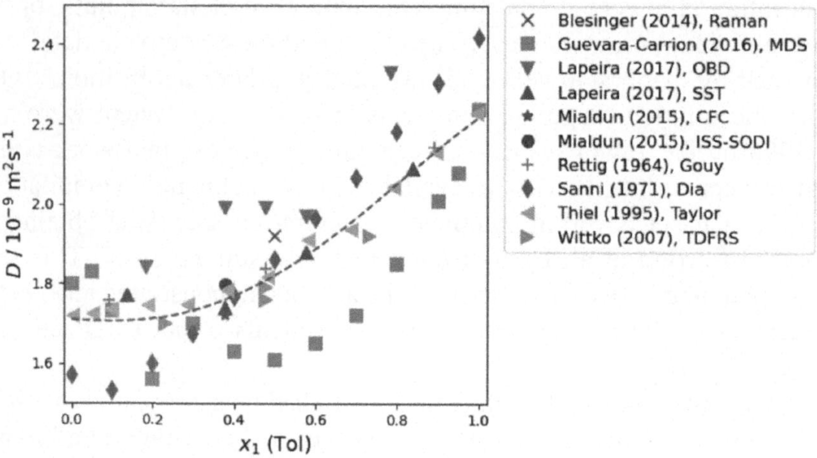

FIGURE 10.4 Mole fraction dependence of the mutual diffusion coefficients of the system toluene–cyclohexane.

To parameterize the concentration dependency with well-known limiting values we apply a classical approach of presenting the property as a sum of the linear and "excess" term,

$$D = D^{lin} + D^{E} = x_1 D_{12}^{\infty} + (1 - x_1) D_{21}^{\infty} + D^{E} \tag{10.22}$$

with following correlation of the concentration dependence of the excess term by applying the Redlich–Kister (RK) type polynomials[59]

$$D^{E} = x_1 x_2 \sum_{n=0}^{N} A_n (x_2 - x_1)^n = x_1 (1 - x_1) \sum_{n=0}^{N} A_n (1 - 2x_1)^n \tag{10.23}$$

Such an approach allows obtaining the concentration dependency with strictly satisfied limiting values. Then, the coefficients of eq 10.23 describing D^{E} can be easily transformed into coefficients of ordinary power series polynomial for D^{E} and then for D as

$$D = \sum_{n=0}^{N} b_n x^n \tag{10.24}$$

More details on the RK polynomials and their transformation to power series can be found elsewhere.[60]

Fitting data by RK polynomials needs in less terms in case the dependency is more symmetric with respect to $x_1 = 0.5$. Thus, a proper choice of the concentration units may result in much simpler parametric expressions. We have tested two obvious options of mass (w) and mole (x) fractions. The test reveals that the preference must be given to the mole fraction, which we will consistently use as the concentration measure over the text.

The next problem to solve is to fix the order of the polynomial which ensures a good fit to the data while limiting the order to a reasonably low number. When looking for the proper polynomial order, we have established our final goal for a future model of parametrization of ternary diffusion using binaries as boundary conditions. In order to do that, we have to keep the same polynomial order for all binary dependencies.

While the dependency for the mixture Tol-Chex is relatively simple, the dependencies of two other mixtures (Tol-MeOH and MeOH-Chex) have imposed a need to apply the RK polynomial of 3rd order, and consequently, power series polynomial of 5th order.

The polynomial coefficients of the corresponding orders found for the mixture are listed in Table 10.4.

TABLE 10.4 Coefficients of RK (*An*) and Power Series (*bn*) Polynomials Describing Concentration Dependencies of *DE* and *D*, Respectively, in the Binary Mixture of Toluene-cyclohexane.

Order n	An / 10^{-9}	$b_n/10^{-9}$
0	−0.57257	1.7100
1	−0.17846	−0.09080
2	0.15024	0.35677
3	–	0.84497
4	–	−0.60094

The dependency outlined by the polynomial is presented in Figure 10.4 by the dashed line. The average absolute deviation (AAD) of all data points from the correlation is 4.3%. And we remind once again that this correlation is fully consistent with the limiting diffusion coefficients established for this mixture in Section. 10.4.1.1.

10.4.2.2 TOLUENE-METHANOL MUTUAL DIFFUSION COEFFICIENTS

One of the first and most complete experimental studies of the mutual diffusion in this binary system was done in 1967 by the diaphragm cell technique.[48] Despite the very wide concentration coverage, the infinite dilution diffusion coefficients were obtained by the authors by extrapolation. This data has been actively cited in later literature, e.g., in Bou-Ali et al.[61] where the original data was interpolated for other concentrations, and in Lusis and Ratcliff, where the data at infinite dilution had been used.[62] Another complete dataset on the mutual diffusion, including the diluted limits, has been obtained by molecular dynamics simulation.[25]

A study with limited data obtained by diaphragm cell at a slightly different mean temperature, 301.15 K, was considered, but, when comparing with other data, this result has appeared completely out of range and was discarded.[63]

Several measurements of concentrated solutions have become available after collaborative work.[50] The work collected data from several different techniques, the sliding symmetric tube technique (isothermal diffusion), as well as optical beam deflection and optical digital interferometry (both utilizing thermodiffusion cells).

Individual concentration measurements were also available, where a single concentration of 0.96 mole fraction of methanol has been measured by Raman spectroscopy[57] and a limiting diffusion coefficient has been measured by the Taylor dispersion technique.[49]

Besides, we present several measurements first time in this work, namely both infinite dilution diffusion coefficients measured in counter-flow cell, and several concentrations assessed by optical digital interferometry.

Also, we found several estimates of the infinite dilution coefficients,[48] being calculated after correlations,[64,65] but since their deviations from the experimental data were all exceeding 10%, we did not take them into account.

The measurements made by the authors as well as the literature data available are presented in Figure 10.5. Limiting values found in literature (10.4.1.2) can be compared with those obtained from the graphic representation in Figure 10.5.

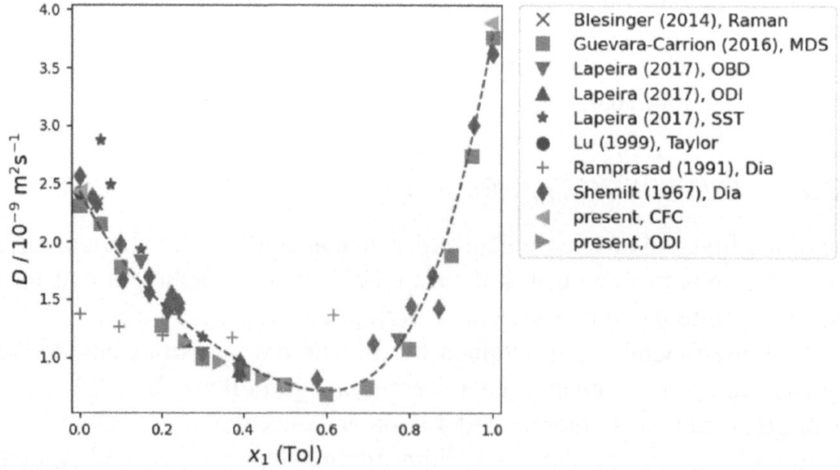

FIGURE 10.5 Mole fraction dependence of the mutual diffusion coefficients of the system toluene–methanol.

The polynomial coefficients of the diffusion coefficient concentration dependency determined for the mixture are listed in Table 10.5.

TABLE 10.5 Coefficients of RK (A_n) and Power Series (b_n) Polynomials Describing Concentration Dependencies of D^E and D, Respectively, in the Binary Mixture of Toluene-methanol.

Order n	A_n / 10^{-9}	b_n / 10^{-9}
0	−9.3682	2.4300
1	5.4412	−7.2010
2	−4.6540	16.315
3	–	−26.350
4	–	18.616

The dependency outlined by the polynomial is presented in Figure 10.5 by the dashed line. The average absolute deviation (AAD) of all data points from the correlation is 11.1%. The deviation is somewhat larger compared with Tol-Chex mixture, probably due to the presence of several definite outliers in the data, clearly visible in the plot. Also, the way we have adopted for the data fit appears insensitive to the presence of the outliers.

10.4.2.3 METHANOL-CYCLOHEXANE MUTUAL DIFFUSION COEFFICIENTS

This binary mixture features a wide miscibility gap at 298.15 K, which results in a relatively limited diffusion data for the system. In the earliest work in 1969, the diffusion and Soret coefficients in the methanol-rich region have been measured at several concentrations using the flow-cell technique.[52] The most extended study of the system was presented in 2017, where the data from two techniques have been combined, the sliding symmetric tubes and counter-flow cell techniques.[50] The same data was later reproduced with the addition of one of the diluted limits.[26]

In Figure 10.6 are presented all the methanol–cyclohexane mutual diffusion coefficients data presented on the literature.

The polynomial coefficients of the diffusion coefficient concentration dependency determined for the mixture are listed in Table 10.6.

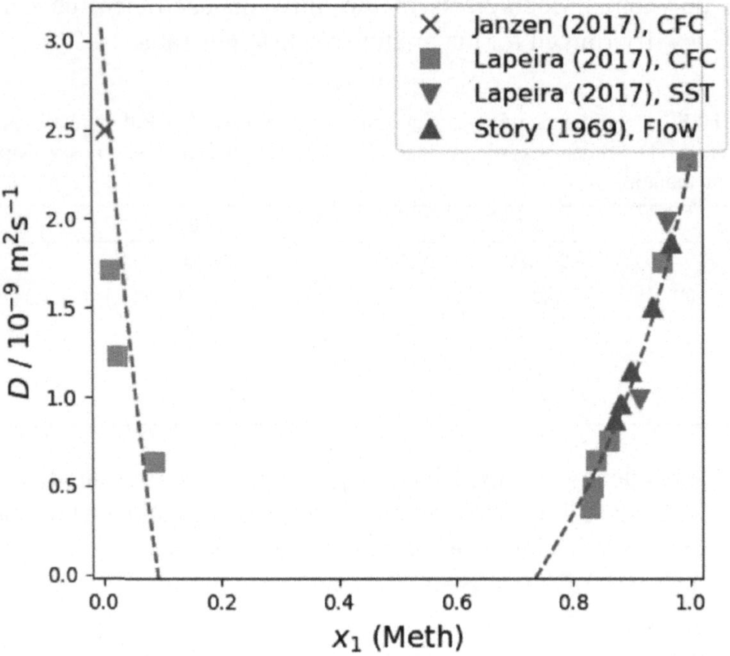

FIGURE 10.6 Mole fraction dependence of the mutual diffusion coefficients of the system methanol–cyclohexane.

TABLE 10.6 Coefficients of RK (A_n) and Power Series (b_n) Polynomials Describing Concentration Dependencies of D^E and D, Respectively, in the Binary Mixture of Methanol-cyclohexane.

Order *n*	An / 10^{-9}	b_n / 10^{-9}
0	−16.010	3.0700
1	−12.877	−43.283
2	−13.625	122.77
3	–	−134.75
4	–	54.499

The dependency outlined by the polynomial is presented graphically in Figure 10.6 by the dashed line. The average absolute deviation (AAD) of all data points from the correlation is 19.7% in this case. The high deviation seemingly arises from the large data scattering specific to the low methanol concentration region.

10.4.3 DIFFUSION COEFFICIENTS IN TERNARY MIXTURES

The Fickian diffusion coefficients for 20 ternary mixtures for the system cyclohexane (1) – toluene (2) – methanol (3), also stated forward as Chex (1) – Tol (2)- MeOH (3), were measured using the Taylor dispersion technique.[10] The third component (methanol) is considered the solvent.

Points under study were chosen along eight paths on the composition diagram, each corresponding to a constant composition of one of the mixture components and originally prepared in mass fraction. The aim was to understand the influence of each one on the transport properties (e.g., to understand the composition dependence on cyclohexane the diffusion measurements were carried out at several compositions with fixed cyclohexane mass fraction of 0.07, 0.2, and 0.4, corresponding respectively to approximate mol fractions 0.04, 0.12, and 0.5). The composition map is represented in Figure 10.7.

TABLE 10.7 Ternary Diffusion Coefficients for the System Cyclohexane (1) – Toluene (2) – Methanol (3) Measured at 298.15 K at a Set of State Points.

x_1	x_2	D_{11}	D_{12}	D_{21}	D_{22}	D_1	D_2	$\|D\|$
0.49	0.19	0.418	−1.257	−0.062	1.813	1.867	0.364	0.680
0.33	0.34	0.859	−0.961	−0.343	1.635	1.940	0.554	1.075
0.17	0.50	1.402	−0.593	−0.785	1.419	2.093	0.728	1.524
0.24	0.11	0.521	−0.136	−0.147	1.911	1.925	0.507	0.976
0.27	0.19	0.555	−0.021	−0.211	1.876	1.879	0.552	1.037
0.05	0.40	2.092	−0.181	−0.417	0.851	2.150	0.793	1.705
0.14	0.32	1.500	−0.582	−0.007	1.600	1.631	1.469	2.396
0.10	0.09	0.801	−0.772	−0.319	1.769	1.978	0.592	1.171
0.53	0.15	0.435	−1.013	−0.013	1.653	1.664	0.424	0.706
0.03	0.08	1.407	−0.245	−0.545	1.476	1.809	1.074	1.943
0.04	0.20	1.325	−0.279	−0.581	1.444	1.791	0.978	1.751
0.06	0.61	1.949	−0.224	−1.141	1.044	2.175	0.818	1.779
0.84	0.04	0.459	−0.842	−0.100	1.806	1.740	0.525	0.913
0.70	0.17	1.201	−0.285	−0.133	1.428	1.540	1.089	1.677
0.20	0.67	1.169	−0.753	−0.207	1.832	2.016	0.985	1.986
0.03	0.02	1.756	−0.216	−0.412	1.304	1.904	1.156	2.201
0.12	0.23	1.408	−0.592	−0.702	1.399	2.048	0.759	1.554
0.21	0.25	1.030	−0.468	−0.177	1.846	1.937	0.939	1.819
0.37	0.30	0.713	−1.059	−0.263	1.724	1.949	0.488	0.951
0.39	0.49	1.432	−1.400	−0.467	1.540	2.296	0.676	1.551

Notes: x_1, x_2 stand for mole fractions of components 1 and 2, D_{11}, D_{12}, D_{21}, and D_{22} are the diffusion coefficients and have units 10^{-9} m^2 s^{-1}, D_1 and D_2 are the calculated eigenvalues and |D| is the determinant.

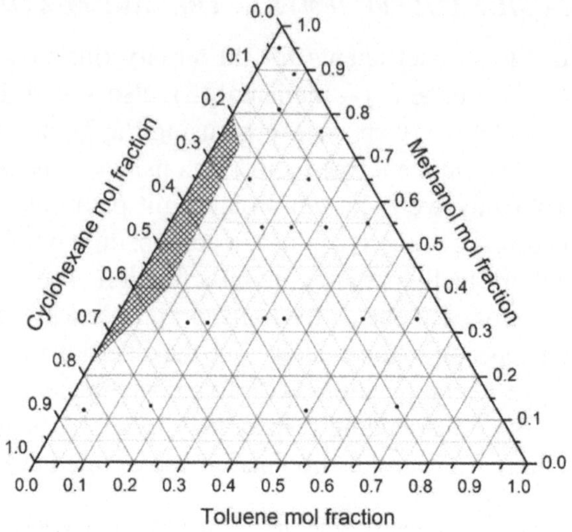

FIGURE 10.7 Map of cyclohexane (1) – toluene (2) – methanol (3) studied mixtures in mol fractions (Dots show selected 20 mixtures; shaded region outlines demixing zone).

Table 10.7 gives the average values of the diffusion coefficients, D_{ij}, for each mixture at 298.15 K, which were determined from four to six replicate dispersion profiles. Additionally, the eigenvalues, D_1 and D_2, of the matrix of mutual diffusion coefficients, and the respective determinants $|D|$ were calculated and are presented in the respective table. The main diffusion coefficients, D_{11} and D_{22}, were normally reproducible within $\pm 0.02 \times 10–^{9}$ $m^2\ s^{-1}$. The cross-coefficients, D_{12} and D_{21}, were, in general, reproducible within about $\pm 0.10 \times 10–^{9}$ $m^2\ s^{-1}$.

From the obtained values it is observed that the results attained for the mutual diffusion coefficients for the mixtures under study show good agreement with the available literature values, measured using the Taylor dispersion technique, by Grossmann and Winkelmann.[23,24] As well as, the eigenvalues are in good agreement to the ones available in the literature.

10.4.3.1 CYCLOHEXANE (1) COMPOSITION DEPENDENCE

Let us first analyze the toluene-rich composition part of the diagram and still far from the demixing zone, that is for small mol fractions of cyclohexane ($x_1 < 0.1$). It can be observed that when toluene mol fraction increases, the diffusion coefficient of cyclohexane D_{11} decreases until $x_2 = 0.20$ and then increases again, while the diffusion coefficient of toluene D_{22} presents the opposite behavior. On the other side, when $x_2 \rightarrow 0$, D_{11} approaches the binary diffusion coefficient for cyclohexane in toluene estimated in Section 10.4.2.1. In this range of composition, the cross diffusion coefficient values, D_{12} and D_{21}, are large and negative, which can lead us to conclude that under these conditions these solutes interact. These possible interactions lead to a constant counter flow of cyclohexane ($D_{12} < 0$) due to the gradient of toluene, in the studied composition range. However, the coupled counter current toluene flow (D_{21}) generated by the gradient of cyclohexane presents negative values larger than D_{12}, increasing when the composition of toluene increases.

For cyclohexane mol fraction $0.1 < x_1 < 0.2$ it is possible to notice that the main diffusion coefficient for cyclohexane D_{11} increases until $x_2 = 0.32$ and then starts to decrease, when the toluene mol fraction increases and $x_2 \rightarrow 1$, while D_{22} present the contrary behavior (Fig. 10.8). This is the opposite to what has been observed for the smaller mol fractions of cyclohexane, at $x_1 < 0.1$. The cross-diffusion coefficient values, D_{12} and D_{21}, in this range of compositions are both large and negative, sign of solute interactions, with an almost constant counter flow of cyclohexane ($D_{12} < 0$) due to the gradient of toluene, in the studied composition range.

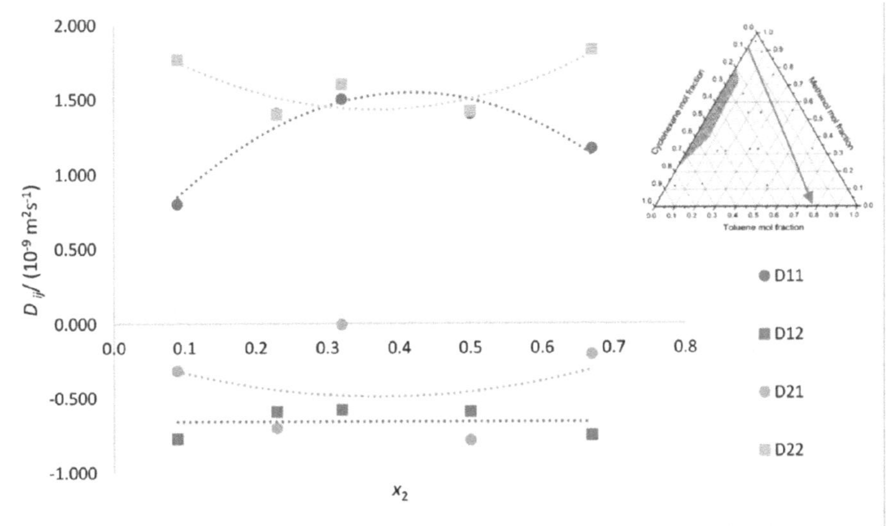

FIGURE 10.8 Mutual diffusion coefficients in the system Chex (1)– Tol (2)– MeOH(3) vs toluene mass fraction at 298.15 K and cyclohexane mol fraction $0.1 < x_1 < 0.2$. Dots represent experimental points; lines are only a guide to the eye.

At higher values of cyclohexane mol fraction ($x_1 > 0.2$) it is possible to observe that when $x_2 \rightarrow 1$, D_{11} and D_{22} present similar behavior than for $0.1 < x_1 < 0$. Also, when $x_2 \rightarrow 0$, D_{11} approaches the binary diffusion coefficient for cyclohexane in toluene calculated in Section 10.4.2.1 and the composition of the system is close to miscibility gap. However, the D_{12} diffusion cross diffusion coefficient values, in the studied range of compositions, are negative and tend to zero when $x_2 \rightarrow 0$, leading to a decrease in the counter flow of cyclohexane ($D_{12} < 0$) due to the gradient of toluene.

Main diffusion coefficient for cyclohexane shows a composition dependence more accentuated with the proximity to the miscibility gap composition. This solute interacts with the remaining toluene and methanol in the mixture, and this leads to counter flows of cyclohexane generated by the toluene gradient. Eigenvalues are positive and seem to be weekly affected by composition at lower cyclohexane mol fractions. At the higher compositions of cyclohexane in the mixture, heading toward the miscibility gap, there is noticeable decrease in the determinant of the diffusion coefficients matrix.

10.4.3.2 TOLUENE (2) COMPOSITION DEPENDENCE

Figure 10.9 represents the data from Table 10.7 for an almost constant composition of toluene (mol fraction $0.1 > x_2 < 0.20$), in the area adjacent to the demixing zone. It is noticeable that the diffusion coefficient for cyclohexane D_{11} shows a strong composition dependence, decreasing until approximately $x_1 = 0.25$ and then increasing, while the main diffusion coefficient for toluene D_{22} shows a behavior closer to ideal, almost independent of the mixture composition, idea reinforced by the fact that D_{21} has small negative values close to zero.

When $x_2 > 0.20$, further away from the demixing zone, the dependence of the diffusion coefficients on the composition of the mixture is even less accentuated, toward a more ideal behavior. When $x_1 \rightarrow 0$, both D_{11} and D_{22} present minor changes and in the limit $x_1 = 0$, D_{22} is close to the corresponding binary diffusion coefficient for toluene in cyclohexane (Section 10.4.2.1). A look at the behavior of the cross diffusion coefficient values, D_{12} and D_{21}, in the studied range of compositions takes us to conclude that there are minor solute interactions affecting the diffusion of the components, and, possible interactions between the solutes leading to a counter flow of cyclohexane due to the gradient of toluene ($D_{12} < 0$) and a counter flow of toluene (D_{21}

<0) generated by the gradient of cyclohexane that ultimately approach zero when $x_1 \rightarrow 0$.

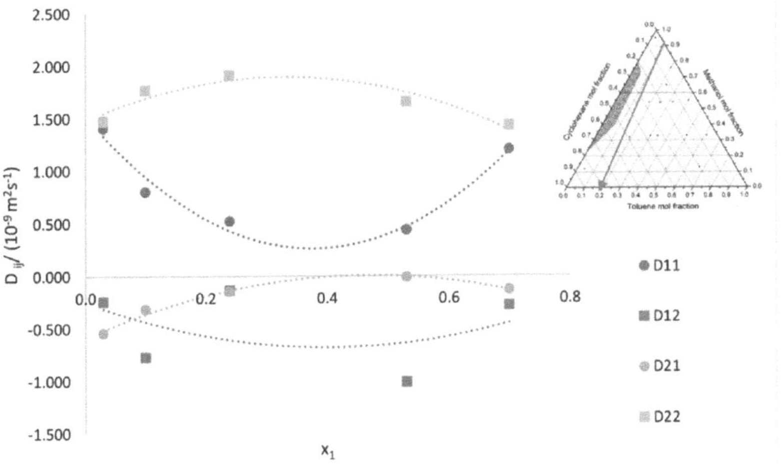

FIGURE 10.9 Mutual diffusion coefficients in the system Chex (1)– Tol (2)– MeOH(3) vs cyclohexane mass fraction at 298.15 K and toluene mol fraction $0.1 < x_1 < 0.2$. Dots represent experimental points; lines are only a guide to the eye.

Main diffusion coefficients for toluene show a behavior close to ideality, independent of the mixture composition, accentuated by the distance to the miscibility gap. Only small solute interactions of toluene with cyclohexane and methanol in the mixture occur, as the counter flow gradients for toluene extinguish when $x_1 \rightarrow 0$. Both eigenvalues and determinant of the diffusion coefficients matrix show a generally constant variation independent of the composition of the mixture.

10.4.3.3 METHANOL (3) COMPOSITION DEPENDENCE

In this study, methanol is considered the reference as the equivalent to a solvent, for the simplicity of the analysis of the results. Nevertheless, its composition dependence is also important to be known when studying the mixture and will be subsequently discussed.

When analyzing the behavior of the studied system in the methanol composition dependence (Fig. 10.10), it is noticeable that when $x_1 \rightarrow 1$, that is following the composition line toward the miscibility gap, the main coefficient for cyclohexane D_{11} shows a composition dependence, while D_{22}

presents a slight increase and then remains almost constant. In what concerns the cross-diffusion coefficient values it is verified that D_{12} and D_{21}, in the studied range of concentrations, are generally large and negative, being this effect more relevant for D_{12}, leading to a counter flow of cyclohexane ($D_{12} < 0$) due to the gradient of toluene, that is almost of the same magnitude of the gradient of concentration generated by the cyclohexane on its own.

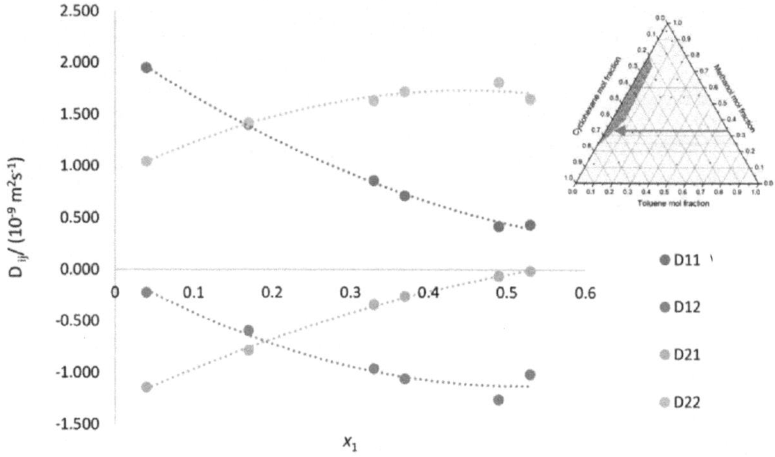

FIGURE 10.10 Mutual diffusion coefficients in the system Chex (1)– Tol (2)– MeOH(3) vs cyclohexane mass fraction at 298.15 K and methanol mol fraction $x_3 = 0.3$. Dots represent experimental points; lines are only a guide to the eye.

From the former analysis to the measured diffusion coefficients, covering most of the composition map of the mixture of cyclohexane (1) + toluene (2) + methanol (3), it was clear that main diffusion coefficients for toluene show a behavior closer to ideal, being almost independent of the mixture composition, while cyclohexane diffusion coefficients show a more nonideal behavior, mainly in mixture compositions close to the miscibility gap, and decreasing drastically when approaching the plait–point composition. This is in line with other authors' findings for this type of mixtures.[66] Moreover, we could observe that cross-diffusion coefficients for cyclohexane, originated by the gradients of toluene, to be as large and negative as main diffusion coefficients in both the cyclohexane-rich region and in the vicinity of the demixing zone, and this is a noticeable behavior, even if we consider the high uncertainty associated with them. At constant methanol mass fractions, the data show also a decrease of |D| when the cyclohexane fraction increases.

That is, when the mixture compositions approach the miscibility gap, along the composition paths for constant mol fractions x_3 equal to 0.1, 0.3, or 0.5, the determinant of the matrix approaches zero.

This study has shown that that in the region of the composition map below toluene mol fraction 0.4 and above cyclohexane mol fraction 0.2 the behavior of the mixture is highly nonideal and diffusion coefficients show a strong composition dependence, while in regions away from the demixing zone diffusion coefficients they present a behavior close to ideality.

10.5 CONCLUSIONS

Diffusion coefficients for the three binary subsystems of the mixture composed by toluene, cyclohexane, and methanol were investigated. A very complete database on diffusion was assembled from experimental results together with an intensive revision on the available literature data. From this database we have defined the values of mutual diffusion coefficients at infinite dilution limits for all three binary mixtures. After fixing the limiting values, correlations for mutual diffusion in concentrated binary solutions, consistent with the diluted limits, were established, providing a model that allows knowledge of the diffusion coefficient in concentration regions not covered by the measurements. Ternary mutual diffusion for the mixture of cyclohexane-toluene-methanol was studied for 20 different compositions, along eight different composition pats, each corresponding to constant composition of one of the mixture components. The associated coupled transport and the magnitude of the diffusion coefficients were analyzed. The results evidenced that the main diffusion coefficients for toluene are generally independent of the mixture composition; however, diffusion of cyclohexane is highly dependent on the mixture composition.

Due to the strong composition dependence shown by main diffusion coefficients and the large and negative cross diffusion coefficients, we can conclude that in the vicinity of the miscibility gap this mixture shows a highly nonideal behavior.

ACKNOWLEDGMENTS

This research was funded by FEDER–European Regional Development Fund through the COMPETE Programme and FCT-Fundação para a Ciência e a Tecnologia, for the KIDIMIX project POCI-01-0145-FEDER-030271, and by "The Coimbra Chemistry Centre" which is supported by the Fundação para a Ciência e a Tecnologia (FCT), Portuguese Agency for Scientific Research, through the programmes UIDB/00313/2020 e UIDP/00313/2020 and COMPETE. V.S. acknowledges support from micro4IIoT (Grant No. KK-2021/00082) from the Basque government and PID2020115086GB-C33 financed by MCIN/AEI of the Spanish Government.

KEYWORDS

- **diffusion coefficient**
- **Taylor dispersion**
- **cyclohexane**
- **toluene**
- **methanol**
- **DCMIX2**

REFERENCES

1. Santos, C. I. A. V.; Mialdun, A.; Legros, J. C.; Shevtsova, V. Sorption Equilibria and Diffusion of Toluene, Methanol, and Cyclohexane in/Through Elastomers. *J. Appl. Polym. Sci.* **2016,** *133*. https://doi.org/10.1002/app.43449
2. Mialdun, A.; Bataller, H.; Bou-Ali, M. M.; Braibanti, M.; Croccolo, F.; Errarte, A.; Ezquerro, J. M.; Fernández, J. J.; Gaponenko, Y.; García-Fernández, L.; Rodríguez, J.; Shevtsova, V. Preliminary Analysis of Diffusion Coefficient Measurements in Ternary mIXtures 4 (DCMIX4) Experiment on Board the International Space Station. *Eur. Phys. J. E* **2019,** *42*. https://doi.org/10.1140/epje/i2019-11851-6
3. Chan, T. C.; Chan, C. H. C.; Chang, N. W.; Lui, Y. Y. Diffusivities of Aromatic Compounds: A New Molecular-Hydrodynamic Model for Nonassociated Pseudoplanar Solutes at Infinite Dilution. *Ind. Eng. Chem. Res.* **2019,** *58*. https://doi.org/10.1021/acs.iecr.9b03859
4. Santos, C. I. A. V.; Shevtsova, V.; Burrows, H. D.; Ribeiro, A. C. F. Optimization of Taylor Dispersion Technique for Measurement of Mutual Diffusion in Benchmark Mixtures. *Microgravity Sci. Technol.* **2016,** *28*. https://doi.org/10.1007/s12217-016-9498-5
5. Callendar, R.; Leaist, D. G. Diffusion Coefficients for Binary, Ternary, and Polydisperse Solutions from Peak-Width Analysis of Taylor Dispersion Profiles. *J. Solution Chem.* **2006,** *35,* 353–379. https://doi.org/10.1007/s10953-005-9000-2
6. Mialdun, A.; Yasnou, V.; Shevtsova, V.; Königer, A.; Köhler, W.; Alonso De Mezquia, D.; Bou-Ali, M. M. A Comprehensive Study of Diffusion, Thermodiffusion, and Soret

Coefficients of Water-isopropanol Mixtures. *J. Chem. Phys.* **2012,** *136*. https://doi.org/10.1063/1.4730306

7. Ribeiro, A. C. F.; Leaist, D. G.; Esteso, M. A.; Lobo, V. M. M.; Valente, A. J. M.; Santos, C. I. A. V.; Cabrai, A. M. T. D. P. V.; Veiga, F. J. B. Binary Mutual Diffusion Coefficients of Aqueous Solutions of β-cyclodextrin at Temperatures from 298.15 to 312.15 K. *J. Chem. Eng. Data.* **2006,** *51*. https://doi.org/10.1021/je060092t
8. Ribeiro, A. C. F.; Sobral, A. J. F. N.; Simões, S. M. N.; Barros, M. C. F.; Lobo, V. M. M.; Cabral, A. M. T. D. P. V.; Veiga, F. J. B.; Santos, C. I. A. V.; Esteso, M. A. Transport Properties of Aqueous Solutions of Sodium Alginate at 298.15K. *Food Chem.* **2011,** 125. https://doi.org/10.1016/j.foodchem.2010.10.036
9. Leaist, D. G. Determination of Ternary Diffusion Coefficients by the Taylor Dispersion Method. *J. Phys. Chem.* **1990,** *94*. https://doi.org/10.1021/j100375a075
10. Santos, C. I. A. V.; Shevtsova, V.; Ribeiro, A. C. F. Isothermal Molecular Diffusion in Mixtures Containing Toluene, Cyclohexane and Methanol. *Eur. Phys. J. E* **2017,** *40*. https://doi.org/10.1140/epje/i2017-11526-4
11. Shevtsova, V.; Santos, C.; Sechenyh, V.; Legros, J. C.; Mialdun, A. Diffusion and Soret in Ternary Mixtures. Preparation of the DCMIX 2 Experiment on the ISS. *Microgravity Sci. Technol.* **2014,** *25*. https://doi.org/10.1007/s12217-013-9349-6
12. Santos, C. I. A. V.; Shevtsova, V.; Ribeiro, A. C. F. Diffusion and Thermodiffusion in Hydrocarbon Mixtures. In *Chemical Science and Engineering Technology*, 2019. https://doi.org/10.1201/9781351048323-17
13. Derlacki, Z. J.; Easteal, A. J.; Edge, A. V. J.; Woolf, L. A.; Roksandic, Z. Diffusion Coefficients of Methanol and Water and the Mutual Diffusion Coefficient in Methanol-Water Solutions at 278 and 298 K. *J. Phys. Chem.* **1985,** *89*. https://doi.org/10.1021/j100270a039
14. Ribeiro, A. C. F.; Lobo, V. M. M.; Natividade, J. J. S. Diffusion Coefficients in Aqueous Solutions of Cobalt Chloride at 298.15 K. *J. Chem. Eng. Data.* **2002,** *47*. https://doi.org/10.1021/je0101808
15. Mialdun, A.; Legros, J. C.; Yasnou, V.; Sechenyh, V.; Shevtsova, V.; Contribution to the Benchmark for Ternary Mixtures: Measurement of the Soret, Diffusion and Thermodiffusion Coefficients in the Ternary Mixture THN/IBB/nC12 with 0.8/0.1/0.1 Mass Fractions in Ground and Orbital Laboratories. *Eur. Phys. J. E* **2015,** *38*. https://doi.org/10.1140/epje/i2015-15027-2
16. Sechenyh, V.; Legros, J. C.; Mialdun, A.; Ortiz De Zárate, J. M.; Shevtsova, V. Fickian Diffusion in Ternary Mixtures Composed by 1,2,3,4-Tetrahydronaphthalene, Isobutylbenzene, and n-Dodecane. *J. Phys. Chem. B* **2016,** *120*. https://doi.org/10.1021/acs.jpcb.5b11143
17. Bochner, N.; Pipman, J. A Simple Method of Determining Diffusion Constants by Holographic Interferometry. *J. Phys. D Appl. Phys.* **1976,** *9*. https://doi.org/10.1088/0022-3727/9/13/003
18. Alizadeh, A.; Nieto de Castro, C. A.; Wakeham, W. A. The Theory of the Taylor Dispersion Technique for Liquid Diffusivity Measurements. *Int. J. Thermophys.* **1980,** *1*. https://doi.org/10.1007/BF00517126
19. Santos, C. I. A. V.; Ribeiro, A. C. F.; Esteso, M. A.; Drug Delivery Systems: Study of Inclusion Complex Formation Between Methylxanthines and Cyclodextrins and Their Thermodynamic and Transport Properties. *Biomolecules* **2019,** *9*. https://doi.org/10.3390/biom9050196

20. Santos, C. I. A. V.; Esteso, M. A.; Lobo, V. M. M.; Cabral, A. M. T. D. P. V.; Ribeiro, A. C. F. Taylor Dispersion Technique as a Tool for Measuring Multicomponent Diffusion in Drug Delivery Systems at Physiological Temperature. *J. Chem. Thermodynamics.* **2015,** *84,* 76–80. https://doi.org/10.1016/j.jct.2014.12.012
21. Legros, J. C.; Gaponenko, Y.; Mialdun, A.; Triller, T.; Hammon, A.; Bauer, C.; Köhler, W.; Shevtsova, V. Investigation of Fickian Diffusion in the Ternary Mixtures of Water-Ethanol-Triethylene Glycol and Its Binary Pairs. *Phys. Chem. Chem. Phys.* **2015,** *17.* https://doi.org/10.1039/c5cp04745e
22. Nagata, I. Liquid-Liquid Equilibria for Four Ternary Systems Containing Methanol and Cyclohexane. *Fluid Phase Equilib.* **1984,** *18.* https://doi.org/10.1016/0378-3812(84)80023-0
23. Großmann, T.; Winkelmann, J. Ternary Diffusion Coefficients of Cyclohexane + Toluene + Methanol by Taylor Dispersion Measurements at 298.15 K. Part 2. Low Toluene Area Near the Binodal Curve. *J. Chem. Eng. Data.* **2009,** *54.* https://doi.org/10.1021/je800477y
24. Grossmann, T.; Winkelmann, J. Ternary Diffusion Coefficients of Cyclohexane + Toluene + Methanol by Taylor Dispersion Measurements at 298.15 K. Part 1. Toluene-Rich Area. *J. Chem. Eng. Data* **2009,** *54.* https://doi.org/10.1021/je800444e
25. Guevara-Carrion, G.; Janzen, T.; Muñoz-Muñoz, Y. M.; Vrabec, J. Mutual Diffusion of Binary Liquid Mixtures Containing Methanol, Ethanol, Acetone, Benzene, Cyclohexane, Toluene, and Carbon Tetrachloride. *J. Chem. Phys.* **2016,** *144.* https://doi.org/10.1063/1.4943395
26. Janzen, T.; Zhang, S.; Mialdun, A.; G. Guevara-Carrion, Vrabec, J.; He, M.; Shevtsova, V. Mutual Diffusion Governed by Kinetics and Thermodynamics in the Partially Miscible Mixture Methanol + Cyclohexane. *Phys. Chem. Chem. Phys.* **2017,** *19.* https://doi.org/10.1039/c7cp06515a
27. Janzen, T.; Vrabec, J. Diffusion Coefficients of a Highly Nonideal Ternary Liquid Mixture: Cyclohexane–Toluene–Methanol. *Indus. Eng. Chem. Res.* **2018,** *57,* 16508–16517. https://doi.org/10.1021/acs.iecr.8b04385
28. Tyrrell, H. J. V. The Origin and Present Status of fick's Diffusion Law. *J. Chem. Educ.* **1964,** *41.* https://doi.org/10.1021/ed041p397
29. Fick, A. Ueber Diffusion. *Ann. Phys.* **1855,** *170.* https://doi.org/10.1002/andp.18551700105
30. Tyrrell, H. J. V.; Harris, K. R. *Diffusion in Liquids: A Theoretical and Experimental Study*; Elsevier, 1984. https://doi.org/10.1016/C2013-0-01002-X
31. Leaist, D. G.; Hao, L. Gravitational stability of Taylor Dispersion Profiles. Revised Diffusion Coefficients for Barium Chloride-Potassium Chloride-Water. *J. Phys. Chem.* **1993,** *97.* https://doi.org/10.1021/j100109a036
32. Deng, Z.; Leaist, D. G. Ternary Mutual Diffusion Coefficients of MgC12 + MgS04+ H20 and Na2S04 + MgS04 + H20 from Taylor Dispersion Profiles. *Can. J. Chem.* **1991,** *69,* 1548–1553
33. Königer, A.; Meier, B.; Köhler, W. Measurement of the Soret, Diffusion, and Thermal Diffusion Coefficients of Three Binary Organic Benchmark Mixtures and of Ethanol-Water Mixtures Using a Beam Deflection Technique. *Philos. Mag.* **2009,** *89.* https://doi.org/10.1080/14786430902814029
34. Zhang, K. J.; Briggs, M. E.; Gammon, R. W.; Sengers, J. V. Optical Measurement of the Soret Coefficient and the Diffusion Coefficient of Liquid Mixtures. *J. Chem. Phys.* **1996,** *104.* https://doi.org/10.1063/1.471355

35. Mialdun, A.; Shevtsova, V. M. Development of Optical Digital Interferometry Technique for Measurement of Thermodiffusion Coefficients. *Int. J. Heat Mass. Transf.* **2008,** *51.* https://doi.org/10.1016/j.ijheatmasstransfer.2007.08.020
36. Platten, J. K. Convection in Liquids. *Int. J. Heat Fluid Flow* **1984,** *5.* https://doi.org/10.1016/0142-727x(84)90068-7
37. Mialdun, A.; Shevtsova, V. Measurement of the Soret and Diffusion Coefficients for Benchmark Binary Mixtures by Means of Digital Interferometry. *J. Chem. Phys.* **2011,** *134.* https://doi.org/10.1063/1.3546036
38. Mialdun, A.; Yasnou, V.; Shevtsova, V. Measurement of Isothermal Diffusion Coefficients in Ternary Mixtures Using Counter Flow Diffusion Cell. *Comptes Rendus Mecanique* **2013,** *341.* https://doi.org/10.1016/j.crme.2013.02.001
39. Komiya, A.; Maruyama, S. Precise and Short-time Measurement Method of Mass Diffusion Coefficients. *Exp. Therm. Fluid Sci.* **2006,** *30.* https://doi.org/10.1016/j.expthermflusci.2005.11.001
40. Torres, J. F.; Komiya, A.; Shoji, E.; Okajima, J.; Maruyama, S. Development of Phase-shifting Interferometry for Measurement of Isothermal Diffusion Coefficients in Binary Solutions. *Opt. Lasers Eng.* **2012,** *50.* https://doi.org/10.1016/j.optlaseng.2012.03.006
41. Rashidnia, N.; Balasubramaniam, R. Measurement of the Mass Diffusivity of Miscible Liquids as a Function of Concentration Using a Common Path Shearing Interferometer. *Exp. Fluids* **2004,** *36.* https://doi.org/10.1007/s00348-003-0735-5
42. Mialdun, A.; Shevtsova, V. Digital Interferometry as a Powerful Tool to Study the Thermodiffusion Effect. *Comptes Rendus Mecanique* **2011,** *339.* https://doi.org/10.1016/j.crme.2011.04.001
43. Sanni, S. A.; Hutchison, H. P.; Fell, C. J. D. Diffusion Coefficients and Densities for Binary Organic Liquid Mixtures. *J. Chem. Eng. Data.* **1971,** *16.* https://doi.org/10.1021/je60051a009
44. Ghai, R. K.; Ertl, H.; Dullien, F. A. L. Liquid Diffusion of Nonelectrolytes: Part I. *AIChE J.* **1973,** *19.* https://doi.org/10.1002/aic.690190502
45. Easteal, A. J.; Woolf, L. A. Solute-solvent Interaction Effects on Tracer Diffusion Coefficients. *J. Chem. Soc. Faraday Trans. 1* **1984,** *80.* https://doi.org/10.1039/F19848001287
46. Thiel, P. Diffusions Messungen in Flussigkeiten Mittels TAYLOR-Dispersions Methode. Ph.D. Thesis, Wittenberg University, 1995.
47. Merzliak, T.; Bartusseck, I.; Stapf, S.; Voda, M. A.; B. Blümich, Pfennig, A. Description of Intra-diffusion in Liquid Mixtures. *Fluid Phase Equilib.* **2006,** *245.* https://doi.org/10.1016/j.fluid.2006.05.001
48. Shemilt, L. W.; Nagarajan, R. Liquid Diffusivities for the System Methanol–Toluene. *Can. J. Chem.* **1967,** *45.* https://doi.org/10.1139/v67-192
49. Lu, J. G.; Kong, R.; Chan, T. C. Effects of Molecular Association on Mutual Diffusion: A Study of Hydrogen Bonding in Dilute Solutions. *J. Chem. Phys.* **1999,** *110.* https://doi.org/10.1063/1.477895
50. Lapeira, E.; Gebhardt, M.; Triller, T.; Mialdun, A.; W. Köhler, Shevtsova, V.; Bou-Ali, M. M. Transport Properties of the Binary Mixtures of the Three Organic Liquids Toluene, Methanol, and Cyclohexane. *J. Chem. Phys.* **2017,** *146.* https://doi.org/10.1063/1.4977078
51. Brun, B.; Peyre, D.; Salvinien, J. Etude Diffusiométrique de L'autoassociation des Alcools. *Comptes Rendus Académie Sci. Sér. C Sci. Chim.* **1968,** *267,* 341–344

52. Story, M. J.; Turner, J. C. R. Flow-cell Studies of Thermal Diffusion in Liquids. Part 5.—Binary Mixtures of CH_3OH with CCl4, Benzene and Cyclohexane at 25°C. *Transac. Faraday Soc.* **1969,** *65*. https://doi.org/10.1039/TF9696501523
53. Sanni, S. A.; Hutchison, P. Diffusivities and Densities for Binary Liquid Mixtures. *J. Chem. Eng. Data.* **1973,** *18*. https://doi.org/10.1021/je60058a028
54. Winkelmann, J. Diffusion in Gases. Liquids and Electrolytes 2018. https://doi.org/10.1007/978-3-662-54089-3
55. Wittko, G.; Köhler, W. Influence of Isotopic Substitution on the Diffusion and Thermal Diffusion Coefficient of Binary Liquids. *Eur. Phys. J. E* **2006,** *21*. https://doi.org/10.1140/epje/i2006-10066-4
56. Wittko, G. Ph.D. Thesis, University Bayreuth, 2007.
57. Blesinger, C.; Beumers, P.; Buttler, F.; Pauls, C.; Bardow, A. Temperature-dependent Diffusion Coefficients from 1d Raman Spectroscopy. *J. Solut. Chem.* **2014,** *43*. https://doi.org/10.1007/s10953-013-0084-9
58. Mialdun, A.; Shevtsova, V. Temperature Dependence of Soret and Diffusion Coefficients for Toluene-Cyclohexane Mixture Measured in Convection-Free Environment. *J. Chem. Phys.* **2015,** *143*. https://doi.org/10.1063/1.4936778
59. Redlich, O.; Kister, A. T. Algebraic Representation of Thermodynamic Properties and the Classification of Solutions. *Ind. Eng. Chem.* **1948,** *40*. https://doi.org/10.1021/ie50458a036
60. Santos, C. I. A. V.; Barros, M. C. F.; Ribeiro, A. C. F.; Bou-Ali, M. M.; Mialdun, A.; Shevtsova, V. Transport Properties of n-Ethylene Glycol Aqueous Solutions with Focus on Triethylene Glycol–Water. *J. Chem. Phys.* **2022,** *156,* 214501. https://doi.org/10.1063/5.0091902
61. Bou-Ali, M. M.; Ecenarro, O.; Madariaga, J. A.; Santamaría, C. M.; Valencia, J. J. Measurement of Negative Soret Coefficients in a Vertical Fluid Layer with an Adverse Density Gradient. *Phys. Rev. E Stat Phys. Plasmas Fluids Relat. Interdiscip. Topics.* **2000,** *62*. https://doi.org/10.1103/PhysRevE.62.1420
62. Ratcliff, G. A.; Lusis, M. A. Diffusion in Simple-Complex Forming Liquid Mixtures. *Ind. Eng. Chem. Fundament.* **1971,** *10*. https://doi.org/10.1021/i160039a021
63. Ramprasad, G.; Das, T. R.; Mukherjee, A. K. Mutual Diffusion Coefficients of Some Binary Liquid Systems. Toluene-n-alkyl Alcohol. *J. Chem. Eng. Japan* **1991,** *24*. https://doi.org/10.1252/jcej.24.389
64. Scheibel, E. G. Correspondence. Liquid Diffusivities. Viscosity of Gases. *Ind. Eng. Chem.* **1954,** *46*. https://doi.org/10.1021/ie50537a062
65. Wilke, C. R.; Chang, P. Correlation of Diffusion Coefficients in Dilute Solutions. *AIChE J.* **1955,** *1*. https://doi.org/10.1002/aic.690010222
66. Clark, W. M.; Rowley, R. L. Ternary Liquid Diffusion Coefficients Near Plait Points. *Int. J. Thermophys.* **1985,** *6*. https://doi.org/10.1007/BF00500335
67. Köhler, W. & Mialdun, Aliaksandr & Bou-Ali, M. & Valentina, Shevtsova. (2023). The Measurement of Soret and Thermodiffusion Coefficients in Binary and Ternary Liquid Mixtures. *International Journal of Thermophysics. 44*. 10.1007/s10765-023-03242-x.

Index

E

F

N

O

P

Q

R

S

T